ADVANCES IN NUTRACEUTICALS AND FUNCTIONAL FOODS

Concepts and Applications

ADVANCES IN NUTRACEUTICALS AND FUNCTIONAL FOODS

Concepts and Applications

Edited by

Sreerag Gopi, PhD

Preetha Balakrishnan, PhD

APPLE ACADEMIC PRESS

First edition published 2022

Apple Academic Press Inc.
1265 Goldenrod Circle, NE,
Palm Bay, FL 32905 USA

4164 Lakeshore Road, Burlington,
ON, L7L 1A4 Canada

CRC Press
6000 Broken Sound Parkway NW,
Suite 300, Boca Raton, FL 33487-2742 USA

2 Park Square, Milton Park,
Abingdon, Oxon, OX14 4RN UK

Apple Academic Press exclusively co-publishes with CRC Press, an imprint of Taylor & Francis Group, LLC

Library and Archives Canada Cataloguing in Publication

Title: Advances in nutraceuticals and functional foods : concepts and applications / edited by Sreerag Gopi, PhD, Preetha Balakrishnan, PhD.

Names: Gopi, Sreerag, editor. | Balakrishnan, Preetha, editor.

Description: First edition. | Includes bibliographical references and index.

Identifiers: Canadiana (print) 20210367172 | Canadiana (ebook) 20210367296 | ISBN 9781774637524 (hardcover) | ISBN 9781774637531 (softcover) | ISBN 9781003277088 (ebook)

Subjects: LCSH: Functional foods. | LCSH: Functional foods—Health aspects. | LCSH: Functional foods—Nutritional aspects.

Classification: LCC QP144.F85 A38 2022 | DDC 613.2—dc23

Library of Congress Cataloging-in-Publication Data

CIP data on file with US Library of Congress

ISBN: 978-1-77463-752-4 (hbk)
ISBN: 978-1-77463-753-1 (pbk)
ISBN: 978-1-00327-708-8 (ebk)

About the Editors

Sreerag Gopi, PhD
Chief Scientific Officer, ADSO Naturals, India;
Vice President, CureSupport, The Netherlands,
Mobile: +91-8594023331,
E-mail: Sreeraggopi@gmail.com

Sreerag Gopi, PhD, is a Chief Scientific Officer at ADSO Naturals, India, and Vice President at CureSupport, The Netherlands. He graduated with a degree in Chemistry from Calicut University, Kerala, India, and an advanced degree from Madras Christian College, Chennai, India. He is a recipient of a prestigious Erasmus Mundus Fellowship from the European Union during his PhD period. He is a materials chemist and nanomaterials scientist and has expertise in nanomaterial synthesis, characterization, biocomposites for natural products, and biomedical and water purification experiments. He has published over 20 peer-reviewed international papers and several book chapters and also has book projects in the works with several publishers, including the Royal Society of Chemistry, Springer, Wiley. He was selected as an Associate Member of the Royal Society of Chemistry in 2018, and he is a chartered member of Royal Australian Chemical Institute.

Preetha Balakrishnan, PhD
Principal Scientist, QA, QC, ADSO Naturals India, and Curesupport, The Netherlands,
Mobile: +91-7025921175
E-mails: preetha@adsonaturals.com

Preetha Balakrishnan, PhD, is the principal scientist, QA, QC, at ADSO Naturals, India, and at CureSupport, Netherlands. She graduated in Chemistry from Calicut University, Kerala, India, and earned her postgraduate degree at Mahatma Gandhi University, Kerala, with a gold medal and first rank. She is a recipient of prestigious INSPIRE Fellowship from the Government of India. She was a postdoctoral researcher in the research group of Professor Sabu Thomas, Vice Chancellor, a renowned scientist in this area who has sustained international acclaims for his work in polymer science and engineering, polymer nanocomposites, elastomers, polymer blends, interpenetrating polymer networks, polymer membranes, green composites and nanocomposites, nanomedicine, and green nanotechnology. She completed her PhD in Chemistry at Mahatma Gandhi University under Dr. Thomas's guidance. She has visited many foreign universities as a part of her research activities and has published over 15 research articles and over 20 book chapters. She has edited ten books with leading publishers, including Elsevier, Springer, Wiley, and the Royal Society of Chemistry. Dr. Balakrishnan has received a number of national and international presentation awards. She also worked as a guest lecturer in chemistry at the Department of Chemistry, Morning Star Home Science College, Angamaly, Kerala, India.

Contents

Contributors .... *ix*
Abbreviations .... *xiii*
Preface .... *xvii*

1. **Introduction to Functional Foods and Nutraceuticals** **1**
Luana Pulvirenti and Angela Paterna

2. **Advanced Nanocarriers for Nutraceuticals Based on Structured Lipid and Nonlipid** **29**
Shafiullah, Syed Wadood Ali Shah, Ismail Shah, Shujat Ali, Aziz Ullah, Samiullah Burki, and Mohammad Shoaib

3. **Nanoparticulate Approaches for Improved Nutrient Bioavailability** .. **59**
Abdul Qadir, Mohd. Aqil, and Dipak Kumar Gupta

4. **Adulteration and Safety Issues in Nutraceuticals and Functional Foods** **79**
Shujat Ali, Syed Wadood Ali Shah, Muhammad Ajmal Shah, Muhammad Zareef, Muhammad Arslan, Md. Mehedi Hassan, Shujaat Ahmad, Imdad Ali, Mumtaz Ali, and Shafi Ullah

5. **Nutraceuticals-Loaded Nano-Sized Delivery Systems: Potential Use in the Prevention and Treatment of Cancer** **105**
Mohammed Jafar, Syed Sarim Imam, Sultan Alshehri, Chandra Kala, and Ameeduzzafar Zafar

6. **Nutrition Nutraceuticals: A Proactive Approach for Healthcare** **123**
Conor P. Akintola, Dearbhla Finnegan, Niamh Hunt, Richard Lalor, Sandra O'Neill, and Christine Loscher

7. **Bioactive Proteins and Peptides as Functional Foods** **173**
Deepa Thomas and M. S. Latha

8. **News and Trends in the Development of Functional Foods: Probiotic Dairy and Non-Dairy Products** **199**
Eliane Maurício Furtado Martins, Wellingta Cristina Almeida do Nascimento Benevenuto, Aurélia Dornelas de Oliveira Martins, Augusto Aloísio Benevenuto Junior, Isabela Campelo de Queiroz, Thainá de Melo Carlos Dias, Daniela Aparecida Ferreira Souza, Daniele De Almeida Paula, and Maurílio Lopes Martins

9. Microencapsulation: An Alternative for the Application of Probiotic Cells in the Food and Nutraceuticals Industries 239

Daniele De Almeida Paula, Carini Aparecida Lelis, and Nataly De Almeida Costa

10. Nutraceuticals-Based Nano-Formulations: An Overview Through Clinical Validations 277

Shelly Singh and Shilpa Sharma

11. Growth Patterns, Emerging Opportunities, and Future Trends in Nutraceuticals and Functional Foods... 311

Asad Ur Rehman, Salman Akram, and Thierry Vandamme

Index.. 347

Contributors

Shujaat Ahmad
Department of Pharmacy, Shaheed Benazir Bhutto University Sheringal, Dir (Upper), Khyber Pakhtunkhwa, Pakistan

Conor P. Akintola
Immune Modulation Group, School of Biotechnology, Dublin City University, Dublin, Ireland

Salman Akram
University of Strasbourg, CNRS 7199, Faculty of Pharmacy, 74 Route du Rhin, CS – 60024, 67401 ILLKIRCH CEDEX, France

Imdad Ali
H.E.J. Research Institute of Chemistry, International Center for Chemical and Biological Sciences, University of Karachi, Karachi – 75270, Pakistan

Mumtaz Ali
Department of Chemistry, University of Malakand, Khyber Pakhtunkhwa – 18800, Pakistan

Shujat Ali
School of Food and Biological Engineering, Jiangsu University, Zhenjiang – 212013, P. R. China; College of Electrical and Electronic Engineering, Wenzhou University, Wenzhou 325035, PR China, E-mail: shujat86@yahoo.com

Sultan Alshehri
Department of Pharmaceutics, College of Pharmacy, King Saud University, Riyadh, Saudi Arabia

Mohd. Aqil
Department of Pharmaceutics, School of Pharmaceutical Education and Research, Jamia Hamdard (Deemed University), M. B. Road, New Delhi – 110062, India, E-mail: maqil@jamiahamdard.ac.in

Muhammad Arslan
School of Food and Biological Engineering, Jiangsu University, Zhenjiang – 212013, P. R. China

Wellingta Cristina Almeida do Nascimento Benevenuto
Federal Institute of Southeast of Minas Gerais, Food Science and Technology Department (DCTA/IF Sudeste MG), Rio Pomba, MG, CEP – 36180-000, Brazil

Samiullah Burki
Department of Pharmacology, Faculty of Pharmacy, Federal Urdu University of Arts, Science, and Technology, Karachi, Pakistan

Nataly De Almeida Costa
Department of Food Technology, Federal University of Viçosa (UFV), P.H. Rolfs Avenue, Campus, Viçosa – 36570-900, MG, Brazil

Thainá de Melo Carlos Dias
Federal Institute of Southeast of Minas Gerais, Food Science and Technology Department (DCTA/IF Sudeste MG), Rio Pomba, MG, CEP – 36180-000, Brazil

Dearbhla Finnegan
Immune Modulation Group, School of Biotechnology, Dublin City University, Dublin, Ireland

Dipak Kumar Gupta
Department of Pharmaceutics, School of Pharmaceutical Education and Research, Jamia Hamdard (Deemed University), M. B. Road, New Delhi – 110062, India

Md. Mehedi Hassan
School of Food and Biological Engineering, Jiangsu University, Zhenjiang – 212013, P. R. China

Niamh Hunt
Immune Modulation Group, School of Biotechnology, Dublin City University, Dublin, Ireland

Syed Sarim Imam
Department of Pharmaceutics, College of Pharmacy, King Saud University, Riyadh, Saudi Arabia

Mohammed Jafar
Assistant Professor, Department of Pharmaceutics, College of Clinical Pharmacy, Imam Abdulrahman Bin Faisal University, P.O. Box – 1982, Dammam – 31441, Saudi Arabia, Mobile: +966502467326, E-mail: jafar31957@gmail.com mjomar@iau.edu.sa

Augusto Aloísio Benevenuto Junior
Federal Institute of Southeast of Minas Gerais, Food Science and Technology Department (DCTA/IF Sudeste MG), Rio Pomba, MG, CEP – 36180-000, Brazil

Chandra Kala
Faculty of Pharmacy, Maulana Azad University, Jodhpur – 342802, Rajasthan, India

Richard Lalor
Fundamental and Translational Immunology Group, School of Biotechnology, Dublin City University, Dublin, Ireland

M. S. Latha
Department of Chemistry, Sree Narayana College, Chathannur, Kollam, Kerala, India; Department of Chemistry, Sree Narayana College, Kollam, Kerala, India, E-mail: lathams2014@gmail.com

Carini Aparecida Lelis
Center for Food Analysis (NAL), Technological Development Support Laboratory (LADETEC), Federal University of Rio de Janeiro (UFRJ), Cidade Universitária, Rio de Janeiro, RJ, 21941-598, Brazil

Christine Loscher
Immune Modulation Group, School of Biotechnology, Dublin City University, Dublin, Ireland

Aurélia Dornelas de Oliveira Martins
Federal Institute of Southeast of Minas Gerais, Food Science and Technology Department (DCTA/IF Sudeste MG), Rio Pomba, MG, CEP – 36180-000, Brazil

Eliane Maurício Furtado Martins
Federal Institute of Southeast of Minas Gerais, Food Science and Technology Department (DCTA/IF Sudeste MG), Rio Pomba, MG, CEP – 36180-000, Brazil, E-mail: eliane.martins@ifsudestemg.edu.br

Maurílio Lopes Martins
Federal Institute of Southeast of Minas Gerais, Food Science and Technology Department (DCTA/IF Sudeste MG), Rio Pomba, MG, CEP – 36180-000, Brazil

Sandra O'Neill
Fundamental and Translational Immunology Group, School of Biotechnology, Dublin City University, Dublin, Ireland, E-mail: Sandra.oneill@dcu.ie

Angela Paterna
Institute of Biophysics, National Research Council, Via Ugo La Malfa – 153,90146, Palermo, Italy

Daniele De Almeida Paula
Federal Institute of São Paulo (IFSP), Campus Avaré - Av. Professor Celso Ferreira da Silva, 1333, Jardim Europa, CEP 18707-150, SP, Brazil, E-mail: daniele.paulaufv@gmail.com

Luana Pulvirenti
Department of Chemical Sciences, University of Catania, Viale Andrea Doria-6, 95125, Catania, Italy [Dipartimento di Scienze Chimiche, Università Degli Studi di Catania, Viale A. Doria – 6,95125, Catania, Italy], E-mail: luanapulvirenti@unict.it

Abdul Qadir
Department of Pharmaceutics, School of Pharmaceutical Education and Research, Jamia Hamdard (Deemed University), M. B. Road, New Delhi – 110062, India

Isabela Campelo de Queiroz
Federal Institute of Southeast of Minas Gerais, Food Science and Technology Department (DCTA/IF Sudeste MG), Rio Pomba, MG, CEP – 36180-000, Brazil

Asad Ur Rehman
University of Strasbourg, CNRS 7199, Faculty of Pharmacy, 74 Route du Rhin, CS – 60024, 67401 ILLKIRCH CEDEX, France; University of Paris Descartes, UTCBS CNRS UMR 8258-INSERM U1267, Faculty of Pharmacy, 4 Avenue de l'Observatoire, Paris – 75006, France

Shafiullah
Department of Pharmacy, University of Malakand, Chakdara, Dir Lower – 18300, Khyber Pakhtunkhwa, Pakistan, E-mail: shafi_ullah34@yahoo.com

Ismail Shah
Department of Pharmacy, Abdulwali Khan University, Mardan – 23200, Khyber Pakhtunkhwa, Pakistan

Muhammad Ajmal Shah
Department of Pharmacognosy, Faculty of Pharmaceutical Sciences, Government College University, Faisalabad, Pakistan

Syed Wadood Ali Shah
Department of Pharmacy, University of Malakand, Chakdara, Dir Lower – 18300, Khyber Pakhtunkhwa – 18800, Pakistan

Shilpa Sharma
Department of Biological Sciences and Engineering, Netaji Subhas University of Technology, Dwarka, New Delhi, India

Mohammad Shoaib
Department of Pharmacy, University of Malakand, Chakdara, Dir Lower – 18300, Khyber Pakhtunkhwa, Pakistan

Shelly Singh
Department of Biological Sciences and Engineering, Netaji Subhas University of Technology, Dwarka, New Delhi, India

Daniela Aparecida Ferreira Souza
Federal Institute of Southeast of Minas Gerais, Food Science and Technology Department (DCTA/IF Sudeste MG), Rio Pomba, MG, CEP – 36180-000, Brazil

Deepa Thomas
Research and Post Graduate Department of Chemistry, Bishop Moore College, Mavelikara, Alappuzha, Kerala, India

Aziz Ullah
Department of Pharmaceutics, Faculty of Pharmacy, Gomal University, D.I. Khan, Khyber Pakhtunkhwa, Pakistan

Shafi Ullah
Department of Pharmacy, University of Malakand, Khyber Pakhtunkhwa – 18800, Pakistan; H.E.J. Research Institute of Chemistry, International Center for Chemical and Biological Sciences, University of Karachi, Karachi – 75270, Pakistan, E-mail: Shafi_ullah34@yahoo.com

Thierry Vandamme
University of Strasbourg, CNRS 7199, Faculty of Pharmacy, 74 Route du Rhin, CS – 60024, 67401 ILLKIRCH CEDEX, France

Ameeduzzafar Zafar
Department of Pharmaceutics, College of Pharmacy, Jouf University, Sakaka, Aljouf, Saudi Arabia

Muhammad Zareef
School of Food and Biological Engineering, Jiangsu University, Zhenjiang – 212013, P. R. China

Abbreviations

ABC	ATP-binding cassette
ACE	angiotensin-converting enzyme
ACE	angiotensin I-enzyme
ADA	American Dietetic Association
Ag	silver
AI	artificial intelligence
ALA	alpha-linoleic acid
ARA	arachidonic acid
Au	gold
AuNPs	gold nanoparticles
BDPP	bioactive dietary polyphenol preparations
CAG	compound annual growth
CAGR	compound annual growth rate
CAPE	caffeic acid phenethyl ester
CCPs	caseinophosphopeptides
CD	Crohn's disease
CdS	cadmium sulfide
CLA	conjugated linoleic acid
CLNA	conjugated α-linolenic acid
CPSC	consumer product safety commission
CST	critical solution temperature
CVD	cardiovascular diseases
DALYs	disability-adjusted life years
DE	dextrose equivalent
DHA	docosahexaenoic acid
DIM	diindolylmethane
DPP4	dipeptidyl peptidase-IV
DSC	differential scanning calorimetry
DTA	differential thermal analysis
EC	European Commission
EE	encapsulation efficiency
EFSA	European Food Safety Authority
EGCG	epigallocatechin gallate
EGFR	epidermal growth factor receptor

EMS	eosinophilia-myalgia-syndrome
Eos	essential oils
EPA	eicosapentaenoic acid
EPA	Environmental Protection Agency
EPS	exopolysaccharides
FAO	Food and Agriculture Organization
FAS	fatty acid synthase
FDA	Food and Drug Administration
FIM	foundation for innovation in medicine
FOSHU	foods for specified health use
FPHs	fish protein hydrolysates
FSA	Food Standards Agency
Gas	glycoalkaloids
GI	gastrointestinal
GIP	glucose-dependent insulinotropic polypeptide
GIT	gastrointestinal tract
GLP-1	glucagon-like peptide-1
GRAS	generally recognized as safe
HCl	hydrochloric acid
HSE	health service executive
IBD	inflammatory bowel disease
IBS	inflammatory bowel syndrome
IGFR	insulin-like growth factor receptor
IL	interleukin
ILSI	International Life Sciences Institutes
IPP	Ile-Pro-Pro
IPP	isopentenyl diphosphate
JECFA	Joint FAO/WHO Expert Committee on Food Additives
LAB	lactic acid bacteria
LCST	lower critical solution temperature
LMP	low methoxyl pectin
LUVs	large unilamellar vesicles
MEP	methylerythritol phosphate
Met-S	metabolic syndromes
MLV	multilamellar vesicles
MMPs	matrix metalloproteinases
MPS	mononuclear phagocyte system
MRP	multidrug resistance protein
MTSG1	mitochondrial tumor suppressor 1

MVA	mevalonic acid
NA	nicotinic acid
NCDs	non-communicable diseases
NEs	nanoemulsions
NIOSH	National Institute for Occupational Safety and Health
NLCs	nanostructured lipid carriers
NPs	nanoparticles
NREA	Nutraceutical Research and Education Act
NSAIDs	non-steroidal anti-inflammatory drugs
O/W	oil-in-water
O/W/O	oil-in-water-in-oil
OCP	office of combination products
OSHA	Occupational Safety and Health Administration
OsLu	lactulose-derived oligosaccharide
PA	palmitic acid
PAA	poly(acrylic acid)
PAAM	poly(acrylamide-co-butyl methacrylate)
PBS	phosphate buffer solution
PC	phosphatidylcholine
PCADK	poly(cyclohexane-1,4-diyl acetone dimethylene ketal)
PCL	polycaprolactone
Pd	palladium
PDEAEM	poly(N,N9-diethylaminoethyl methacrylate)
PDEAM	poly(N,N-diethylacrylamide)
PDMAEMA	poly[2-(dimethylamino)ethyl methacrylate]
PE	phosphotidyl ethanolamine
PECs	polyelectrolyte complexes
PEG	poly(ethylene glycol)
PGA	poly(glycolic acid)
PK	polyketals
PLA	polylactic acid
PLGA	poly(lactic-co-glycolic acid)
PNIPAM	poly(N-isopropylacrylamide)
PNPs	polymeric nanoparticle systems
PPARγ	peroxisome proliferator-activated receptors
Pt	platinum
PUFAs	polyunsaturated fatty acids
PVCL	poly(N-vinylcaprolactam)

PVCL-PVA-PEG	polyvinyl caprolactam-polyvinyl acetate-polyethylene glycol
RA	rheumatoid arthritis
RES	reticuloendothelial system
RESS	rapid expansion of supercritical solution
RNA	ribonucleic acid
ROS	reactive oxygen species
SERS	surface-enhanced Raman spectroscopy
siRNA	small interfering RNA
SLN	solid lipid nanoparticles
SMEDDS	self-micro emulsifying drug delivery system
SOD	superoxide dismutase
SUVs	small unilamellar vesicles
T2DM	type 2 diabetes mellitus
TEM	transmission electron microscopy
TJs	tight junctions
TMC	N-trimethyl chitosan
TNF	tumor necrosis factor
TNF-α	tumor necrosis factor-alpha
UC	ulcerative colitis
UCST	upper critical solution temperature
US	United States
USA	United States of America
USDA	US Department of Agriculture
USPTO	US Patent and Trademark Office
VEGF	vascular endothelial growth factor
VPP	val-pro-pro
W/O/W	water-in-oil-in-water
WHO	World Health Organization

Preface

In recent years there is a growing interest in nutraceuticals, which provide health benefits and are alternative to modern medicine. Nutrients, herbals, and dietary supplements are significant constituents of nutraceuticals which make them instrumental in maintaining health, act against various disease conditions, and thus promote the quality of life. The explosive growth, research developments, lack of standards, marketing zeal, quality assurance, and regulation will play a vital role in its success or failure.

The demand for foods with a positive impact on human health and wellness has exploded globally over the past two decades. This growth is driven by socioeconomic and scientific factors, including increases in population, disposable income, life expectancy, and healthcare costs. Advancements also enhance the market for healthier foods in our understanding of dietary bioactive ingredients and their effects on various aspects of human health at a systems and molecular level. This book examines the rapidly growing field of functional foods to prevent and manage chronic and infectious diseases. It attempts to provide a unified and systematic account of functional foods by illustrating the connections among the different disciplines needed to understand foods and nutrients, mainly: food science, nutrition, pharmacology, toxicology, and manufacturing technology. Advances within and among all these fields are critical for the successful development and application of functional foods. Chapters in the present volume explore the varied sources, biochemical properties, metabolism, health benefits, and safety of bioactive ingredients. Special emphasis is given to linking the molecular and chemical structures of biologically active components in foods to their nutritional and pharmacological effects on human health and wellness. In addition to discussing scientific and clinical rationales for different sources of functional foods, the book also explains in detail the scientific methodologies used to investigate the functionality, effectiveness, and safety of bioactive ingredients in food.

CHAPTER 1

Introduction to Functional Foods and Nutraceuticals

LUANA PULVIRENTI[1] and ANGELA PATERNA[2]

[1]Department of Chemical Sciences, University of Catania, Viale Andrea Doria-6, 95125, Catania, Italy

[2]Institute of Biophysics, National Research Council, Via Ugo La Malfa – 153,90146, Palermo, Italy

ABSTRACT

Nowadays, the term functional food gained more attention, especially by the younger generations, since are certainly more informed about the increasingly close correlation between food and health. This term was first used in Japan in 1980 and since that time it has been possible to record a growing interest from the scientific community around the world, in order to clarify their potential role in the prevention of chronic diseases and in the maintenance of good health of a population with a longer life expectancy than in the past. In this context, this chapter aims to offer a simple and comprehensive overview about definitions and classifications of functional food. Furthermore, attention was focused on the close relationship that exists between the chemical composition of a food in terms of 'functional' chemical compounds known today as nutraceuticals, and the ability of the food to play a functional role.

1.1 INTRODUCTION

Today, the common thought is that foods together with a good lifestyle may be able to prevent diseases or physiological disorders. This belief is actually much older than might think, and even Hippocrates about 2500 years ago claimed, "Let food be thy medicine and medicine be thy food." The aforementioned concept is particularly felt mostly by the younger generations, who represent a new class of consumers, of course, more health-conscious than before. More in general people take more into account the strict relationship between diet and health. The reason is that they are more informed about it, thanks also to many scientific and popular magazines, tv programs, social media posts, and blogs which often deal with topics concerning the content of bioactive chemical substances in foods and their potential activity as chemopreventive agents of degenerative diseases. Therefore, these foods defined 'functional foods' are considered desirable in a good diet. The term 'functional foods,' used for the first time in Japan in 1980 [1, 2], includes every food or food ingredients exerting a nutritional function but at the same time express promising healthy effect when eaten regularly in a varied diet [3]. The above consideration allows to enclose in this group not processed foods such as fruit and vegetable, but also foods formulated with a specific health purpose.

In recent years a renew attention has been registered on functional foods from researchers in the world working in different fields of science due to a growing global interest for these foods. Indeed, the fields of investigation involving functional foods are manifold and often linked together; for example, using on Scopus.com the index term 'functional food' about 61.000 documents (articles, chapters, and books) were found, published between 1980 and 2019, with an increasing number of publication year by year, confirming the growing scientific interest. Furthermore, it is noteworthy that analyzing quickly the results of search by subject area, it is possible to observe that this topic involves many scientific fields (Figure 1.1).

Therefore, it is not surprising that the functional foods development has required interconnection with related field like food chemistry, biology, nutrition, pharmacology, and statistics [4].

In this context, important contributions have been made by many epidemiological studies reported in literature with the purpose to evaluate the relationship between diet habits and the risk of contracting a large share of the global diseases, through conditions such as high blood pressure and elevated blood glucose and cholesterol levels. Two famous examples are the Mediterranean-style diet and high-fat diets, also well known as the French paradox, incorporating moderate red wine intake are reported to benefit to human health [5]. In particular French paradox refers to the lower risk of the French people towards cardiovascular diseases (CVD), despite their high fat diet, attributed to their habitual but moderate consumption of red wine. Also, the Mediterranean-style diet, expressed by a reduced drinking of alcohol, a balanced eating of meat and its subproducts, an increased ingestion of fruits, vegetables, and extra-virgin olive oil, is widely recognized to have beneficial effects on CVD.

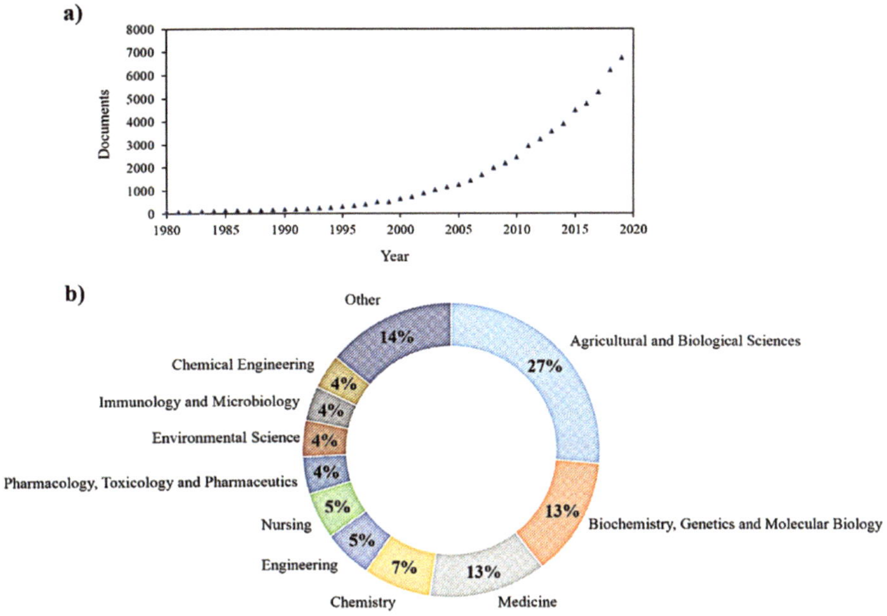

FIGURE 1.1 Total number of articles (a) by year; and (b) subject area.

Source: Published from: 1980 to 2019 at Scopus.com using the index term 'functional food.'

Nowadays, is commonly approved that the helpful outcomes of functional foods can be attributed to the chemical substances' characteristic of their composition, well known with the term 'nutraceuticals,' for which it was registered an increased interest corroborated by the growth of the nutraceutical trend aiding the growth of the global market. The global nutraceutical market size was valued to grow from about \$209 billion in 2017 to \$373 billion in 2025, predicting a spread at a CAG (compound annual growth) rate of 8.3% over the estimated period [6].

The growing scientific interest in the development of functional foods and nutraceuticals also goes hand in hand with the increased life expectancy average at the global level (around 70–80 years), and with the necessity in maintaining a good overall health status in the years. Therefore, both are considered a valid and safe help that together with a healthy lifestyle they can prevent chronic diseases, very frequent in the elderly.

The purpose of this chapter is to offer a comprehensive overview of functional foods and nutraceuticals, briefly mentioning about definitions, classifications, and their potential role in the chemoprevention of chronic diseases such as diabetes, obesity, CVD, and cancer.

1.2 FUNCTIONAL FOODS

1.2.1 CONCEPTUAL DEFINITION AND CLASSIFICATION

Since 1980, when the term "functional food" was used for the first time [1, 2], inaugurating a new sector of food sciences around the world, researchers, government agencies, and national and international organizations have tried to formulate their own definition. This circumstance certainly created little clarity and many opportunities to generate confusion due to the large number of definitions and their large variations of meaning that make it difficult to provide industry partners with solid information on market trends and potential or adequately protect consumers through legislation.

In this context, the lack of an official or commonly accepted formal meaning for functional foods has promoted the opportunity of an international debate involving many researchers. Scientists have accepted the challenge of trying to make a concise analysis of all the definitions in the literature, trying to clarify at least a conceptual level of which type of food should be considered "functional" and which scientific requirements should support them.

The monograph of International Life Sciences Institutes [7], and the work done by Doyon and Labrecqueri [8], respectively, were really important in the formulation of the conceptual meaning of functional foods, and have allowed to summarize the most important phases of their development. Table 1.1 offers only a quick overview of the most important definitions found in literature for "functional food" formulated over the past 30 years from different countries and health institutions. Therefore, it is not surprising that there is ambiguity among the highest government offices, public health professionals, and of course the population.

Furthermore, it is noteworthy to highlight that currently, Japan is the unique country that has defined a precise regulatory recommendation for functional foods practices well known with the acronym FOSHU (foods for specified health use) [9], while in the rest of the world the boundary between conventional and functional foods remains undeciphered and make trouble also experts such as nutritionists and health experts. It is obvious to think that the lack of specific legislative regulation was, on the one hand, a limitation for the definition of coherent guidelines to be followed in their development and, on the other hand, did not allow the release of health claims that were regulated.

TABLE 1.1 Various Definitions of Functional Foods

Year	Definition of Functional Food	Source/Reference(s)
1991	'Foods which are, based on the knowledge between foods or food components and health, expected to have certain health benefits, and have been licensed to bear a label claiming that a person using them for specified health use may expect to obtain the health use through the consumption thereof.'	FOSHU, Japanese Ministry of Health and Welfare
1994	'Foods that encompass potentially healthful products, including any modified food or food ingredient that may provide a health benefit beyond that of the traditional nutrients it contains.'	National Academy of Sciences Food and Nutrition Board
1999	'Food which could be regarded as 'functional' as being one that has been satisfactorily demonstrated to beneficially affect one or more functions in the body, beyond adequate nutritional effects, in a way which is relevant to either an improved state of health and well-being and/or a reduction of risk. It is consumed as part of a normal food pattern. It is not a pill, a capsule or any form of dietary supplement.'	The European Commission Concerted Action Group on Functional Food Science in Europe (FUSOSE), International Life Sciences Institute (ILSI) [10]
1999	'Modified foods or food ingredients that provide health benefits beyond their traditional nutrients.'	Adelaja and Schilling
2000	'Foods or food components that may have health benefits that reduce the risk of specific diseases or other health concerns.'	National Institute of Nutrition
2002	'A food component (being a nutrient or not) which affects one or a limited number of function(s) in the body in a targeted way so as to have positive effects that may justify health claims.'	Roberfroid
2003	'Functional foods serve naturally primarily the supply of nutrients, but additionally they offer a special advantage for the health.'	European food information council
2004	'Substances that provide essential nutrients often beyond quantities necessary for normal maintenance, growth, and development, and/or other biologically active components that impart health benefits or desirable physiological effects.'	Institute of Food Technologists (IFT)

TABLE 1.1 *(Continued)*

Year	Definition of Functional Food	Source/Reference(s)
2006	'Foods that may provide health benefit beyond basic nutrition.'	International Food Information Council (IFIC) (IFIC Foundation (2006)
2006	'A functional food is a conventional food or a food similar in appearance to a conventional food, it is part of a regular diet and has proven health-related benefits and (or) reduces the risk of specific chronic diseases above its basic nutritional functions.'	Health Canada
2014	'Natural or processed foods that contains known or unknown biologically-active compounds; which, in defined, effective non-toxic amounts, provide a clinically proven and documented health benefit for the prevention, management, or treatment of chronic disease.'	The Functional Food Center (FFC)

The analysis of these definitions is not the purpose of this chapter, but briefly, it is clear that the main characteristics that can be extrapolated in order to define a food as "functional" include the type of the food and the relationship with their potential health benefits and consumption pattern.

It is widely recognized that the general perception is that a functional food is any healthy food consumed regularly during the daily life, and is declared to possess a physiological advantage such as health-promoting or disease-preventing properties beyond the basic function of supplying nutrients. On this basis, functional foods can be classified in different classes depending on the origin, modification, and their potential biological activities.

In the following are reported practical examples of foods that should be considered "functional foods" classified based on their possible modification [11]:

- Not processed natural food (or conventional food), such as fruit, vegetable, or fish, well known for their promising content of bioactive natural products;
- A natural food which may be modified during plant breeding or other technological procedures in order to improve desired characteristics;
- A modified food by adding a bioactive component;
- A modified food by removing or reducing component.

As regards the composition of functional foods often substances were eliminated or incorporated into foods targeted for specific group of consumers.

In the development and production of this kind of food products, food technology plays a key role, considering always palatability and convenience as essential requirement for the success of the product on the reference market. Among the technologies that allow the modification of the food composition, fortification, and extraction are very important. In food technology, the term fortification is used to indicate the enhancement of a product with a specific nutrient before to be processed. Whereas, the simply use of the term "enrichment" indicates a product in which a component, not normally present in the food in the original composition, is added. Extraction and purification are techniques used in the field of food technology with two main purposes: obtaining bioactive substances from plant, food or waste materials possessing special activity related to health and well-being, in order to add them to food products; eliminating a component that interferes with the optimal nutritional value of food product. Therefore, an assortment of functional foods can be developed and classified [11] as reported in Table 1.2.

TABLE 1.2 Different Types of Functional Foods

Types of Functional Foods	Description	Practical Example
Fortified products	Product with an increased content of a component.	Vitamins in juice
Enriched products	Product with a new component in its content.	Spread with added phytosterols
Altered products	Product in which a component is removed, and replaced with a beneficial one.	Fiber in meat products
Enhanced products	Product in which the nutrient composition is altered by raw commodities.	Vitamins in fruit and vegetables

Thus, the term functional foods can include traditional foods like fruit, vegetable, fish, meat, and derivatives and modified foods (fortified, enriched, altered, and enhanced) which have proven nutritional and preventive qualities. The consumption could therefore improve well-being, prolong existence, and prevent the development of chronic diseases.

Many of conventional foods considered functional for their beneficial qualities are known since the tradition in which they were used together with spices, medicinal herbs, and roots for the preparation of recipes deemed capable of treating and preventing health-related ailments. Nowadays, thanks to the scientific progress, it is possible to demonstrate the role of food on human health and not just hypothesize it.

A growing attention has been registered in particular on plant based functional food such as fruit, vegetables, and spices, also supported by the health claims issued by the most authoritative health organizations such as World Health Organization (WHO). In fact, among the WHO guidelines, to guarantee a healthy diet it is necessary to introduce minimum five portions of fruit and vegetables daily [12]. It was also highlighted an increased demand for plant-based foods from consumers more attracted to the opportunity to have health benefits in a natural way. A wide range of examples can be cited in this regard (Table 1.3). Among foods possessing a "functional" aspect in the body, pomegranate (*Punica granatum*), a fruit widely consumed as fresh fruit and juice, is known since the traditional medicine for its therapeutic qualities in the treatment of diarrhea, diabetes, hemorrhage, and inflammation [13]. In recent years, studies both *in vitro* and *in vivo* demonstrated its antioxidant, antidiabetic, hypolipidemic, and shows antibacterial, anti-inflammatory, antiviral, and anticarcinogenic activities [14]. Many studies have suggested a beneficial potential on health in dietary grape consumption; in particular cardiovascular benefits and cancer chemopreventive are only some of the potential disease prevention activities of grape [15]. Grapes are also the raw material used to obtain red wine, whose cardioprotective potential has been widely studied in people who consume it in a habitual but moderate way to accompany main meals [5]. Special mention must be made for *Citrus* fruit, for which the annual total global production is estimated to be about 120 million tons [16]. The interest in consumption of these fruit is certainly favored by the delicious flavors, but more importantly, their nutritional value and the health promotion effects are of considerable impact. The human health-promoting by *Citrus* fruit has been the object of many scientific investigations that highlighted the efficacy against various diseases including cardiovascular, cancer, and inflammatory diseases [17]. There are several works in literature focused on characterizing the intrinsic health-promoting potential of blueberry, today recognized by media as "superfruit" for its several health benefits that include maintenance of blood sugar levels, reduction of oxidative stress, anti-inflammatory effect, prevention of CVD, antimicrobial, and antitumor activity [18]. Another example of functional food much cited in the literature is tomato (*Lycopersicon esculentum*) that represents an important and significant part of the human diet, and its regular consumption has been related with a risk reduction to various types of cancers and CVD [19]. In particular, there is scientifically supported epidemiological evidence, which suggests a reduction in the risk of susceptibility to certain types of cancer, in more exposed subjects such as smokers [20]. Furthermore, is noteworthy a recent study published on "Public Health Nutrition" associates a lower risk of cancer mortality increasing the consumption of tomato [21]. Since the 2000s was denoted a growing interest in ginger (9.247 document results using scopus.com as source), known for its beneficial properties for health already in traditional oriental medicine to treat different illnesses. Today is widely used as spice in foods and beverages and strongly recommended as a functional ingredient in our daily diet for its nutraceutical attributes that include digestive stimulant action, anti-inflammatory influence, and anticancer effect [22]. It is interesting to note that although ginger performs a digestive action capable of promoting the absorption of nutrients, in conditions of high-fat diet this spice is able to suppress the accumulation of cholesterol and fats in the body suggesting a potential role in weight management and hence in preventing risk of obesity [23] and CVD. Among the foods that have promising properties in the prevention of the risk of chronic diseases such as diabetes, obesity, and CVD we can also mention broccoli (*Brassica oleracea*) [24], recently studied by Aranaz et al. [25] that provided new knowledge about their potential role in the prevention of metabolic syndrome. Most of the foods plant-based discussed until now are low in calories, with the exception of Hass Avocado (1.7 kcal/g) very popular as the main ingredient of the avocado-based guacamole. Hass Avocado is a tropical nutrient-dense fruit that has a high oil content composed of highly digestible unsaturated fatty acids; due to these nutritional characteristics, it could be a valid substitute for not very healthy high-calorie snacks carrying out a protective and preventive action against CVD. In fact, an avocado-rich diet has been shown to reduce blood cholesterol, preserving the level of high-density lipoproteins, significantly reducing low density lipoproteins [26] and diminishing the risk of metabolic syndrome [27]. Taking into account beverages, tea infusion is one of the most popular widely consumed worldwide, known also to be linked to the promising activity of prevention of many types of cancer and to reduce the risk of chronic diseases [28]; for all the health-promoting qualities, reviewed in the last 30 years, tea is recognized as functional food by many authors in literature. Moreover, a recent epidemiologic study on the Japanese population showed that a higher consumption of green tea is associated with lower risk of mortality for heart and cerebrovascular diseases, and a moderate consumption decreased the risk of total cancer and respiratory disease mortality [29]. Another spice for which there has been greater interest in recent years is cinnamon; its introduction into weight-reducing diets has proved positive, improving the weight reduction effect, moreover, many studies have supported its activity as an antiobesity [30]. These are just a few examples of foods that could have a functional role if inserted in a balanced diet.

TABLE 1.3 List of Conventional Foods and Their Beneficial Qualities

Food Plant-Based	Beneficial Qualities	References
Pomegranate (*Punica granatum L.)*	Antioxidant, antidiabetic, hypolipidemic, antibacterial, anti-inflammatory, antiviral, anticarcinogenic	[14]
Grapes (*Vitis vinifera*)	Cardioprotective, decreased platelet aggregation, antihypertensive, anticancer, antioxidant	[15]
Citrus fruits (orange, tangerine, lime, lemon, grapefruit)	Cardioprotective, anti-inflammatory, anticancer	[17]
Blueberry	Anti-hyperglycemic, antioxidant, anti-inflammatory, cardioprotective, antimicrobic, anti-mutagenic, antitumoral	[18]
Tomato (*Lycopersicon esculentum*)	Antioxidant, cardioprotective, anticancer	[19]
Ginger (*Zingiber officinale*)	Antibacterial, antiviral, analgesic, antipyretic, carminative, anti-inflammatory, immunomodulator, antitumorigenic, antihyperglycemic, anti-lipidemic, antidiabetic	[22]
Broccoli (*Brassica oleracea*)	Antioxidant, cardioprotective, anticancer	[24]
Hass avocado (*Persea Americana*)	Antioxidant, cardioprotective, LDL-oxidation, immune system, diabetes, cancer	[26]
Green tea (*Camellia sinensis*)	Antioxidant, cardioprotective, anticancer, anti-inflammatory, antiarthritic, antibacterial, antiangiogenic, antioxidative, antiviral, neuroprotective	[28]
Cinnamon (*Cinnamomum zeylanicum*)	Stringent, carminative, antiseptic, antifungal, antiviral, digestive, antihyperlipidemic, antihyperglycemic, antiobesity	[30]

In this chapter, we wanted to highlight the close correlation between healthy food and the prevention of chronic degenerative diseases. In this context, however, it should not be forgotten that the opposite is also true, namely that bad eating habits and unregulated lifestyles are among the main causes of twentieth-century diseases. Especially in economically developed western countries, the diseases with a higher incidence are obesity, metabolic syndrome, and diabetes, in which weight management represents a key strategy in prevention. In particular, obesity is today considered one of the main public health problems worldwide [31] due to its high incidence and represents an important risk factor for diseases such as type 2 diabetes, CVD, and tumors [32]. Therefore, functional foods that have a preventive action in this sense, such as the above-cited foods, are considered desirable in the diet also in view of the pandemic impact of chronic diseases that are becoming the leading causes of global morbidity and mortality. Finally, the most pragmatic and widely scientifically-supported recommendation for populations in general is a balanced diet, with an emphasis on fruit and vegetables and increased physical activity.

1.3 NUTRACEUTICALS

1.3.1 DEFINITION AND REGULATORY ASPECTS

Today it is known that functional foods owe their beneficial qualities to the biologically active chemicals that make them up; in particular, edible plants and fruit together with their agro-industrial waste are considered promising sources of potential chemopreventive agents for degenerative diseases. Among the food sciences, food chemistry, thanks also to the development of analytical techniques, has invested many efforts on the evaluation and characterization of the molecular composition of foods. Despite, nowadays, the chemical composition of most of these foods is known, the study of potential biological activities has become the object of interest. In this regard, a lot of efforts has been made to isolate and characterize the chemical compounds believed to be beneficial to health in order to perform *in vitro* and *in vivo* studies aiming to demonstrate their potential role on human health.

Natural products, both isolated compounds from foods and comestible plants, able to achieve a "functional" role in the body are called "nutraceuticals." Currently, there is a growing global interest in nutraceuticals due to the recognition that they may play a major role in health enhancement and they are considered to be a promising source of potential chemotherapeutic agents. Many studies support the hypothesis that they are capable to counteract pathologies such as inflammation, obesity, diabetes, carcinogenesis, or neurodegenerative disorders; moreover, many of those perform a very effective antioxidant activity (radical scavengers). For this reason, these natural compounds have also been exploited in drug discovery, developing new synthetic analogs with the purpose to improve their biological activity, bioavailability, the route of administration, etc. A variety of examples could be cited in this regard. In 1989 Stephen De Felice, Chairman of the Foundation for Innovation in Medicine (FIM), coined the term nutraceutical combining the terms "Nutrition" and "Pharmaceutical," to highlight the strict connection between some groups of food molecular components and their capability to act with effects attributable to drugs. Therefore, it can be considered nutraceutical any substances, part of a food, able to contribute to well-being as well as prevention and treatment of diseases. In summary, the concept of nutraceutics integrates, in its definition, the union between food (or rather nutrition), understood as a generic intake of substances that allow the body to function, and the pharmaceutical that has to do with substances or components synthesized or isolated for therapeutic purposes, going for the first time to sanction a fundamental concept that sees food as a container of potentially biologically active substances. According to this principle, food can be medicine but at the same time poison depending on how they are composed.

As a consequence of misinformation and maybe the lack of specific regulations, there is a lot of confusion regarding the boundary between nutraceuticals and pharmaceuticals. From a practical point of view, recognized by many scientists, the difference between these two product groups is patent coverage that supports the pharmaceutical function of drugs [33]. Indeed, both pharmaceuticals and nutraceuticals have the ability to cure or prevent diseases; only pharmaceuticals have the authoritative approval. The use of nutraceuticals with a potential therapeutic effect has also met the interest of pharmaceutical companies that are more incentivized in the development of this type of product, which requires a lower basic investment if compared to pharmaceutical ones. This data is also confirmed by the size of the global nutraceutical market, which has been estimated to be $230.9 billion in 2018 and should reach $336.1 billion by 2023 with a compound annual growth rate (CAGR) of 7.8% [34]. According to the Food and Drug Administration (FDA) regulations, nutraceuticals in the Unites States would be recognized as "dietary supplements" comprising vitamins, minerals, herbs, and extracts which are able to provide nutrients [35]. Instead, the safety assessment and regulation of nutraceuticals in the European Community are special product category regulated by the European Food and Safety Authority (EFSA) called "food supplements" concentrated sources of nutrients or other substances with beneficial nutritional effects (Directive 2002/46/EC) [36]. Thanks to constant progress in scientific research, today we are able to accurately assess the influence of different nutraceuticals on the normal physiological functions of the body. However, it was highlighted that most part of nutraceuticals can have a multiple therapeutic effect, and more in general, they can have a role in the protection against obesity, diabetes, metabolic syndrome, etc. Therefore, it is not so surprising that since prehistoric times, humans have been able to draw most part of their medicines from foods and plants in order to treat multiple pathologies. At this regard, the Ebers Papyrus (1550 BC) is rich of examples. Nutraceuticals can be classified differently, according to the needs of the discussion; thus, we could classify them on the basis of the potential biological activity and the potential sector of use, but this would make the discussion more complicated since, as already said, most of these substances can have a multiple biological effect. For that reason, the classification that from our point of view is more exemplary is that which takes into account the biogenetic origin and therefore the chemical class to which they belong.

1.3.2 NUTRACEUTICALS AND CHEMICAL CLASSIFICATION

All the living organisms such as plants, animals, and microorganisms require a wide number of organic compounds to live, grow, and reproduce. The metabolic pathways are responsible, through a complex system of enzyme-mediated chemical reactions, to provide the essentially molecules carbohydrates, proteins, fats, and nucleic acids. The synthesis, transformation, and degradation of these compounds is called primary metabolism and the molecules originated are primary metabolites. Moreover, in organisms occur the secondary metabolism contributing to the production of the most pharmacologically active natural compounds described as secondary metabolites. These bioactive compounds are distributed and sometimes confined in specific organisms as a consequence of the environmental context, defense, or resistance against insects and are classified according to their biosynthetic pathways. The classification of these compounds comprises phenolic compounds, terpenoids, alkaloids, and fatty acids [37].

1.3.2.1 POLYPHENOLS

Polyphenols, belonging to an important family of natural products, are compounds that should be found in fruits, vegetables, herbs; moreover, in foods and beverages derived from them, have been the subject of studies indicating their role in the chemoprevention of degenerative diseases, such as

CVD, cancer or Alzheimer's disease. Some of these polyphenols are considered nutraceuticals because they are constituents of foods and drinks acting as and beverages able to play a 'functional' role in the body. Several studies support the hypothesis that polyphenols, derived from natural sources, are potent antioxidants (radical scavengers) and are able to counteract pathologies such as inflammation, diabetes, carcinogenesis, or neurodegenerative disorders. Phenolic compounds are widespread mainly in the Plant Kingdom and include more than 8000 known compounds. Their role in the plant is presumably defensive, but they may also have other biological activities in interspecies relationships. This group of compounds is one of the most studied worldwide, and many publications report beneficial effects of polyphenols on various aspects of human health and well-being [38].

The growing interest in (poly)phenolic compounds and their exploitation in the fields of agro-food, cosmetic, and drug industry has led to a broader (and sometimes inappropriate) use of the term 'polyphenols' with respect to the original definition of 'plant polyphenols,' later expanded by E. Haslam [39], and recently by S. Quideau [40]. Originally, the 'plant polyphenols' were substantially equivalent to 'vegetable tannins,' with reference to the tanning action of some plant extracts that had been employed for centuries in the leather-making process. However, this definition has subsequently been broadened in the common use to include low-molecular-weight phenolic molecules as well, not necessarily water-soluble or exerting a 'tanning' action. Consequently, the common feature of polyphenols has been reconfigured with regard to their biosynthetic origin, thus including phenolic metabolites biosynthetically derived through the shikimate and/or the acetate/malonate pathways. Scheme 1.1 briefly reviews the biosynthesis of phenolic compounds, mainly through the shikimate pathway (Scheme 1.1) [41, 42]. Some examples of bioactive polyphenols are reported below.

Resveratrol is the widely recognized polyphenol, should be found in grapes and red wine, considered cardioprotective and anticarcinogenic, which has become very popular due to the so-called French paradox, already discussed above. J. Pezzuto, in a recent review, cites about 512 references on its ability to prevent cancer [15]. A further well-known phenolic compound is Genistein, an isoflavone present in soybean (Glycine soja), with estrogen-like activity able to relieve menopause symptoms and prevent some estrogen-dependent cancers, such as breast cancer [43]. Tannins are another class of natural polyphenols known for their several biological activities related to their antioxidant [44], antiviral [45], host-mediated antitumor activities [46, 47], moreover, recently they are reported for their promising antidiabetic properties [48]. Many polyphenols are esters or amides of phenolic acids such as CAPE (caffeic acid phenethyl ester), found in substance produced by bees, known as propolis; this compound is capable of acting as a potent antioxidant, reported also for its promising antitumor properties [49].

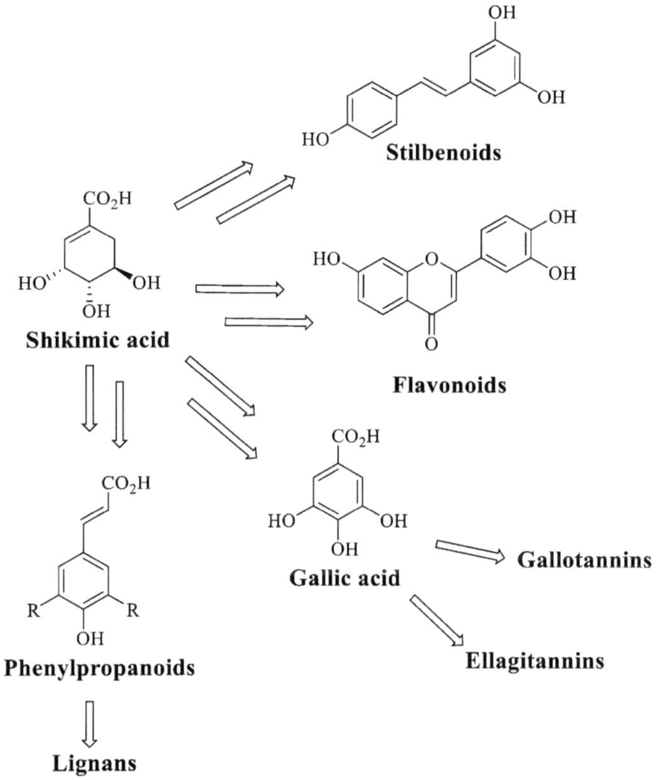

Scheme 1.1 Biosynthesis of phenolic compounds; shikimate pathway.

1.3.2.2 TERPENOIDS

Terpenoids represent a wide and diversified class of secondary metabolites derived from C5 isoprene units joined together. Classification of terpenoids is based on the number of carbon skeleton $(C5)_n$ linked through a linear head-to-tail organization leading to monoterpenes (C10), sesquiterpenes (C15), diterpenes (C20), sesterterpenes (C25), triterpenes (C30) and tetraterpenes (C40). Two different metabolic pathways should be involved in the synthesis of the isoprenoids units-dimethylallyl diphosphate (DMAPP) and isopentenyl diphosphate (IPP)-the mevalonic acid (MVA) pathway and the 2-*C*-methyl-D-erythritol 4-phosphate (methylerythritol phosphate: MEP) pathway (Figure 1.2).

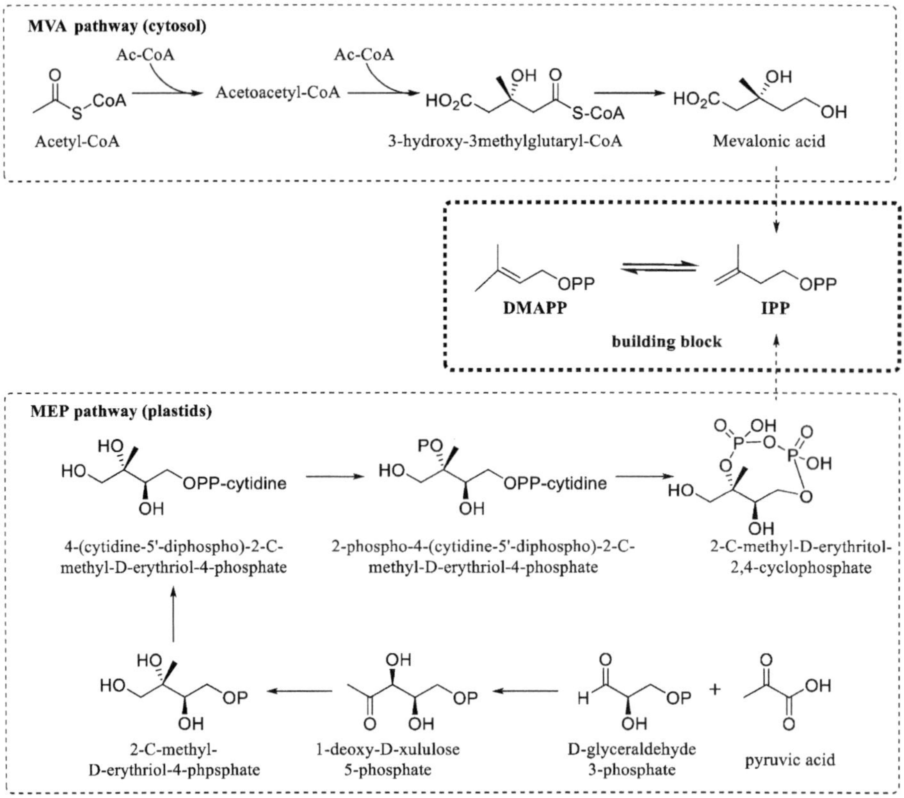

FIGURE 1.2 Terpenoids biosynthesis, mevalonic acid pathway, and methylerythritol phosphate pathway.

MAV and MEP pathways are responsible for providing the isoprene units for the biosynthesis of particular classes of terpenes; in particular animals and fungi utilize the mevalonate pathway exclusively, instead MEP is presents in plants, algae, and bacteria. In plants, both pathways are present and compartmentalized, MAV in the cytosol and MEP in the plastids. Thus, triterpenes, and sesquiterpenes are formed by the mevalonate pathway, mono-di-, and tetraterpenes are MEP derived [50].

Among the natural products scaffolds, terpenoids play a crucial role in a wide variety of therapeutic indications. Essential oils (EOs) are secondary metabolites produced by plants, consist of a mixture of volatiles terpenoids, phenylpropanoids, and short-chain aliphatic hydrocarbon derivatives containing a major constituent up to 85%, while other constituents are present in traces. They are usually extracted by steam distillation from natural sources (flowers, seeds, leaves, bark, herbs, wood, fruits, and roots) and have been used since ancient times because of their perfumes, flavors, and preservatives features [51]. EOs are characterized by high chemical diversity and biochemical specificity being responsible for their biological activities. EOs extracted from *Origanum vulgare*, *Thymus vulgaris* and *Rosmarinus officinalis* have been shown to possess antibacterial activity against *Staphylococcus aureus* and *Listeria monocytogenes* [52]. EOs of the fresh leaves, unripe, and ripe fruit peels of *Citrus reshni* have been displayed potential antiviral activity against avian influenza virus A (H5N1 subtype) [53]. Moreover, EOs extracted from plants exhibited antifungal [54], insecticidal [55, 56] and anticancer activities [57]. In drug discovery programs, diterpenes represent another important class of terpenoids. Macrocyclic diterpenes presenting lathyrane and jatropha scaffolds, extracted from *Euphorbia* species, displayed a potential anticancer activity as MDR inhibitors in multidrug resistance, acting through P-gp modulation [58–61]. Moreover, *Rosmarinus officinalis* extracts exhibited high antioxidant properties due to the presence of phenolic diterpenes [62].

1.3.2.3 ALKALOIDS

Alkaloids are compounds characterized for their basicity and the presence of nitrogen atoms in the molecule. Morphine were the first alkaloid discovered obtained from plants, consequently the early definition of alkaloids included these three characteristics nitrogen-containing, basicity, and plant origin. Successively, the theory of being derived from amino acids was added, together with the idea that the nitrogen should be in a heterocyclic ring [63].

Alkaloids are frequently classified based on the structure, such as the presence of the nitrogen in the ring. Some different amino acid precursors are involved in alkaloid biosynthesis (Figure 1.3).

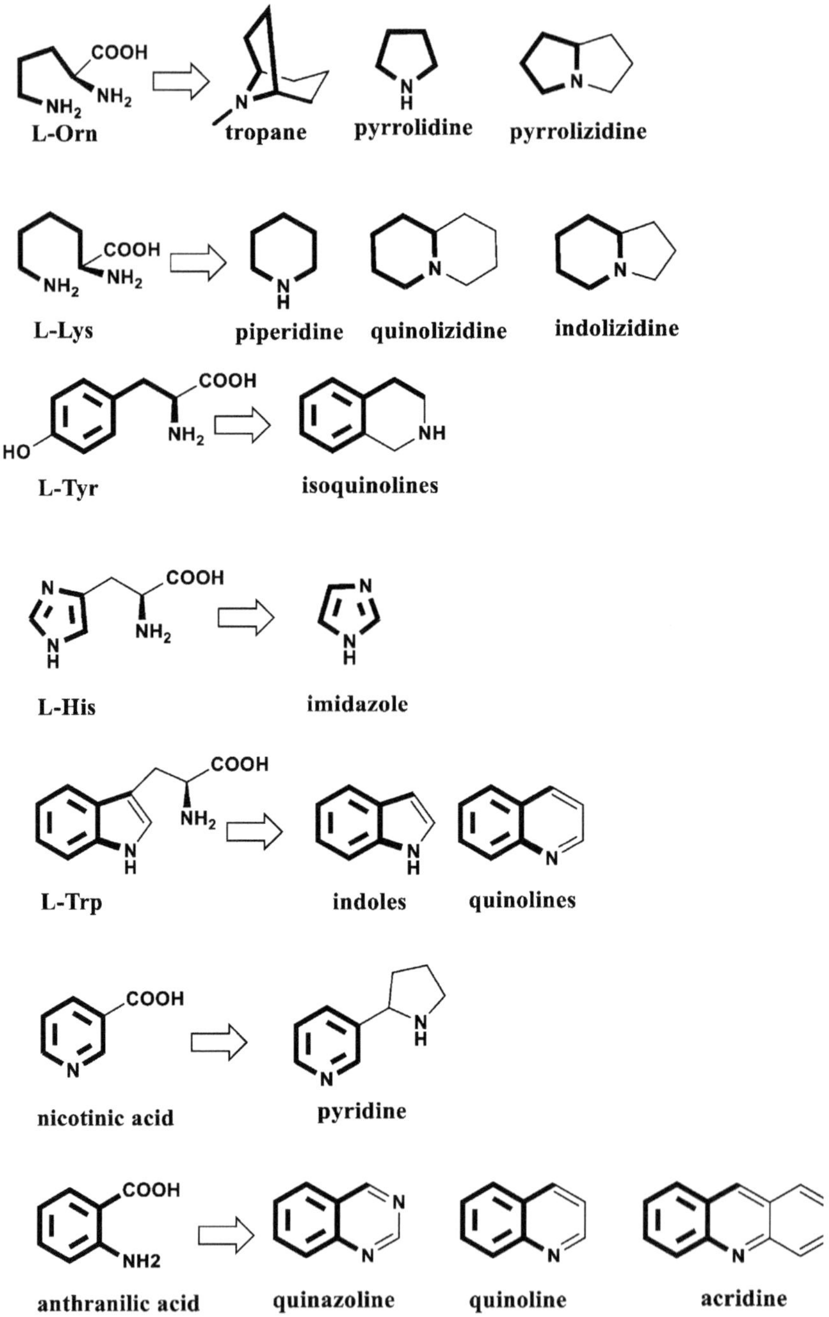

FIGURE 1.3 Alkaloids biosynthesis and basic structures.

Nevertheless, a large group of alkaloids is found to acquire their nitrogen atoms via transamination reactions, incorporating only the nitrogen from an amino acid, whilst the rest of the molecule may be derived from acetate or shikimate; others may be terpenoid or steroid in origin [64].

Indole alkaloids are plant-derived compounds, comprising over 3000 members, characterized by a wide range of biological activities (Figure 1.4), including cytotoxic and anti-inflammatory [63, 65, 66]. Vincristine and vinblastine, clinically important anticancer agents, are examples of useful bioactive indole alkaloids from Apocynaceae [67]. Plant species belonging to the Apocynaceae family are important source of these secondary metabolites. Rubiaceae, Loganiaceae, and Nyssaceae families are also known for synthesize bioactive indole alkaloids (Figure 1.4) [68–71].

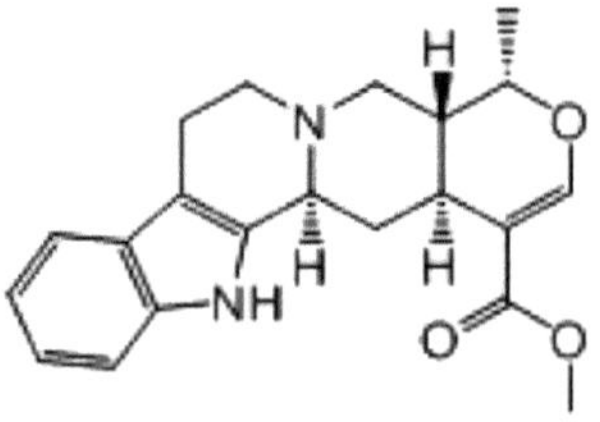

Ajmalicine
(Antihypertensive)

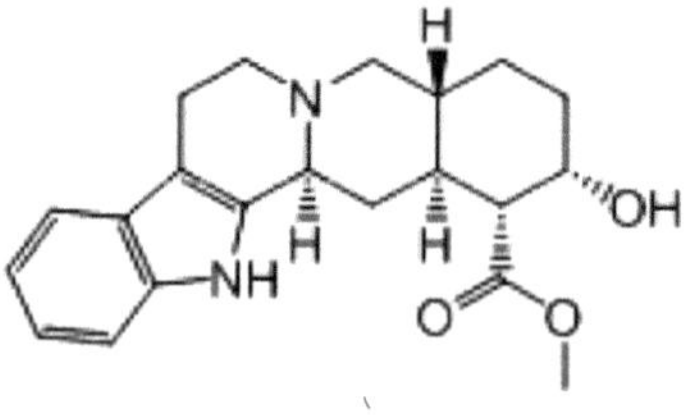

Yohimbine
(Adrenergic receptor blocker)

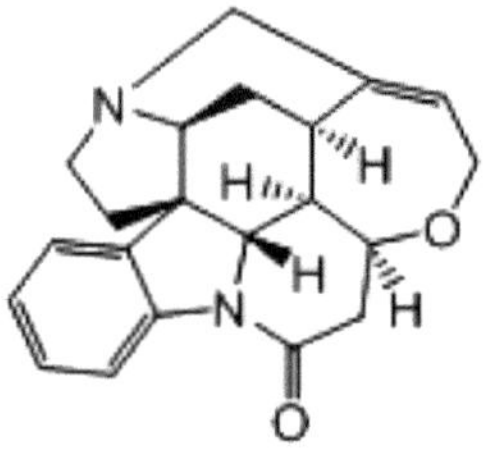

Strychnine
(Convulsant)

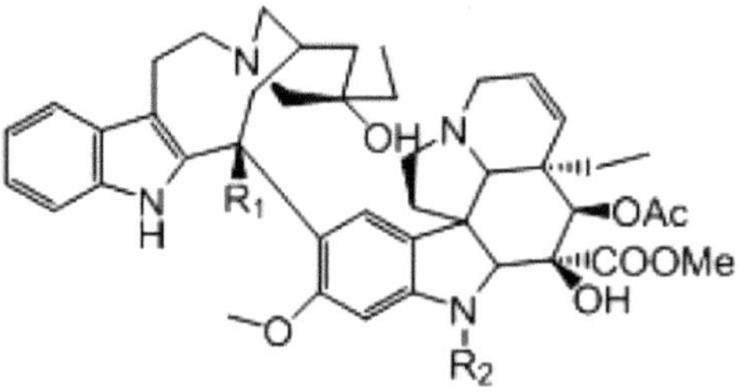

Vincristine R_1= COOMe; R_2= CHO
Vinblastine R_1= COOMe; R_2= H
(Anti-cancer)

FIGURE 1.4 Representative terpene indole alkaloids, with the corresponding biological function.

Moreover, *Capsicum* genus fruits (chili peppers) are food ingredients and additives used widespread. The principal secondary metabolite present is capsaicin, a phenylalanine derived alkaloid well-known for the mucosal irritant peculiarity and beyond food flavoring possess multiple health benefits like obesity, cardiovascular, and gastrointestinal (GI) disorders cancer [72] and *in vivo* antioxidant activity [73].

Glycoalkaloids (GAs), are secondary metabolites synthesized by plant belonging to the *Solanaceae* family (i.e., tomato, potato, and eggplant). The two major GAs present in potato (*Solanum tuberosum*) are α-solanine and α-chaconine, tomato plants (*S. lycopersicum*) present α-tomatine and dehydrotomatine and in eggplant fruits (*S. melongena*) are found solanine and solamargine. These bioactive compounds, besides have antifungal, antimicrobial, and insecticidal properties as a protective activity against several insects, pests, and herbivores; several studies reported their potential anticancer activity [74].

1.3.2.4 FATTY ACIDS

The acetate pathway is involved in the biosynthesis of fatty acids, another crucial class of nutraceuticals. Fatty acids are classified in saturated and unsaturated; their formation is catalyzed by the enzyme fatty acid synthase (FAS) and natural saturated fatty acids may contain from 4 (butyric acid) to 30 (melissic acid), or even more, carbon atoms. The unsaturated fatty acids, usually containing one or more double bonds in a non-conjugated pattern, occurs in animals and plants. Omega 3 and omega 6 are polyunsaturated fatty acids (PUFAs), also called essential fatty acids because of our organism is not able to synthesize them *ex Novo* and have beneficial effects in human health. Is important to assume them through the diet. PUFAS are well known to take part in cellular physiology, are involved in energy storage and are structural components of cell membrane conferring fluidity, thickness, stability, and permeability [75]. Moreover, PUFAs like arachidonic acid (ARA, C20:4n–6), eicosapentaenoic acid (EPA, C-20:5n–3) and docosahexaenoic acid (DHA, C-22:6n–3) are precursors of specific lipid mediators with a potent pro- and anti-inflammatory activity. The inflammation resolution pathways need numerous biochemical signals fundamental to achieve the inflammatory response and the lipid mediators' synthesis, including prostaglandins, leukotrienes, resolvins, lipoxins, maresins, and protectins. All these compounds are mono-, di-, and tri-hydroxylated and epoxidized derivatives of PUFAs. PUFAs and the corresponding derived lipid mediators possess a strategic function as potential pharmaceutical and nutraceutical targets in the prevention and treatment of several chronical immune diseases [76].

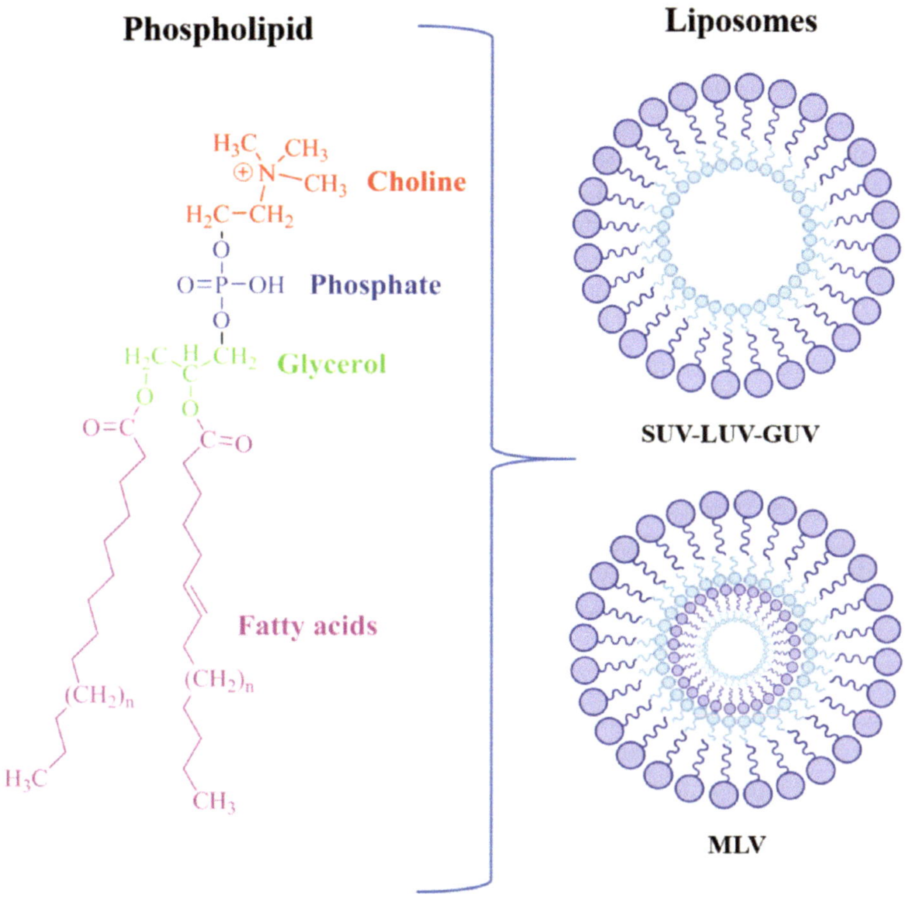

FIGURE 1.5 Illustrative description of phospholipid and liposomes.

1.3.3 NUTRACEUTICALS AND NANOTECHNOLOGY

As mentioned above, nutraceuticals possess a wide array of chemical and physical properties, features which often do not allow to use these products directly in their pure form. Nowadays, nanotechnology techniques are tools used by food and nutraceuticals industries to develop new products with improved characteristics. Encapsulation methods are useful to separate and entrap the biologically active compound from the outer part conferring many benefits. These benefits include mix incompatible ingredients, reduce toxicity, enhance solubility and stability, improve effectiveness, prevent chemical degradation in food matrix, increase bioavailability and lead to targeted delivery; moreover, in food can change color, smell, and flavor [77]. Among all the encapsulation techniques, the one which fascinate researchers and still capture attention in the literature is the use of liposomes. Liposomes are spherical-shaped structures consisting of a central aqueous compartment enclosed by one or more concentric phospholipid layers. The phospholipid vesicles are characterized by hydrophilic (aqueous cavity) and hydrophobic (within lipidic bilayer) elements consenting to amphiphilic bioactive substances to be incorporated within these structures. Liposomes were usually classified on the basis of their size and number of bilayers, the unilamellar vesicles are identified in small, large, and giant (SUV: 20–100 nm; LUV: > 100 nm; GUV: > 1000 nm) and multilamellar vesicles (MLV): > 500 nm (Figure 1.5) [78].

Food industries exploit nutraceuticals entrapped in liposomes to produce functional foods, reinforcing the nutritional efficacy, enhancing the health of buyers, and reducing the risk of some diseases. Encapsulation of phenolic compounds from Pistachio green hulls of *Ahmad aghaei* variety, possessing antioxidant and antimicrobial properties, could enhance the bioavailability of the extract [79]. Vitamins are another vast class of nutrients involved in several biochemical functions such as preventing cancer and cardiovascular disorders and improving the immune system. These bioactive compounds are produced in few amounts by our body leading to a necessary implementation of vitamins through food supplements. Vitamins are classified in two groups water-soluble and fat-soluble, encapsulation of vitamins in liposomes is a suitable technique to increase their stability and solubility [80]. PUFAs are another class of nutraceuticals widely used and are susceptible to autoxidation reactions which is considered the key disadvantage to overcome. Encapsulation of PUFAs get better foods in terms of sensory parameters, like smell and flavor, and in terms of stability [81]. Furthermore, betalains are natural pigments extensively used as colorants for food products, and despite this use, some studies disclosed about the antioxidant, anti-inflammatory, anticancer, and antidiabetic properties. The weakness of betalains is the limited oral bioavailability, encapsulation in liposomes increased its stability and enhanced the antioxidant activity [82]. Turmeric plant (*Curcuma longa*) is the source of curcumin, a yellow phenolic compound, known worldwide for its antioxidant and anti-inflammatory properties and used in Asian traditional medicine for the multiple health benefits. Although curcumin exhibit numerous therapeutic effects possess some disadvantages limiting the real effectiveness, has not a good solubility in water, low bioavailability and is quickly metabolized and eliminated. Several studies proved that encapsulation of curcumin in liposomes increase anticancer activity by improving pharmacokinetics and pharmacodynamics and reducing the dose required [83].

1.4 CONCLUSION

Although the use of plants in traditional medicine to prevent and treat several diseases have been known and employed by indigenous people since ancient times; just a while ago, the effectiveness of nutraceuticals has been supported by scientific reports. The enlarged scientific interest in nutraceuticals and functional foods led food industries in a growing attention to develop health-promoting ingredients. Despite regulation of nutraceuticals and functional foods is still ambiguous and varies from country to country, an expanded global market of nutraceuticals and functional foods has been documented together with a recent customer conscience and careful to choose healthy food and a constant search for high dietary intake of nutraceuticals. An important challenge and still a topic of discussion concerns improving the characteristics of this bioactive compound and promote their biological efficacies. Nanotechnology techniques are employed to enhance stability, solubility, bioavailability, targeted delivery, smell, color, and flavor.

KEYWORDS

- **cardiovascular diseases**
- **compound annual growth rate**
- **European Food and Safety Authority**
- **isopentenyl diphosphate**
- **mevalonic acid**
- **World Health Organization**

REFERENCES

1. Hardy, G., (2000). Nutraceuticals and functional foods: Introduction and meaning. *Nutrition, 16*(7, 8), 688–689.

2. Kwak, N. S., & Jukes, D. J., (2001). Functional foods. Part 1: The development of a regulatory concept. *Food Control, 12*(2), 99–107.
3. Martirosyan, D. M., & Singh, J., (2015). A new definition of functional food by FFC: What makes a new definition unique? *Funct. Foods Heal. Dis., 5*(6), 209–223.
4. Granato, D., Nunes, D. S., & Barba, F. J., (2017). An integrated strategy between food chemistry, biology, nutrition, pharmacology, and statistics in the development of functional foods: A proposal. *Trends Food Sci. Technol., 62*, 13–22.
5. Ndlovu, T., Van, J. F., & Caleb, O. J., (2019). French and Mediterranean-style diets: Contradictions, misconceptions, and scientific facts: A review. *Food Res. Int., 116*, 840–858.
6. Jorge, I. P. F., (2019). *Nutraceuticals Market Assessment.* https://pt.scribd.com/document/437146584/2019-Nutraceuticals-Assessment-Executive-Summary-pdf (accessed on 19 June 2021).
7. Ashwell, M., (2003). ILSI Europe concise monograph on concepts of functional foods. *Int. Life Sci.* Institute, Washington, DC.
8. Doyon, M., & Labrecque, J., (2008). Functional foods: A conceptual definition. *Br. Food J.*
9. Shimizu, T., (2003). Health claims on functional foods: The Japanese regulations and an international comparison. *Nutr. Res. Rev., 16*(2), 241–252.
10. Action, E. C., (1999). Scientific concepts of functional foods in Europe: Consensus document. *Br. J. Nutr., 81*(1), 1–27.
11. Spence, J. T., (2006). Challenges related to the composition of functional foods. *J. Food Compos. Anal., 19*, S4–S6.
12. World Health Organization, (2020). *Healthy Diet.* https://www.who.int/news-room/fact-sheets/detail/healthy-diet (accessed on 19 June 2021).
13. Wu, S., & Tian, L., (2017). Diverse phytochemicals and bioactivities in the ancient fruit and modern functional food pomegranate (*Punica granatum*). *Molecules, 22*(10), 1606.
14. Viuda-Martos, M., Fernández-López, J., & Pérez-Álvarez, J. A., (2010). Pomegranate and its many functional components as related to human health: A review. *Compr. Rev. Food Sci. Food Saf., 9*(6), 635–654.
15. Pezzuto, J. M., (2008). Grapes and human health: A perspective. *J. Agric. Food Chem., 56*(16), 6777–6784.
16. Putnik, P., Barba, F. J., Lorenzo, J. M., Gabrić, D., Shpigelman, A., Cravotto, G., & Bursać, K. D., (2017). An integrated approach to mandarin processing: Food safety and nutritional quality, consumer preference, and nutrient bio-accessibility. *Compr. Rev. food Sci. Food Saf., 16*(6), 1345–1358.
17. Surampudi, P., Enkhmaa, B., Anuurad, E., & Berglund, L., (2016). Lipid-lowering with soluble dietary fiber. *Curr. Atheroscler. Rep., 18*(12), 75.
18. Silva, S., Costa, E. M., Veiga, M., Morais, R. M., Calhau, C., & Pintado, M., (2020). Health-promoting properties of blueberries: A review. *Crit. Rev. Food Sci. Nutr., 60*(2), 181–200.
19. Borguini, R. G., & Ferraz, D. S. T. E., (2009). Tomatoes and tomato products as dietary sources of antioxidants. *Food Rev. Int., 25*(4), 313–325.
20. Weisburger, J. H., (1999). Mechanisms of action of antioxidants as exemplified in vegetables, tomatoes, and tea. *Food Chem. Toxicol., 37*(9, 10), 943–948.

21. Mazidi, M., Ferns, G. A., & Banach, M., (2020). A high consumption of tomato and lycopene is associated with a lower risk of cancer mortality: Results from a multi-ethnic cohort. *Public Health Nutr., 23*(9), 1569–1575.
22. Srinivasan, K., (2017). Ginger rhizomes (*Zingiber officinale*): A spice with multiple health beneficial potentials. *Pharma Nutrition, 5*(1), 18–28.
23. Wang, J., Li, D., Wang, P., Hu, X., & Chen, F., (2019). Ginger prevents obesity through regulation of energy metabolism and activation of browning in high-fat diet-induced obese mice. *J. Nutr. Biochem., 70*, 105–115.
24. Raiola, A., Errico, A., Petruk, G., Monti, D. M., Barone, A., & Rigano, M. M., (2018). Bioactive compounds in *Brassicaceae* vegetables with a role in the prevention of chronic diseases. *Molecules, 23*(1), 15.
25. Aranaz, P., Navarro-Herrera, D., Romo-Hualde, A., Zabala, M., López-Yoldi, M., González-Ferrero, C., Gil, A. G., et al., (2019). Broccoli extract improves high fat diet-induced obesity, hepatic steatosis, and glucose intolerance in Wistar rats. *J. Funct. Foods, 59*, 319–328.
26. Yahia, E. M., Ornelas-Paz, J. D. J., & Gonzalez-Aguilar, G. A., (2011). Nutritional and health-promoting properties of tropical and subtropical fruits. In: *Postharvest Biology and Technology of Tropical and Subtropical Fruits* (pp. 21–78). Elsevier.
27. Fulgoni, V. L., Dreher, M., & Davenport, A. J., (2013). Avocado consumption is associated with better diet quality and nutrient intake, and lower metabolic syndrome risk in US Adults: Results from the national health and nutrition examination survey (NHANES) 2001–2008. *Nutr. J., 12*(1), 1.
28. Chacko, S. M., Thambi, P. T., Kuttan, R., & Nishigaki, I., (2010). Beneficial effects of green tea: A literature review. *Chin. Med., 5*(1), 13.
29. Abe, S. K., Saito, E., Sawada, N., Tsugane, S., Ito, H., Lin, Y., Tamakoshi, A., et al., (2019). *Green Tea Consumption and Mortality in Japanese Men and Women: A Pooled Analysis of Eight Population-Based Cohort Studies in Japan*. Springer.
30. Yazdanpanah, Z., Azadi-Yazdi, M., Hooshmandi, H., Ramezani-Jolfaie, N., & Salehi-Abargouei, A., (2020). Effects of cinnamon supplementation on body weight and composition in adults: A systematic review and meta-analysis of controlled clinical trials. *Phytother. Res., 34*(3), 448–463.
31. Mechanick, J. I., Hurley, D. L., & Garvey, W. T., (2016). Adiposity-based chronic disease as a new diagnostic term: The American association of clinical endocrinologists and American college of endocrinology position statement. *Endocr. Pract.*
32. Field, A. E., Coakley, E. H., Must, A., Spadano, J. L., Laird, N., Dietz, W. H., Rimm, E., & Colditz, G. A., (2001). Impact of overweight on the risk of developing common chronic diseases during a 10-year period. *Arch. Intern. Med., 161*(13), 1581–1586.
33. Rajasekaran, A., Sivagnanam, G., & Xavier, R., (2008). Nutraceuticals as therapeutic agents: A review. *Res. J. Pharm. Technol., 1*(4), 328–340.
34. Natraj, P., (2017). *Nutraceuticals: Global Markets to 2023*. https://www.bccresearch.com/market-research/food-and-beverage/nutraceuticals-global-markets.html (accessed on 19 June 2021).
35. Lewis, C. A., Jackson, M. C., & Bailey, J. R., (2019). *Chapter 15: Understanding Medical Foods Under FDA Regulations*. Nutraceutical and Functional Food Regulations in the United States and Around the World. Academic Press.

36. Directive 2002/46/EC of the European Parliament and of the Council of 10 June 2002 on the approximation of the laws of the member states relating to food supplements, (2002). *Off. J. Eur. Communities,* 183/57-57.
37. Dewick, P. M., (2009). Secondary metabolism: The building blocks and construction mechanisms. In: *Medicinal Natural Products: A Biosynthetic Approach* (3rd edn., pp. 7–38). John Wiley & Sons, L., Ed.
38. Umar, L. S., & Xia, W., (2005). Food phenolics, pros and cons: A review. *Food Rev. Int., 21*(4), 367–388.
39. Haslam, E., & Cai, Y., (1994). Plant polyphenols (vegetable tannins): Gallic acid metabolism. *Nat. Prod. Rep., 11*, 41–66.
40. Quideau, S., Deffieux, D., Douat-Casassus, C., & Pouységu, L., (2011). Plant polyphenols: Chemical properties, biological activities, and synthesis. *Angew. Chemie Int. Ed., 50*(3), 586–621.
41. Crozier, A., Clifford, M. N., & Ashihara, H., (2008). *Plant Secondary Metabolites: Occurrence, Structure and Role in the Human Diet.* John Wiley & Sons.
42. Fraga, C. G., (2009). *Plant Phenolics and Human Health: Biochemistry, Nutrition and Pharmacology* (Vol. 1). John Wiley & Sons.
43. Van, P. C. L., Olivotto, I. A., Chambers, G. K., Gelmon, K. A., Hislop, T. G., Templeton, E., Wattie, A., & Prior, J. C., (2002). Effect of soy phytoestrogens on hot flashes in postmenopausal women with breast cancer: A randomized, controlled clinical trial. *J. Clin. Oncol., 20*(6), 1449–1455.
44. Cerdá, B., Tomás-Barberán, F. A., & Espín, J. C., (2005). Metabolism of antioxidant and chemopreventive ellagitannins from strawberries, raspberries, walnuts, and oak-aged wine in humans: Identification of biomarkers and individual variability. *J. Agric. Food Chem., 53*(2), 227–235.
45. Martinez, J. P., Sasse, F., Brönstrup, M., Diez, J., & Meyerhans, A., (2015). Antiviral drug discovery: Broad-spectrum drugs from nature. *Nat. Prod. Rep., 32*(1), 29–48.
46. Quideau, S., (2009). *Chemistry and Biology of Ellagitannins: An Underestimated Class of Bioactive Plant Polyphenols.* World Scientific.
47. Buzzini, P., Arapitsas, P., Goretti, M., Branda, E., Turchetti, B., Pinelli, P., Ieri, F., & Romani, A., (2008). Antimicrobial and antiviral activity of hydrolysable tannins. *Mini-Reviews Med. Chem., 8*(12), 1179.
48. Cardullo, N., Muccilli, V., Pulvirenti, L., Cornu, A., Pouységu, L., Deffieux, D., Quideau, S., & Tringali, C., (2020). *C-glucosidic ellagitannins* and galloylated glucoses as potential functional food ingredients with antidiabetic properties: A study of α-glucosidase and α-amylase inhibition. *Food Chem., 313*, 126099.
49. Dorai, T., & Aggarwal, B. B., (2004). Role of chemopreventive agents in cancer therapy. *Cancer Lett., 215* (2), 129–140.
50. Dewick, P. M., (2009). The mevalonate and methylerythritol phosphate pathways: Terpenoids and steroids. In: *Medicinal Natural Products* (3rd edn., pp. 187–310). John Wiley & Sons, Ltd.
51. Tariq, S., Wani, S., Rasool, W., Shafi, K., Bhat, M. A., Prabhakar, A., Shalla, A. H., & Rather, M. A., (2019). A comprehensive review of the antibacterial, antifungal, and antiviral potential of essential oils and their chemical constituents against drug-resistant microbial pathogens. *Microb. Pathog., 134*, 103580.
52. Pesavento, G., Calonico, C., Bilia, A. R., Barnabei, M., Calesini, F., Addona, R., Mencarelli, L., et al., (2015). Antibacterial activity of oregano, *Rosmarinus* and *thymus*

essential oils against Staphylococcus aureus and listeria monocytogenes in beef meatballs. *Food Control, 54*, 188–199.
53. Nagy, M. M., Al-Mahdy, D. A., Abd, E. A. O. M., Kandil, A. M., Tantawy, M. A., & El Alfy, T. S. M., (2018). Chemical composition and antiviral activity of essential oils from citrus Reshni hort. ex Tanaka (*Cleopatra mandarin*) cultivated in Egypt. *J. Essent. Oil Bear. Plants, 21*(1), 264–272.
54. Hu, F., Tu, X. F., Thakur, K., Hu, F., Li, X. L., Zhang, Y. S., Zhang, J. G., & Wei, Z. J., (2019). Comparison of antifungal activity of essential oils from different plants against three fungi. *Food Chem. Toxicol., 134*, 110821.
55. Ma, S., Jia, R., Guo, M., Qin, K., & Zhang, L., (2020). Insecticidal activity of essential oil from *Cephalotaxus Sinensis* and its main components against various agricultural pests. *Ind. Crops Prod., 150*, 112403.
56. Ikbal, C., & Pavela, R., (2019). Essential oils as active ingredients of botanical insecticides against aphids. *J. Pest Sci., 92*(3), 971–986.
57. Sugier, D., Sugier, P., Jakubowicz-Gil, J., Winiarczyk, K., & Kowalski, R., (2019). Essential oil from *Arnica montana* L. achenes: Chemical characteristics and anticancer activity. *Molecules, 24*(22), 4158.
58. Reis, M. A., Ferreira, R. J., Serly, J., Duarte, N., Madureira, A. M., Santos, J. V. A. D., Molnar, J., & Ferreira, M. J. U., (2012). Colon adenocarcinoma multidrug resistance reverted by euphorbia diterpenes: Structure-activity relationships and pharmacophore modeling. *Curr. Med. Chem. Agents, 12*(9), 1015–1024.
59. Reis, M. A., Ahmed, O. B., Spengler, G., Molnár, J., Lage, H., & Ferreira, M. J. U., (2017). Exploring Jolkinol D derivatives to overcome multidrug resistance in cancer. *J. Nat. Prod., 80*(5), 1411–1420.
60. Reis, M. A., Paterna, A., Ferreira, R. J., Lage, H., & Ferreira, M. J. U., (2014). Macrocyclic diterpenes resensitizing multidrug-resistant phenotypes. *Bioorganic Med. Chem., 22*(14).
61. Reis, M. A., Paterna, A., Mónico, A., Molnar, J., Lage, H., & Ferreira, M. J. U., (2014). Diterpenes from *Euphorbia piscatoria*: Synergistic interaction of lathyranes with doxorubicin on resistant cancer cells. *Planta Med., 80*(18).
62. Petiwala, S. M., & Johnson, J. J., (2015). Diterpenes from rosemary (*Rosmarinus officinalis*): Defining their potential for anticancer activity. *Cancer Lett., 367*(2), 93–102.
63. Cordell, G. A., Quinn-Beattie, M. L., & Farnsworth, N. R., (2001). The potential of alkaloids in drug discovery. *Phytother. Res., 15*(3), 183–205.
64. Dewick, P. M., (2009). Alkaloids. *Med. Nat. Prod. A Biosynthetic Approach* (3rd edn., pp. 311–420).
65. De Sa, A., Fernando, R., Barreiro, E. J., Fraga, M., & Alberto, C., (2009). From nature to drug discovery: The indole scaffold as a 'privileged structure.' *Mini Rev. Med. Chem., 9*(7), 782–793.
66. Ishikura, M., Abe, T., Choshi, T., & Hibino, S., (2013). simple indole alkaloids and those with a non-rearranged monoterpenoid unit. *Nat. Prod. Rep., 30*(5), 694–752.
67. Cragg, G. M., & Newman, D. J., (2005). Plants as a source of anticancer agents. *J. Ethnopharmacol., 100*(1, 2), 72–79.
68. Liang, S., He, C. Y., Szabó, L. F., Feng, Y., Lin, X., & Wang, Y., (2013). Gelsochalotine, a novel indole ring-degraded monoterpenoid indole alkaloid from *Gelsemium elegans*. *Tetrahedron Lett., 54*(8), 887–890.

69. Lopes, S., Von, P. G. L., Kerber, V. A., Farias, F. M., Konrath, E. L., Moreno, P., Sobral, M. E., et al., (2004). Taxonomic significance of alkaloids and iridoid glucosides in the tribe Psychotria (Rubiaceae). *Biochem. Syst. Ecol., 32*(12), 1187–1195.
70. Ramani, S., Patil, N., Nimbalkar, S., & Jayabaskaran, C., (2013). Alkaloids derived from tryptophan: Terpenoid indole alkaloids. In: *Natural Products* (pp. 575–604). Springer.
71. Paterna, A., Gomes, S. E., Borralho, P. M., Mulhovo, S., Rodrigues, C. M. P., & Ferreira, M. J. U., (2016). Vobasinyl-iboga alkaloids from *Tabernaemontana elegans*: Cell cycle arrest and apoptosis-inducing activity in hct116 colon cancer cells. *J. Nat. Prod., 79*(10).
72. Patowary, P., Pathak, M. P., Zaman, K., Raju, P. S., & Chattopadhyay, P., (2017). Research progress of capsaicin responses to various pharmacological challenges. *Biomed. Pharmacother, 96*, 1501–1512.
73. Chaudhary, A., Gour, J. K., & Rizvi, S. I., (2019). Capsaicin has potent antioxidative effects *in vivo* through a mechanism which is non-receptor mediated. *Arch. Physiol. Biochem.,* 1–7.
74. Siddique, M. A. B., & Brunton, N., (2019). Food glycoalkaloids: Distribution, structure, cytotoxicity, extraction, and biological activity. In: *Alkaloids-Their Importance in Nature and Human Life.* IntechOpen.
75. Mostofsky, D. I., Shlomo, Y., & N. S. J., (2001). *Fatty Acids Physiological and Behavioral Functions*. Springer Science Business Media: New York.
76. Bennett, M., & Gilroy, D. W., (2017). Lipid mediators in inflammation. Myeloid cells heal. *Dis. A Synth.,* 343–366.
77. Chang, T. M. S., (2013). *Cell Encapsulation Technology and Therapeutics*. Springer Science & Business Media.
78. Laouini, A., Jaafar-Maalej, C., Limayem-Blouza, I., Sfar, S., Charcosset, C., & Fessi, H., (2012). Preparation, characterization, and applications of liposomes: State of the art. *J. Colloid Sci. Biotechnol., 1*(2), 147–168.
79. Rafiee, Z., Barzegar, M., Sahari, M. A., & Maherani, B., (2017). Nanoliposomal carriers for improvement the bioavailability of high-valued phenolic compounds of pistachio green hull extract. *Food Chem., 220*, 115–122.
80. Ko, S., & Lee, S. C., (2010). Effect of nanoliposomes on the stabilization of incorporated retinol. *African J. Biotechnol., 9*(37), 6158–6161.
81. Ghorbanzade, T., Jafari, S. M., Akhavan, S., & Hadavi, R., (2017). Nano-encapsulation of fish oil in nano-liposomes and its application in fortification of yogurt. *Food Chem., 216*, 146–152.
82. Amjadi, S., Ghorbani, M., Hamishehkar, H., & Roufegarinejad, L., (2018). Improvement in the stability of betanin by liposomal nanocarriers: Its application in gummy candy as a food model. *Food Chem., 256*, 156–162.
83. Feng, T., Wei, Y., Lee, R. J., & Zhao, L., (2017). Liposomal curcumin and its application in cancer. *Int. J. Nanomedicine, 12*, 6027.

CHAPTER 2

Advanced Nanocarriers for Nutraceuticals Based on Structured Lipid and Nonlipid

SHAFIULLAH,[1] SYED WADOOD ALI SHAH,[1] ISMAIL SHAH,[2] SHUJAT ALI,[3,6] AZIZ ULLAH,[4] SAMIULLAH BURKI,[5] and MOHAMMAD SHOAIB[1]

[1]*Department of Pharmacy, University of Malakand, Chakdara, Dir Lower – 18300, Khyber Pakhtunkhwa, Pakistan*

[2]*Department of Pharmacy, Abdulwali Khan University, Mardan – 23200, Khyber Pakhtunkhwa, Pakistan*

[3]*School of Food and Biological Engineering, Jiangsu University, Zhenjiang – 212013, P. R. China*

[4]*Department of Pharmaceutics, Faculty of Pharmacy, Gomal University, D.I. Khan, Khyber Pakhtunkhwa, Pakistan*

[5]*Department of Pharmacology, Faculty of Pharmacy, Federal Urdu University of Arts, Science, and Technology, Karachi, Pakistan*

[6]*College of Electrical and Electronic Engineering, Wenzhou University, Wenzhou 325035, PR China*

ABSTRACT

Nanocarriers-based therapeutics are gaining greater attention and research trends in the biomedical field because they have excellent commercialization potential. In this context, an upsurge in nanocarriers based on commercially available products has been observed. Nanotechnology basically deals with the engineering of particles at molecular levels, and this term was originally the subject of building nanoscale devices and machines. In recent years, nanotechnology has broached in various unfamiliar fields of Pharmaceuticals and nutrition. Nutraceuticals have immense importance among both consumers and researchers because of the growing interest in alternative medicinal sources. Nutraceuticals are foods, or foods' parts, that provide health or medical benefits, including the treatment and prevention of diseases. Nutraceuticals is a broad term where foods, i.e., dietary supplements, antioxidants, dairy products (fortified), and minerals, vitamins, herbals, citrus fruits, cereals, and milk comes under this umbrella. Due to the proven healthcare and fitness benefits of nutraceuticals, researchers as well as ordinary people are engrossed towards these natural dietary agents. The increasing interest in such products is also due to the innate biological activities (i.e., antioxidant activities), biocompatibility, and non-toxic nature of these phytochemicals. The World Health Organization (WHO) has started a worldwide strategy to cope with traditional medicines-related issues due to this growing public interest in nutraceuticals. Nutraceuticals from various sources have shown to suppress pro-inflammatory pathways for treating cancer and similarly in other ailments; however, their low *in vivo* bioavailability limits their use. Research on nano-encapsulation of nutraceuticals has been the recent trend to resolve the limitations associated with them and improve their health benefits. This chapter gives an overview of nutraceuticals, different types of nanocarriers, recent developments in the field of nanocarriers based nutraceuticals delivery, their absorption mechanisms, and major challenges in the way of commercialization of nanocarriers based nutraceuticals.

2.1 INTRODUCTION

Natural products are the main sources of bioactives and are regarded as the key discovery of modern medicine [1, 2]. Search for novel bioactive compounds and pharmacophores has always been in regular practice because still organic and synthetic medicinal chemists yet have to find replacement for many natural compounds [3]. The field of drug delivery is revolutionized by nanotechnology-based drugs where bioactive compounds are encapsulated in nanocarriers for addressing various bioactive related issues. Several phytochemicals face the issue of poor bioavailability due to their low intrinsic solubility. Encapsulating such phytochemicals and bioactive compounds in appropriate nanocarriers can enhance their bioavailability by virtue of altered pharmacokinetics and biodistribution profiles in nanosized delivery carriers [4]. Similarly, it has also been shown that using nanocarriers based delivery systems, the drug is localized to specific tissues, and the therapeutic index of drugs is increased [5, 6]. The availability of nanocarriers based products in the market strongly suggests that they have commercialization potentials. At present, researchers as well as the ordinary population is attracted towards agents from natural dietary sources because of their proven effectiveness in fitness and healthcare [7]. As an example, due to the proven health benefits of the cactus plant (having taurine as the key constituent) such as anti-diabetic, anti-viral, and anticancer potentials, it has become an active constituent of nutraceuticals. WHO has started a worldwide strategy to resolve issues related to traditional medicines keeping in view the growing researchers and scientists' interest in nutraceuticals. Similarly, the European Commission (EC) has also decided to put the disease risk reduction at priority in future plans. Research on encapsulation and delivery of nutraceuticals in nanocarriers-based systems has been the subject of interest for many researchers to address the issues associated with nutraceuticals and enhance their healthcare outcomes. This chapter describes nutraceuticals; different types of nanocarriers, recent developments in the field of nanocarriers based nutraceuticals delivery, their absorption mechanisms, and major challenges commercialization of nanocarriers based nutraceuticals.

2.2 NUTRACEUTICALS

The word nutraceutical is the combination of two terms, i.e., nutrition, and pharmaceuticals. Nutraceuticals are food or food products offering nutritional as well as pharmaceutical benefits, i.e., give nutrients to the body, provide resistance against various diseases, and also help in treating certain ailments [8]. Long ago, people having knowledge and working in the field of medicine thought to develop and search such foods that may be served as medicine for treatment and prevention of diseases. Ultimately, such sparkling ideas led to the development of nutraceuticals, whereas the term nutraceuticals were first coined by Dr. Stephen DeFelice in 1989 by combining nutrition and pharmaceuticals [9]. There are three main classes of nutraceuticals, i.e., functional foods, dietary supplements, and functional beverages. The dietary supplements can be further sub-divided into mineral supplements, protein supplements, vitamins, herbal supplements, and plant extracts.

Probiotics and omega fatty acid foods fall in the category of functional foods, while functional beverages are sub-classified into fortified juices, sports drinks, and energy drinks. Dietary supplements, functional food, multi-functional food, etc., are some of the common words used as synonyms or related to nutraceuticals. Functional foods are just the basic foods; however, some special or specific ingredients are incorporated in them for providing health benefits to the body along with nutrients [9]. Functional foods exclusively designed for promoting good health in human beings is made possible due to recent technological advancements in the field of food technology. Isolation, identification, characterization, and purification are some basic requirements to be considered while incorporating food components in functional foods. Moreover, characterization of incorporated food components in terms of medicinal values, nutritional values, etc., are also important. Our body need primary food components composed of proteins, carbohydrates, and lipids for normal energy and proper body functioning. Some secondary food elements, i.e., vitamins, are commonly not produced in the human body, and are necessary for proper body function, thus these components must be taken in food diet. Nutraceuticals are also minor food elements that improve the body function by virtue of fighting against some chronic disease conditions [10].

The therapeutic efficacy of any drug, food product or nutraceuticals is dependent on its bioavailability. In pharmacological perspectives, bioavailability is the rate and extent to which the drug reaches systemic circulation or its site of action, while in terms of nutritional concept-bioavailability means some nutrients in food are partially available. When orally administered, certain parameters, e.g., low solubility and/or permeability within the gastrointestinal tract (GIT), less gastric residence time and instability in GIT or under food processing conditions limit their activity. The increasing popularity and ever-growing public interests in nutraceuticals as preventive medicine demands and put pressure on manufacturers and regulators of health-related products to address their bioavailability-related issues [11].

2.3 NANOCARRIERS FOR NUTRACEUTICALS DELIVERY

Nanotechnology-based delivery systems, e.g., nanoparticles (NPs), vesicles, hydrogels or microparticles are attaining promising place in pharmaceutical industry and food technology because they offer tools for enhancing therapeutic efficacy drugs and nutraceuticals. Nanotechnological tools are mainly applied to those drugs or nutraceuticals having poor aqueous solubility, low bioavailability, and GI stability problems. As an example, the stability of flavonoids and anthocyanidins depends upon GI pH whereas nearly 60% of probiotic bacteria cannot survive in GI environment. Thus, to protect nutraceuticals or drugs from harsh GI environment and to improve their bioavailability, their delivery in suitable carrier-based system is highly demanding [12].

Nanotechnology-based products possess great commercialization potential and a multifold increase of such products in the market is expected in coming years because such products overcome many of the limitations associated with them [13]. Nevertheless, the safety of newly developed nanocarriers-based systems must be ensured prior to their incorporation in commercial food products. Various desirable features of nanocarriers-based systems must be considered prior to the construction of such systems. Firstly, as the size of the drug/nutraceuticals is reduced to nano-size range, the behavior of this system will be different from the conventional particles in GIT [14, 15].

Toxicity issues may occur if the degradation products of the nanocarriers-based delivery system is different from conventional particulate matter. Thus, to ensure safety, it is imperative to evaluate toxicity profiles to ensure the safety of such food-grade nanoscale devices. For the fabrication and construction of nanocarriers-based delivery systems, the application of food-grade materials is highly preferable. In addition, these nano-formulations must be strong enough to withstand storage conditions, economically feasible as well as have strong potentials for practical applications. Moreover, the quality of products should not be adversely affected by the incorporation of such materials into final food products. Figure 2.1 shows the advantages of using nanocarriers'-based delivery systems over other conventional dosage forms. However, the nanotechnology-based nutraceuticals delivery systems should be:

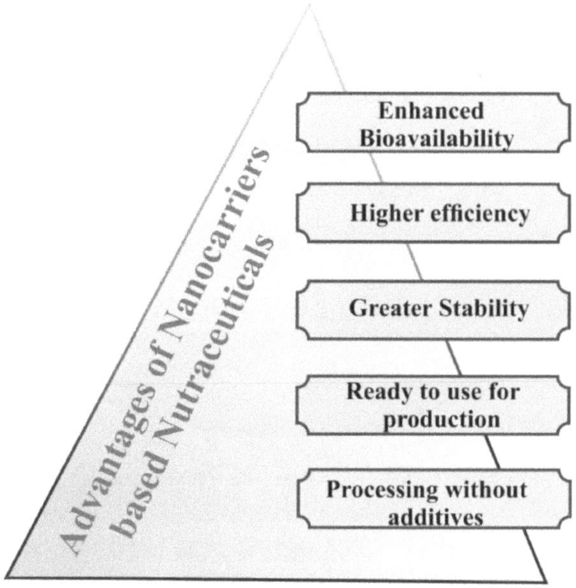

FIGURE 2.1 Advantages of using nanocarriers'-based delivery systems over other conventional dosage forms.

- Physicochemically stable to environmental conditions and should preserve its functional properties [16];
- Capable of improving GI stability of labile bioactive components;
- Capable of maintaining constant dose levels in systemic circulation;
- Capable of facilitating lymphatic transport in case of highly lipophilic compounds;
- Capable of extending gastric retention times [17].

Several researchers have reported the application of nanotechnology for nutraceuticals over the last few years. The increase in oral bioavailability and absorption and thus the nutraceutical effects of certain phenolic compounds have been reported [18, 19]. The advantages of nanocarriers-based delivery systems for nutraceuticals is summarized in Figure 2.2. The nanocarriers-based nutraceuticals have been mainly prepared using polymers (natural or synthetic polymers, polysaccharides, proteins) or lipids-based systems (liposomes, solid lipid nanoparticles (SLN), nanoemulsions (NEs)). These polymers and lipids-based delivery systems for nutraceuticals are discussed in the following subsections.

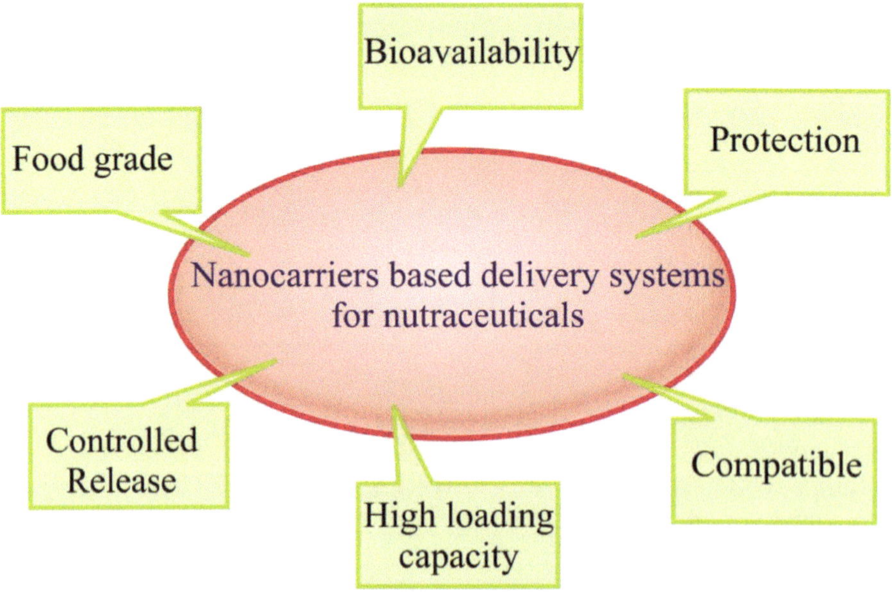

FIGURE 2.2 The advantages of nanocarriers-based delivery systems for nutraceuticals.

2.3.1 POLYMERS

In targeted drug delivery, polymers are extensively employed as delivery vehicles due to their specialized flexible structures. In addition, polymers prevent nutraceuticals from severe conditions of the GI tract due to their entrapment of nutraceuticals. Polymers ideally deliver nutraceuticals to the targeted site either in natural or synthetic form, usually in nano or micro-sized particles [20].

2.3.1.1 BIODEGRADABLE POLYMERS

Biodegradable synthetic or natural polymers have been extensively used in the field of tissue engineering and drug delivery as nanocarriers for nutraceuticals/drug delivery because of their unique characteristics such as biodegradability, biocompatibility, and flexible nature [21]. Polymers with biodegradability characteristics certainly degrade via normal biological and chemical processes in the body. Specifically, in most cases, the synthetic polymers are engineered in such a way that they biodegrade by hydrolysis phenomena. It should also be kept in mind that the degradation product of a synthetic biodegradable polymer should also be biocompatible and biodegradable. These biodegradable polymers encapsulate the nutraceuticals and are tailored in such a way that the encapsulated product is released upon degradation of the polymer in the desired site.

2.3.1.2 NATURALLY OCCURRING BIODEGRADABLE POLYMERS

Though, the naturally occurring biodegradable polymers have certain limitation such as structural complexities, poor biomechanical characteristics, etc., still they are considered as nanocarriers for numerous attractive reasons like their commercial availability, capability of structural modification and outstanding biocompatible and biodegradation potentials. Moreover, the natural polymers based nanocarriers could easily be metabolized by the host successfully and cleared from the body. The important naturally occurring biodegradable polymers include proteins such as collagen, gelatin, albumin, elastin, globulin, zein, gliadin, and casein. Chitosan, carrageenan, chitin, alginate, dextran, and hyaluronic acid are classified as naturally occurring polysaccharides-based polymers [22].

2.3.1.2.1 PROTEINS BASED NATURAL POLYMERS

Proteins are long-chain macro-molecules comprised of one or more amino acids chains and are considered vital for basic biological and normal physiological functions of living systems. Proteins are macromolecules not only important for nutritional point of view, but also extensively used in drug/nutraceuticals delivery systems, especially as an encapsulating and coating materials for nutraceuticals. Among animal proteins; gelatin [23, 24], casein [25], collagen [26–28], albumin [29, 30] and whey proteins [31, 32] have been extensively used as delivery vehicles, while glycinin [33], zein [34] and wheat gliadin [35] are proteins obtained from plant sources that have been investigated for drug and nutraceuticals delivery to the target sites.

Owing to non-immunogenic and biodegradable nature of collagen and gelatin, they have been extensively used in drugs/nutraceuticals delivery and tissue engineering. Several delivery systems of collagen in the form of nano-films, nano-gels, and nano-sponges have been investigated for the delivery of nutraceuticals and drugs (antibiotics anti-inflammatory drugs) [36]. The applicability of collagen as a nanocarrier has not been too much attractive, and this might be due to difficulty in entrapping the bioactive molecules in collagen. However, more widespread applications of gelatin (a derivative of collagen) have been reported for encapsulation of bioactives as nanotechnology-based systems [37]. Another animal protein; albumin is also widely used as a natural polymer in drugs/nutraceuticals delivery systems due to its outstanding non-toxic nature, and acceptable non-immunogenicity, biocompatibility, and biodegradability features. Examples of albumin-based nanoparticulate systems that have been investigated for the delivery of key anticancer drugs such as doxorubicin for chemotherapy of breast cancer and noscapine have been reported. Similarly, anti-inflammatory drugs like SB202190 delivery in albumin-based NPs have also been reported for inhibition of p38 mitogen-activated protein kinase and multiple inflammatory cytokines' secretion [38, 39].

Oral-based nano-drug delivery systems have also been constructed from Gliadin and Zein proteins. These are prolamin proteins used by plants for storage purposes. Key examples of prolamin proteins extensively employed and investigated as ideal nanocarriers are Zein from corn, hordein from barley and gliadin from wheat. Beside these proteins; the antioxidant proteins-inhibiting catalase, superoxide dismutase (SOD), and scavenging free radicals have also been regarded as applicable in delivery systems. On the other hand, these antioxidant proteins are prone to degradation in the harsh GI environment because of its low GI pH and the presence of protein degrading enzymes, i.e., trypsin and pepsin. The antioxidant proteins coated with gliadin, zein or other prolamin proteins successfully protect them from harsh GIT environment [40].

2.3.1.2.2 POLYSACCHARIDES BASED NATURAL POLYMERS

These are long-chain molecules composed of monosaccharide units. Primarily their function in plants is related to storage and provides firm structure to the plants. Polysaccharide polymers are usually obtained from both plants' sources such as inulin fiber, pectin, and starch and animals' sources like chondroitin sulfate, glycogen, and chitosan. The biotransformation or breakdown of polysaccharide polymers can be accomplished in different parts of GI by normal bacterial flora. Nutraceuticals are protected from harsh GI conditions when encapsulated and delivered in polysaccharides-based drug delivery systems. When the polysaccharides-based nanocarriers reach the colon portion of GIT, the polysaccharides are hydrolyzed, and the encapsulated nutraceuticals are released in the colon. Probiotics like bifidobacteria and lactobacilli delivery is mainly accomplished with the application of polysaccharides-based nanocarriers [41].

2.3.1.3 SYNTHETIC BIODEGRADABLE POLYMERS

For the delivery of nutrients/nutraceuticals, the synthetic biodegradable polymers can be used to enhance the encapsulation efficiency (EE), and they can also release the nutraceuticals/bioactives for a couple of days to weeks. Comparatively, polymers from natural sources have a relatively short period of time for drug release and commonly confined by the use of organic solvents and need comparatively harder formulation conditions. Synthetic polymers are able enough to control or sustainly release the bioactives for a couple of days to weeks as compared to natural polymers. Nevertheless, synthetic polymers could potentially lead to chronic inflammation and toxicity. Thus, if synthetic polymers have to be used for nano-encapsulation of nutraceuticals, their immunogenicity and toxicity should be taken into consideration and should be properly evaluated [42].

2.3.2 ESTER, ANHYDRIDE, AND AMIDE FUNCTIONAL GROUPS' BASED POLYMERS

Synthetic polymers which are biodegradable include polylactic acid (PLA), polyglycolic acid (PGA), poly(lactic-co-glycolic) acid (PLGA), polyanhydrides, polyorthoesters, and polyamide. Ester-based polymers like PLGA, PGA, and PLA have been extensively used in the field of nanotechnology [43]. These polymers have a number of useful applications in the field of biomedical science as drug delivery carriers, artificial-tissue materials, and resorbable sutures. PLGA is a copolymer composed of lactic acid and glycolic acid in which a different make-up of two monomers represents the polymer properties. PGA is a hard, tough, and crystalline polymer due to glycolic acid composition only. PGA has excellent fiber-forming properties but is not soluble in almost all those common polymer solvents that affect its application for drug carriers, as it cannot be made into films, rods, or capsules. PLA is a thermoplastic biodegradable polymer and is degraded by hydrolysis because of lactic acid composition only. PLGA has been used as a nanocapsule to increase the hydrophilicity, bioavailability, and anticancer property of lipophilic molecules such as lutein [44].

For controlled delivery of nutraceuticals/drugs, different nano-systems like microspheres, coatings, tubes, and disks can be constructed from biocompatible poly-anhydrides. The antibiotics like ampicillin or non-steroidal anti-inflammatory drugs (NSAIDs; e.g., salicylic acid) has been encapsulated in polyanhydrides and the payload was released as the polymers degraded [45]. At room temperature, the polyorthoesters based polymers are stable in dry conditions. NSAIDs were encapsulated in Polyorthoesters and they released the active drug by surface erosion. These polyorthoester polymers have the advantage of prolonged drug release rates from nano-/microspheres formulation from a couple of days up to months [46]. Even polyamide is biodegradable; however, its application has been confined due to its immunogenicity and poor mechanical characteristics.

2.3.3 SMART/STIMULI-RESPONSIVE POLYMERS

2.3.3.1 TEMPERATURE-SENSITIVE POLYMERS

Temperature-sensitive polymers are those that show a drastic change in their physical properties (mostly solubility) with temperature. The phase change behavior of such polymers can be due to the disruption of intra- and intermolecular interactions at their critical solution temperature (CST) causing expansion or collapse in the polymer within the aqueous solution. Temperature sensitive polymers with a lower critical solution temperature (LCST) will show phase separation (e.g., precipitation) above a specific temperature, while those with an upper critical solution temperature (UCST) will display phase separation (e.g., precipitation) below a specific temperature. These polymers are mostly used for the construction of hydrogels, and thus, hydrogels formulated from such polymers are called smart gels that show sole-gel transition at a specific temperature [47–49]. Polymeric hydrogels form a loosely cross-linked three-dimensional polymeric network, which absorbs a sufficient quantity of water by hydration as they have a large number of hydrophilic groups. Sol is regarded as a stable colloidal suspension (0.1–1 mm) of solid particles or polymers in a liquid.

Poly(N-isopropylacrylamide) (PNIPAM) is one of the temperature-sensitive polymers which is currently used mostly in hydrogels. Below its LCST, PNIPAM is soluble in water and this solubility is due to the change in phase of PNIPAM from a hydrated swollen state to a dehydrated shrunken state upon temperature change [50]. In this case when the hydrogel is in solution form, the volume of hydrogel changes up to 90% at approximately 32°C (LCST for PNIPAM) [51]. PNIPAM can be used for such drugs or nutraceuticals where the release of the payload is dependent on the temperature of the body or specific tissue. Hence, it can be used for drugs or nutraceuticals for inflammation and cancer where specific body part or tissue temperature will be comparatively higher, and the drug/nutraceuticals will be released at the target site. As the nutraceuticals/drugs that are encapsulated in PNIPAAM will be stable above LCST; however, the release of nutraceuticals/drugs will be started at a temperature above LCST due to swelling of the polymer/hydrogel [52]. Examples of thermo-sensitive polymers with LCST consist of poly[2-(dimethylamino)ethyl methacrylate] (PDMAEMA), poly(N-vinylcaprolactam) (PVCL), and poly(N,N-diethylacrylamide) (PDEAM). Thermo-sensitive polymers with UCST have opposite phase transition compared to LCST hydrogels. Hydrogels with UCST swell and are soluble in water above their UCST. Below UCST, the solution of such hydrogels will give a cloudy appearance. Examples of UCST hydrogels are polyacrylamide (PAAM), poly(acrylamide-co-butyl methacrylate) or poly(acrylic acid) (PAA). Smart hydrogels not merely show response to a change in temperature but will also show response to other conditions like change in pH, enzyme, or electrical potential depending on polymer and hydrogel nature [53, 54].

2.3.3.2 PH-SENSITIVE POLYMERS

Almost all polymers with pH-sensitivity characteristics have either basic (ammonium salts) or acidic (sulfonic or carboxylic acid) functional groups which are responsible for changes in environmental pH [53, 54]. Poly(sulfonic acid) and PAA polymers are the examples of polyanions polymers employed in drug delivery. At alkaline, neutral, or high pH, the acidic functional groups containing polymers are dissolved due to ionization polymers swelling or complete dissolution. Whereas; polycationic polymers, e.g., poly(N,N9-diethylaminoethyl methacrylate) (PDEAEM) swell or dissolve at low pH [55]. Polycationic base polymer hydrogels have been mainly used for drug delivery in the stomach because of their swelling in acidic environments. Such type of hydrogels can best be used for antibiotics (e.g., metronidazole, amoxicillin, etc.), delivery to the stomach for the treatment of stomach infections such as *Helicobacter pylori* infection. Many nutraceuticals, e.g., Phytosterols [56], carotenoids, vitamins, and lutein efficient delivery through such polymers have also been reported [57].

Polyketals (PK), is a new class of synthetic acid-responsive polymers having ketal linkages in their backbone. For drug/nutraceuticals delivery, they are designed to hydrolyze by macrophages in the acidic environment of phagosome after phagocytosis. Thus, this class of polymers can successfully be used for enhancing intracellular delivery of therapeutic drugs/nutraceuticals. As an example, the poly(cyclohexane-1,4-diyl acetone dimethylene ketal) (PCADK) based microparticles significantly enhanced the activity of SOD; superoxide scavenging enzyme) to scavenge reactive oxygen species (ROS) produced by macrophages [58]. Polyketal PK3 is another polymer that has been efficiently used for the delivery of tumor necrosis factor-alpha (TNF-α)-small interfering RNA (siRNA) to Kupffer cells *in vivo* and successfully inhibited gene expression in the liver [59].

2.3.4 LIPOSOMES

Liposomes are lipid bilayer vesicles consisting of phosphatidylcholine (PC) based phospholipids. A liposome may also comprise of lipid or lipid chains like phosphatidylethanolamine (PE), cholesterol, sphingolipids, and long-chain fatty acids. Liposome exist in spherical shape with an inner aqueous core and hydrophobic membrane layer in the middle of the bilayer [60, 61]. *In vivo*, liposomes are cleared by the reticuloendothelial system (RES) of the liver and mononuclear phagocyte system (MPS) very rapidly, and due to this limitation, the application of liposomes is slightly hampered. However, this selective clearance of liposomes by MPS can certainly be exploited to target cells of the MPS, particularly macrophages and carry drugs to MPS with high proficiency. Liposomes can be pegylated to reduced their clearance by the RES and extend their systemic circulating times. Consequently, pegylation enhances the efficiency of liposomes for drugs' delivery. Similarly, liposomes can also be fabricated with a targeting species including a ligand or an antibody to target specific kinds of tissue or cells [62].

Liposomes have been applied for the delivery of a number of therapeutic drugs for central nervous system illnesses such as brain infection, ischemia, and brain tumors [63, 64]. Drugs encapsulated within liposomal vesicles can be administrated through different routes, i.e., intracerebrally, intraventricularly, or intravenously. Liposomes are among the most widely explored delivery systems owing to their biocompatibility and low immunogenicity characteristics; however, they face few limitations such as high cost of production, low stability, short shelf life and rapid removal by RES after intravenous injection [65].

2.3.5 SOLID LIPID NANOPARTICLES (SLNS)

SLNs are aqueous surfactant solutions or colloidal dispersions of lipids in water [66]. They have several advantages such as excellent stability of the encapsulated material, controlled release, easy to scale-up, protection of incorporated molecules from the external environment, and ability to carry both hydrophilic as well as lipophilic drugs. These lipid-based systems can also serve to improve functional and organoleptic properties. Furthermore, these systems contain the species with generally recognized as safe (GRAS) status [67]. Shortcomings such as aggregation, flocculation, increased particle size, comparatively high-water content and compound release may happen during storage [68].

2.3.6 NANOEMULSIONS (NES)

Emulsions are fundamentally bi-phasic structures which are comprised of an outer phase, i.e., continuous phase and an inner phase, i.e., dispersed phase, whereas, surfactant molecules make an interphase. NEs are emulsions which have very small sizes and appear translucent or transparent. They occur in much smaller size as compared to the conventional emulsions, i.e., size range from 50 to 200 nm [69]. Generally, a micelle is 5 nm or more in diameter and a surfactant molecule have a length of 2 nm. However, incorporation of oil phase into micellar core may result in increase in its size occasionally to a large range [4]. NE is a good choice to incorporate nutraceuticals with poor solubility into food matrix, and it is understood that maximum biologically active phytochemicals are either lipophilic in nature or having low solubility. Systemic bioavailability of these bioactive phytochemicals is significantly influenced by their low solubility, because their characteristics including lipophilicity, partition coefficient, solubility, etc., dictate their way of transport, administration, and target sites. Modification of these bioactives into NEs can offer the advantages such as increase surface area (the small particle size of NEs), thereby resulting in improved epithelium cell permeability, rapid diffusion across mucus membrane and enhanced digestion rates [17, 70, 71].

Furthermore, as may protect chemically reactive compounds from oxidation, and hence resulting in minimum degradation in the GIT and increased shelf life [72, 73]. Several reports have been published on the entrapment of bioactives into NEs, and current drifts have revealed the application of food-grade NEs [74]. Carrier oil is a significant constituent in the synthesis of food-grade NEs, as it regulates the bioavailability of encapsulated components [75, 76]. However, the carrier oil must have the ability to form mixed micelles, should be fully digestible and have a high solubilization capacity for active components [77]. Different types of nanocarriers commonly used as nutraceuticals/drug delivery vehicles are given in Figure 2.3.

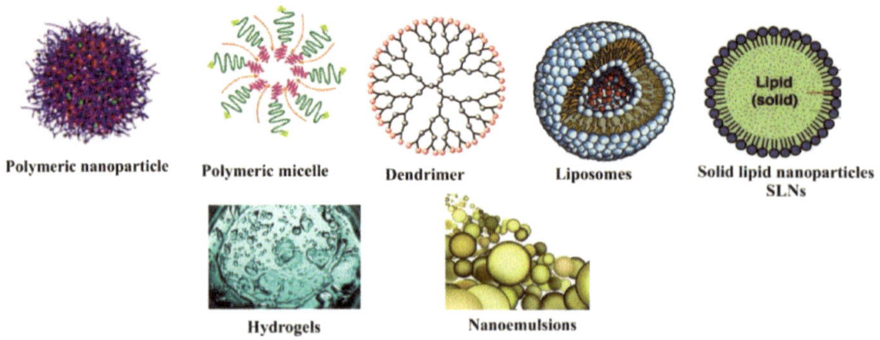

FIGURE 2.3 Different types of nanocarriers-based delivery systems for nutraceuticals/drugs.

2.4 EXAMPLE OF NANOCARRIERS BASED NUTRACEUTICALS DELIVERY

Probiotics keep the digestive system healthy by controlling microbial balance. Though, most of the probiotic bacteria (about 60%) cannot endure in the gastric environment. Only a limited number of bacterial species ($0–10^3$ CFU/mL) can exist in the stomach (due to the low intragastric pH). The main bacterial species in the stomach are Enterobacteriaceae, lactobacilli, staphylococci, streptococci, and yeasts. Most of bacterial species, about 500 strains live in the intestinal microbiota. More probiotic bacteria can live in the small intestine and their number further increase in the duodenum ($0–10^5$ CFU/g) to ileum (10^8 CFU/g) and colon ($10^{10}–10^{12}$ CFU/g) [78]. Hence, a delivery system for nutraceuticals is crucial to protect nutraceuticals or probiotic bacteria from the severe gastric environment. Micro- or nano-encapsulation of probiotics can keep safe these biological cells in an unpleasant environment.

By using different gel-forming approaches, probiotic bacteria have been entrapped in the gel matrix. Probiotic cells can be encapsulated by emulsion, extrusion, and spray drying approaches. Extrusion is a conventional method for probiotic formulation. Using an extrusion method for alginate capsule, cell suspension is obtained by the addition of probiotic cells into hydrocolloid solution. Then, to produce droplets, the cell suspension is passed through the syringe needle and are directly dropped into the hardening solution comprising cations such as calcium. In the hardening solution, the alginate polymers in the cell suspension are crosslinked by the cations which result in the alginate capsule. Finely, the as-prepared alginate capsule is obtained and dried by applying a suitable approach.

Emulsion is a suitable means for the encapsulation of lactic acid bacteria (LAB). In this approach, polymer slurry and a dispersed phase that contain a small volume of cells are emulsified into a continuous phase that having a large volume of vegetable oil such as sunflower, soy oil, light paraffin, and corn oil. The gel formation of emulsion is done by various cross-linking approaches including interfacial, enzymatic, and ionic polymerization.

During the drying process, starches, and gums tend to produce sphere-shaped microparticles. In spray drying approach, dissolved polymer matrix and probiotic cells of starches or gum Arabic is obtained. The spray drying approach can successfully produce microparticles; but, during the drying process, probiotic cells may be damaged because of physical injury to microparticles and heat generation. In order to minimize the damage of probiotic cells, the outlet and inlet heating for spray drying must be adjusted and appropriate cryoprotectant must be applied during freeze-drying.

Probiotic bacteria are thought to influence the immune system; hence their delivery is important. To efficiently deliver probiotic bacteria, suitable encapsulating materials and encapsulation methods should be used. The probiotic bacteria should be released to changes of osmotic force, time, temperature, environmental pH, enzymatic activity, and mechanical stress. During the formation, several parameters such as heat generation should also be regulated to improve the viability of probiotic bacteria.

According to the Food Standards Agency (FSA), titanium nitride, nano-clay on silver and fumed silica are the nanomaterials that are allowed to be applied in food if they follow the pertinent legislation. Center for Food Safety has generated a database which registers about 300 foodstuff contact products that use nanotechnology [79]. In addition, Chinese nano tea, nano-gold, and nanosilver have been applied as mineral supplements. Similarly, carotenoid NPs have been used in fruit drinks and patented "Nanodrop" delivery structures have been applied for encapsulation of vitamins, etc., and Nanoclusters or Nano cages have been applied in nanoceutical foodstuffs such as chocolate drink, thus, giving sweetness without addition of any sweeteners or sugar [80].

2.5 ABSORPTION MECHANISMS OF NANOCARRIERS BASED NUTRACEUTICALS

The small intestine portion of GIT is regarded as the major site for absorption of nanocarriers/nutraceuticals. The wall of the intestine is an active and complex structure that regulates nutrients absorption, immunes system and interactions between intestinal microflora, thus guarantees intestinal equilibrium [81]. A nanocarrier/nutraceuticals must first diffuse through the thick intestinal mucus layer prior to reaching the endothelial cells. The goblet cells synthesize this mucus layer and is composed of lipids and glycoproteins combination [82]. Various parameters, e.g., charge, size, and viscosity influence the passage of nutrients/drug to pass through this layer. The interaction of this layer with mucoadhesive hydrophobic materials or large molecules may reduce their permeability. Permeation of nutraceuticals through gut endothelia may be restricted; however, lipophilic components (e.g., and kaempferol and resveratrol) can pass through the mucus layer when mixed with bile salts, free fatty acids, and phospholipids [83]. Alternatively, nutraceuticals can adhere or pass-through mucus layer when encapsulated in polymeric nanocarriers, e.g., chitosan, lectin, polyethylene glycol (PEG) and gelatin thus can allow the uptake of nutraceuticals [84]. Nanocarriers with negative surface charges are repelled by the mucus layer, and their cellular uptake is decreased because of less residence time in the epithelial cells. Estradiol intracellular uptake was increased when it was encapsulated in positively charged poly(lactic-co-glycolic acid) (PLGA) NPs as compared to neutral or negatively NPs [85]. As the mucus layer barrier is overcome by nanocarrier/nutraceuticals, they may then cross the epithelium barrier via either through paracellular (through tight junctions) or transcellular (including M-cell-mediated transport) transport routes.

2.5.1 PARACELLULAR ROUTE

Passive transport of materials (drugs/nutraceuticals) via passive diffusion occurs through inter-cellular spaces of epithelial cells of the intestine [86]. Epithelial tight junctions (TJs) are present at the intercellular spaces, and these TJs are composed of proteins (e.g., claudin, and occludin and claudin), making it a complex structure [87]. These TJs regulates intestinal permeability of substances, intercellular adhesion, paracellular transport, mediate passage of molecules from lumen to lamina propria and impede in the access of microbes to host cells and tissues [88]. Polar molecules and small water-soluble molecules such as sugars, water, amino acids, ions, and peptides having molecular weight less than 500 Da can pass through these TJs [89]. The function of TJs is increased by certain agents (e.g., polyphenols) while others, e.g., caprylic acid can decrease the function of the TJ barrier thus enhances the small molecules' uptake [90]. Paracellular transport can potentially decrease intracellular metabolism, which is important for nutraceuticals. However, it has also been reported that polyphenols (i.e., quercetin, chrysin, caffeic acid, rutin, gallic acid, and resveratrol) are poorly transported through passive diffusion both in Caco-2 cells monolayer and similar artificial membrane permeability assays [91]. Moreover, TJs cannot open more than 20 nm; thus, the transport of nanocarriers across intestinal epithelium through paracellular route is very low, and this space impedes most of the nanocarriers based delivery systems [92].

Nano-systems with less than 20 nm size can adversely affect TJs and release the payload to systemic circulation. Later, the TJs restore their function to their original regular position. In addition, positively charged particles can easily be transported via paracellular transport because the negatively charged membrane surface will attract them. Nanosystems with cationic chitosan have the ability to open TJs and thus induce paracellular transport. As an example, the tea catechins delivery and transport is enhanced when encapsulated in poly(glycolic acid) (PGA) and chitosan-based NPs and cross the intestinal barrier through paracellular transport [93].

2.5.2 TRANSCELLULAR ROUTE

The transcellular absorption mechanism is dependent on the active or passive transport of molecules through cells via endocytosis. Most of the nutraceuticals are believed to be absorbed simply by passive transport without the involvement of carrier or receptor via transcellular route [94]. This mechanism has been proposed for non-polar polyphenol aglycones and carotenoids [90]. For instance, Guri reported the curcumin transport through passive diffusion in Caco-2 cells when loaded in SLNs [95]. In contrast, some charged and polar biomolecules bind to a specific receptor (receptor-mediated transport) or naturally-occurring membrane protein transporter (carrier-mediated transport) located in the apical cell membrane. These molecules are then transported against concentration gradient within the intestinal cells with expenditure of energy; a phenomenon known as active transport [96], rather, they might not cross the cell membrane [97]. Such receptors and membrane carriers are vital for the uptake of numerous nutraceuticals. For example, fatty acids, vitamin C, and some peptides are carried via fatty acid-binding proteins, sodium vitamin C co-transporter and proton-coupled peptide transporters, respectively of [83]. Moreover, the capacity of stimulation or inhibition of membrane transporters may be affected by many polyphenols and the stimulation or inhibition potential depend on the polyphenol concentration, form, exposure time, etc., [90].

Molecules that attach to specific carriers at the apical cell membrane or cell membrane receptors are internalized to the cell by endocytosis mechanisms (including pinocytosis or phagocytosis) [98]. Entry to M-cells of the Peyer's patch (specialized in antigen sampling) is mainly based on phagocytosis (M-cells mediated transport), thus offering a supposed route for nanocarriers-based delivery systems [96, 99]. When the expression of M-cells comes under less than 1% of total intestine area, then transport through these cells becomes very difficult [100]. Nanocarriers can also be internalized by pinocytosis mechanism where they bind to complementary cell surface receptors (like lectins, lactoferrin, and α5β1 integrin) [101]. Some nanocarriers designed for nutraceuticals have been reported using specific ligands on their surface for specific receptors to achieve enhanced intracellular delivery both in M-cells and enterocytes [101]. For example, the gambogic acid transport in lactoferrin-based NPs resulted in enhanced transport through cell membrane because of the presence of lactoferrin receptor [102]. It is worthy to mention that some nutraceuticals are excreted back into the lumen of GIT after their absorption by efflux pumps (efflux transporter) present lipid bilayers of the cell membrane, thus, limit the bioavailability of such nutraceuticals [103]. Therefore, the knowledge and understanding of various transport mechanisms across GIT is vital for the successful development of nanocarriers-based nutraceuticals delivery system.

2.6 RELEASE MECHANISMS OF NUTRACEUTICALS FROM NANOCARRIER

The knowledge and understanding of release mechanisms are very important for the development of controlled and tailored nano-based delivery systems. On the basis of a good understanding of release mechanisms, one can predict ways for better protection of the payload in the nanocarrier, their absorption as well as optimize their release from the system [104]. Release of the encapsulated material from the carrier can occur through various processes depending on the nature of the encapsulated molecule, composition of the carrier, loaded amount, the release media, and the particle's geometry. The release of encapsulated drug/nutraceutical from the carrier may certainly fall in one of the following four main mechanisms:

1. **Diffusion:** The drug/nutraceutical molecules simply diffuse out from the intact non-biodegradable biopolymers to the surrounding medium. This diffusion can take place via homogeneous matrix, water-filled pores, or via an external shell from an internal reservoir. The overall rate of mass transfer is dependent on the solubility of drug/nutraceuticals in the matrix, geometry, and size of the carrier as well as on its diffusion coefficient through the matrix. The diffusion coefficient in turn is affected by various environmental and particle parameters like porosity (porosity $\propto$ diffusion coefficient), tortuosity (tortuosity $1/\propto$ diffusion coefficient) and temperature (temperature $\propto$ diffusion coefficient).

2. **Erosion:** Another mechanism for payload release from nanocarriers is the erosion. The encapsulated drug/nutraceuticals are released to the medium either via homogenous (occurring in the bulk volume of the nanocarrier; or heterogeneous erosion (occurring at the nanocarrier surface. Enzymatic and/or chemical processes can induce the erosion process. Bulk erosion is the process in which the nanocarriers' size remains almost the same where the external fluid goes inside the nanocarrier via breaking of the physical or chemical bonds. In contrast, the surface erosion is the process where the nanocarrier (usually a biopolymer) size is gradually reduced through erosion at the external surface [105]. The rate of erosion is dependent on various parameters, i.e., physicochemical stability, polymer molecular weight (erosion $1/\propto$ molecular weight), size (erosion $1/\propto$ size) and the release medium [106].
3. **Swelling-Shrinkage Mechanism:** In this phenomenon, when the drug or other payload dimensions (e.g., size) is higher than the nanocarriers' pore size, they are entrapped within the nanosystem. Then the nano-system conditions are changed through different triggers (e.g., temperature, water activity, ionic strength, or pH) that cause the swelling of the nanosystem leading to increase in the nanocarriers' pore size and ultimately release of the entrapped payload takes place. On the other hand, in shrinkage-induced release mechanism, the payload is entrapped in the nanosystem initially upon its swelling and then released upon shrinkage via altering the solution conditions [107].
4. **Fragmentation:** In this case, the entrapped drug/nutraceuticals are released through physical disruption of the nanocarrier to medium. The physical disruption of the nanocarriers may be either fragmentation or fracturing through shear or compression mechanisms in the mouth and gastrointestinal (GI) environments or during processing [108].

It is noteworthy that diffusion is always involved in each of the above mechanisms. For the design of the efficient delivery system with desired EE and release profiles, mathematical modeling is also of prime importance [109]. A preliminary understanding of the drug release mechanism is required for the selection of an appropriate release model. For a full understanding of the release mechanisms, various parameters are considered such as nanocarrier size, concentration, solubility of the entrapped substance in the release medium and nanocarrier matrix, the porosity, pore size distribution along with the effective diffusion coefficients. Such parameters can often be found for drug delivery systems; however, only limited data have been published for nutraceuticals delivery systems in recent years [110, 111].

2.7 REGULATORY ASPECTS

Regulatory issues for medicinal products should have to be resolved regarding quality, efficacy, safety, testing, and marketing authorization processes for nutraceutical products claiming medicinal benefits [112]. The application and designing of nano-delivery systems not only for fresh foods but also for healthier foods has been the recent trend, however several of such products can pose serious threats to peoples' safety [113]. Several governing bodies like the European Food and Safety Authority (EFSA), United States Food and Drug Administration (FDA), Environmental Protection Agency (EPA), Occupational Safety and Health Administration (OSHA), National Institute for Occupational Safety and Health (NIOSH), US Department of Agriculture (USDA), US Patent and Trademark Office (USPTO) and Consumer Product Safety Commission (CPSC) regulates the application of nanosystems in food [114].

FDA published guidance documents about nanotechnology in 2012. According to FDA guidelines, the dietary supplements are considered a a category of food, so these documents have no specific mentions for dietary supplements. Rather, they talk about food and cosmetics. Moreover, FDA guidance papers have mentioned that if chemical or physical properties of food substance is changed its bioavailability will also be changed. Additionally, such physical or chemical changes in food products can potentially lead to toxicity. In a later FDA draft guidance for new dietary ingredient, the agency pointed nanotechnology as a process that construct new dietary ingredients thus should notify FDA properly. However, nanoceuticals can be brought to market with little or no safety verification because they are not properly regulated. The FDA anticipated that nanocarriers-based products should come under the jurisdiction of the Office of Combination Products (OCP) [115]. When nanotechnology-based products are aimed for food applications, such products should be designed from non-toxic, mycotoxins, and heavy metals free materials as per EC Food Law Regulation [116]. The directive 89/107/EEC further states that nanomaterials for food packaging application should first be as a direct food additive [117].

2.8 CONCLUSION

Nutraceuticals are foods or food parts that provide health or medical benefits, including basic nutrition as well as the treatment and prevention of chronic disease conditions. Medicinally important nutraceuticals include natural antioxidant foods or essential minerals, vitamins, functional foods, pre/probiotics, and phytochemicals. For efficient delivery to target tissue or systemic circulation, nutraceuticals should be encapsulated in a biocompatible, safe, and targeted delivery system. Polymers offer unique characteristics as delivery vehicles that cannot be attained by using any other material. Particularly, stimuli-responsive, and biodegradable polymers have been the subject of interest for controlled drug delivery systems. On the other hand, nanocarriers made of amphiphilic materials such as liposomes and micelles possess lower serum stability. Such systems have also been extensively employed for the delivery of small molecules, siRNA, antisense nucleotides, and small proteins.

Production of nanotechnology-based products via eco-friendly processes is quite a promising research area for the development of various food products. Though, to a large extent, important goals have been reached in achieving food products with controlled release characteristics, the cost of such products production is still the overriding factor that hinders the introduction of more sophisticated controlled release technologies in food technology. The potential health benefits of probiotics and nutraceuticals are very well known. Thus, the addition of nutraceutical ingredients and nanocarriers for maintaining the stability of these materials will justify the additional cost of nanoencapsulation technology. Nanocarriers-based delivery systems will be more commercially available in the markets than in the past, as indicated from the published literature. It looks like these new technologies are feasible and promising tools for the food product industries and convince and persuade manufacturers to introduce nanocarriers-based ingredients into their food products as a part of their marketing strategy. Nanocarriers-based technologies can minimize various unique problems such food products via safeguarding their stability and preserving safety, appeal (texture, color, odor, and taste), stability, low cost, and nutritional value. Thus, it is concluded from published literature that nanocarriers-based delivery systems will have more commercial status in the market in the near future.

KEYWORDS

- **critical solution temperature**
- **European Commission**
- **gastrointestinal tract**
- **lower critical solution temperature**
- **non-steroidal anti-inflammatory drugs**
- **polyglycolic acid**
- **solid lipid nanoparticles**

REFERENCES

1. Molinski, T. F., (1993). Developments in marine natural products. Receptor-specific bioactive compounds. *Journal of Natural Products, 56*(1), 1–8.
2. Grabley, S., & Thiericke, R., (1999). Bioactive agents from natural sources: Trends in discovery and application. In: *Thermal Biosensors, Bioactivity, Bioaffinitty* (pp. 101–154). Springer.
3. Leach, A. R., et al., (2010). Three-dimensional pharmacophore methods in drug discovery. *Journal of Medicinal Chemistry, 53*(2), 539–558.
4. Huang, Q., Yu, H., & Ru, Q., (2010). Bioavailability and delivery of nutraceuticals using nanotechnology. *Journal of Food Science, 75*(1), R50–R57.
5. Riehemann, K., et al., (2009). Nanomedicine: Challenge and perspectives. *Angewandte Chemie International Edition, 48*(5), 872–897.
6. Ferrari, M., (2005). Cancer nanotechnology: Opportunities and challenges. *Nature Reviews Cancer, 5*(3), 161–171.
7. Amin, A. R., et al., (2009). Perspectives for cancer prevention with natural compounds. *Journal of Clinical Oncology, 27*(16), 2712.
8. Trottier, G., et al., (2010). Nutraceuticals and prostate cancer prevention: A current review. *Nature Reviews Urology, 7*(1), 21.
9. Kalra, E. K., (2003). Nutraceutical-definition and introduction. *Aaps. Pharmsci., 5*(3), 27, 28.
10. McClements, D., (2012). Requirements for food ingredient and nutraceutical delivery systems. In: *Encapsulation Technologies and Delivery Systems for Food Ingredients and Nutraceuticals* (pp. 3–18). Elsevier.
11. Rapaka, R. S., & Coates, P. M., (2006). Dietary supplements and related products: A brief summary. *Life Sciences, 78*(18), 2026–2032.
12. Lee, S., (2017). Strategic design of delivery systems for nutraceuticals. In: *Nanotechnology Applications in Food* (pp. 65–86). Elsevier.
13. Augustin, M. A., & Hemar, Y., (2009). Nano-and micro-structured assemblies for encapsulation of food ingredients. *Chemical Society Reviews, 38*(4), 902–912.
14. Tiede, K., et al., (2008). Detection and characterization of engineered nanoparticles in food and the environment. *Food Additives and Contaminants, 25*(7), 795–821.

15. Brower, V., (1998). Nutraceuticals: Poised for a healthy slice of the healthcare market? *Nature Biotechnology, 16*(8), 728–731.
16. McClements, D. J., (2012). Advances in fabrication of emulsions with enhanced functionality using structural design principles. *Current Opinion in Colloid & Interface Science, 17*(5), 235–245.
17. Ting, Y., et al., (2014). Common delivery systems for enhancing *in vivo* bioavailability and biological efficacy of nutraceuticals. *Journal of Functional Foods, 7*, 112–128.
18. Rein, M. J., et al., (2013). Bioavailability of bioactive food compounds: A challenging journey to bio-efficacy. *British Journal of Clinical Pharmacology, 75*(3), 588–602.
19. Munin, A., & Edwards-Lévy, F., (2011). Encapsulation of natural polyphenolic compounds: A review. *Pharmaceutics, 3*(4), 793–829.
20. Vorhies, J. S., & Nemunaitis, J. J., (2009). Synthetic vs. natural/biodegradable polymers for delivery of shRNA-based cancer therapies. In: *Macromolecular Drug Delivery* (pp. 11–29). Springer.
21. Nicolas, J., et al., (2013). Design, functionalization strategies and biomedical applications of targeted biodegradable/biocompatible polymer-based nanocarriers for drug delivery. *Chemical Society Reviews, 42*(3), 1147–1235.
22. Joshi, J. R., & Patel, R. P., (2012). Role of biodegradable polymers in drug delivery. *Int. J. Curr. Pharm. Res., 4*(4), 74–81.
23. Payne, R. G., et al., (2002). Development of an injectable, in situ cross-linkable, degradable polymeric carrier for osteogenic cell populations: Part 1. Encapsulation of marrow stromal osteoblasts in surface crosslinked gelatin microparticles. *Biomaterials, 23*(22), 4359–4371.
24. Franz, J., et al., (1998). Adjuvant efficacy of gelatin particles and microparticles. *International Journal of Pharmaceutics, 168*(2), 153–161.
25. Latha, M., et al., (1995). Bioavailability of theophylline from glutaraldehyde cross-linked casein microspheres in rabbits following oral administration. *Journal of Controlled Release, 34*(1), 1–7.
26. Swatschek, D., et al., (2002). Microparticles derived from marine sponge collagen (SCMPs): Preparation, characterization, and suitability for dermal delivery of all-trans-retinol. *European Journal of Pharmaceutics and Biopharmaceutics, 54*(2), 125–133.
27. Alex, R., & Bodmeier, R., (1990). Encapsulation of water-soluble drugs by a modified solvent evaporation method. I. Effect of process and formulation variables on drug entrapment. *Journal of Microencapsulation, 7*(3), 347–355.
28. Rössler, B., Kreuter, J., & Scherer, D., (1995). Collagen microparticles: Preparation and properties. *Journal of Microencapsulation, 12*(1), 49–57.
29. Chen, L., Remondetto, G. E., & Subirade, M., (2006). Food protein-based materials as nutraceutical delivery systems. *Trends in Food Science & Technology, 17*(5), 272–283.
30. Tomlinson, E., & Burger, J., (1985). [3] Incorporation of water-soluble drugs in albumin microspheres. In: *Methods in Enzymology* (pp. 27–43). Elsevier.
31. Rosenberg, M., & Young, S., (1993). Whey proteins as microencapsulating agents. Microencapsulation of anhydrous milkfat-structure evaluation. *Food Structure, 12*(1), 4.
32. Beaulieu, L., et al., (2002). Elaboration and characterization of whey protein beads by an emulsification/cold gelation process: Application for the protection of retinol. *Biomacromolecules, 3*(2), 239–248.
33. Lazko, J., Popineau, Y., & Legrand, J., (2004). Soy glycinin microcapsules by simple coacervation method. *Colloids and Surfaces B: Biointerfaces, 37*(1, 2), 1–8.

34. Chen, S., et al., (2020). Co-delivery of curcumin and piperine in zein-carrageenan core-shell nanoparticles: Formation, structure, stability, and *in vitro* gastrointestinal digestion. *Food Hydrocolloids, 99*, 105334.
35. Wu, W., et al., (2020). Fabrication and characterization of resveratrol-loaded gliadin nanoparticles stabilized by gum Arabic and chitosan hydrochloride. *LWT*, 109532.
36. Friess, W., (1998). Collagen-biomaterial for drug delivery. *European Journal of Pharmaceutics and Biopharmaceutics, 45*(2), 113–136.
37. Fathi, M., Donsi, F., & McClements, D. J., (2018). Protein-based delivery systems for the nanoencapsulation of food ingredients. *Comprehensive Reviews in Food Science and Food Safety, 17*(4), 920–936.
38. Sebak, S., et al., (2010). Human serum albumin nanoparticles as an efficient noscapine drug delivery system for potential use in breast cancer: Preparation and *in vitro* analysis. *International Journal of Nanomedicine, 5*, 525.
39. Bae, S., et al., (2012). Doxorubicin-loaded human serum albumin nanoparticles surface-modified with TNF-related apoptosis-inducing ligand and transferrin for targeting multiple tumor types. *Biomaterials, 33*(5), 1536–1546.
40. Lee, S., Alwahab, N. S. A., & Moazzam, Z. M., (2013). Zein-based oral drug delivery system targeting activated macrophages. *International Journal of Pharmaceutics, 454*(1), 388–393.
41. Kwiecień, I., & Kwiecień, M., (2018). Application of polysaccharide-based hydrogels as probiotic delivery systems. *Gels, 4*(2), 47.
42. Coelho, J. F., et al., (2010). Drug delivery systems: Advanced technologies potentially applicable in personalized treatments. *EPMA Journal, 1*(1), 164–209.
43. Makadia, H. K., & Siegel, S. J., (2011). Poly lactic-co-glycolic acid (PLGA) as biodegradable controlled drug delivery carrier. *Polymers,* 3(3), 1377–1397.
44. Arunkumar, R., et al., (2015). Biodegradable poly (lactic-co-glycolic acid)-polyethylene glycol nanocapsules: An efficient carrier for improved solubility, bioavailability, and anticancer property of lutein. *Journal of Pharmaceutical Sciences, 104*(6), 2085–2093.
45. Griffin, J., et al., (2011). Salicylic acid-derived poly (anhydride-ester) electrospun fibers designed for regenerating the peripheral nervous system. *Journal of Biomedical Materials Research Part A, 97*(3), 230–242.
46. Engesæter, L. B., Sudmann, B., & Sudmann, E., (1992). Fracture healing in rats inhibited by locally administered indomethacin. *Acta Orthopaedica Scandinavica, 63*(3), 330–333.
47. Shim, W. S., et al., (2007). pH-and temperature-sensitive, injectable, biodegradable block copolymer hydrogels as carriers for paclitaxel. *International Journal of Pharmaceutics, 331*(1), 11–18.
48. Park, T. G., (1999). Temperature modulated protein release from pH/temperature-sensitive hydrogels. *Biomaterials, 20*(6), 517–521.
49. Wang, B., et al., (2008). Synthesis and properties of pH and temperature-sensitive P (NIPAAm-co-DMAEMA) hydrogels. *Colloids and Surfaces B: Biointerfaces, 64*(1), 34–41.
50. Lutz, J. F., Akdemir, Ö., & Hoth, A., (2006). Point by point comparison of two thermosensitive polymers exhibiting a similar LCST: Is the age of poly (NIPAM) over? *Journal of the American Chemical Society, 128*(40), 13046–13047.
51. Plunkett, K. N., et al., (2006). PNIPAM chain collapse depends on the molecular weight and grafting density. *Langmuir, 22*(9), 4259–4266.

52. Rejinold, N. S., et al., (2014). Dual drug encapsulated thermo-sensitive fibrinogen-graft-poly (N-isopropyl acrylamide) nanogels for breast cancer therapy. *Colloids and Surfaces B: Biointerfaces, 114*, 209–217.
53. Traitel, T., Goldbart, R., & Kost, J., (2008). Smart polymers for responsive drug-delivery systems. *Journal of Biomaterials Science, Polymer Edition, 19*(6), 755–767.
54. Alexander, C., (2006). Temperature-and pH-responsive smart polymers for gene delivery. *Expert Opinion on Drug Delivery, 3*(5), 573–581.
55. Qiu, Y., & Park, K., (2001). Environment-sensitive hydrogels for drug delivery. *Advanced Drug Delivery Reviews, 53*(3), 321–339.
56. Fujiwara, G. M., et al., (2013). Production and characterization of alginate-starch-chitosan microparticles containing stigmasterol through the external ionic gelation technique. *Brazilian Journal of Pharmaceutical Sciences, 49*(3), 537–547.
57. Arunkumar, R., Prashanth, K. V. H., & Baskaran, V., (2013). Promising interaction between nanoencapsulated lutein with low molecular weight chitosan: Characterization and bioavailability of lutein *in vitro* and *in vivo*. *Food Chemistry, 141*(1), 327–337.
58. Lee, S., et al., (2007). Polyketal microparticles: A new delivery vehicle for superoxide dismutase. *Bioconjugate Chemistry, 18*(1), 4–7.
59. Lee, S., et al., (2009). Solid polymeric microparticles enhance the delivery of siRNA to macrophages *in vivo*. *Nucleic Acids Research, 37*(22), e145–e145.
60. Van, M. G., Voelker, D. R., & Feigenson, G. W., (2008). Membrane lipids: Where they are and how they behave. *Nature Reviews Molecular Cell Biology, 9*(2), 112–124.
61. Bozzuto, G., & Molinari, A., (2015). Liposomes as nanomedical devices. *International Journal of Nanomedicine, 10*, 975.
62. Immordino, M. L., Dosio, F., & Cattel, L., (2006). Stealth liposomes: Review of the basic science, rationale, and clinical applications, existing and potential. *International Journal of Nanomedicine, 1*(3), 297.
63. Chakraborty, C., et al., (2009). Future prospects of nanoparticles on brain targeted drug delivery. *Journal of Neuro-Oncology, 93*(2), 285–286.
64. Blasi, P., et al., (2009). Lipid nanoparticles for drug delivery to the brain: *In vivo* veritas. *Journal of Biomedical Nanotechnology, 5*(4), 344–350.
65. Corvo, M. L., et al., (2002). Superoxide dismutase entrapped in long-circulating liposomes: Formulation design and therapeutic activity in rat adjuvant arthritis. *Biochimica et Biophysica Acta (BBA)-Biomembranes, 1564*(1), 227–236.
66. Weber, S., Zimmer, A., & Pardeike, J., (2014). Solid lipid nanoparticles (SLN) and nanostructured lipid carriers (NLC) for pulmonary application: A review of the state of the art. *European Journal of Pharmaceutics and Biopharmaceutics, 86*(1), 7–22.
67. Severino, P., et al., (2012). Current state-of-art and new trends on lipid nanoparticles (SLN and NLC) for oral drug delivery. *Journal of Drug Delivery, 2012*.
68. Das, S., & Chaudhury, A., (2011). Recent advances in lipid nanoparticle formulations with solid matrix for oral drug delivery. *AAPS Pharmscitech, 12*(1), 62–76.
69. Solans, C., et al., (2005). Nano-emulsions. *Current Opinion in Colloid & Interface Science, 10*(3, 4), 102–110.
70. Sivakumar, M., Tang, S. Y., & Tan, K. W., (2014). Cavitation technology-a greener processing technique for the generation of pharmaceutical nanoemulsions. *Ultrasonics Sonochemistry, 21*(6), 2069–2083.

71. Yu, H., & Huang, Q., (2013). Bioavailability and delivery of nutraceuticals and functional foods using nanotechnology. *Bio-Nanotechnology: A Revolution in Food, Biomedical and Health Sciences*, 593–604.
72. Augustin, M. A., et al., (2011). Effects of microencapsulation on the gastrointestinal transit and tissue distribution of a bioactive mixture of fish oil, tributyrin and resveratrol. *Journal of Functional Foods, 3*(1), 25–37.
73. Frede, K., et al., (2014). Stability and cellular uptake of lutein-loaded emulsions. *Journal of Functional Foods, 8*, 118–127.
74. Liu, X., et al., (2018). Nanoemulsion-based delivery systems for nutraceuticals: Influence of long-chain triglyceride (LCT) type on *in vitro* digestion and astaxanthin bio-accessibility. *Food Biophysics, 13*(4), 412–421.
75. Zheng, J., et al., (2014). Improving intracellular uptake of 5-demethyltangeretin by food-grade nanoemulsions. *Food Research International, 62*, 98–103.
76. Qian, C., et al., (2012). Nanoemulsion delivery systems: Influence of carrier oil on β-carotene bioaccessibility. *Food Chemistry, 135*(3), 1440–1447.
77. Li, Y., Xiao, H., & McClements, D. J., (2012). Encapsulation and delivery of crystalline hydrophobic nutraceuticals using nanoemulsions: Factors affecting polymethoxyflavone solubility. *Food Biophysics, 7*(4), 341–353.
78. Zilberstein, B., et al., (2007). Digestive tract microbiota in healthy volunteers. *Clinics, 62*(1), 47–54.
79. Bernela, M., et al., (2018). Nano-based delivery system for nutraceuticals: The potential future, In: *Advances in Animal Biotechnology and its Applications* (pp. 103–117). Springer.
80. Paul, S., & Dewangan, D., (2015). Nanotechnology and neutraceuticals. *Int. J. Nanomater. Nanotechnol. Nanomed., 1*, 30–33.
81. Davitt, C. J., & Lavelle, E. C., (2015). Delivery strategies to enhance oral vaccination against enteric infections. *Advanced Drug Delivery Reviews, 91*, 52–69.
82. Boegh, M., & Nielsen, H. M., (2015). Mucus as a barrier to drug delivery-understanding and mimicking the barrier properties. *Basic & Clinical Pharmacology & Toxicology, 116*(3), 179–186.
83. Gleeson, J. P., Ryan, S. M., & Brayden, D. J., (2016). Oral delivery strategies for nutraceuticals: Delivery vehicles and absorption enhancers. *Trends in Food Science & Technology, 53*, 90–101.
84. Mansuri, S., et al., (2016). Mucoadhesion: A promising approach in drug delivery system. *Reactive and Functional Polymers, 100*, 151–172.
85. Hariharan, S., et al., (2006). Design of estradiol loaded PLGA nanoparticulate formulations: A potential oral delivery system for hormone therapy. *Pharmaceutical Research, 23*(1), 184–195.
86. Daugherty, A. L., & Mrsny, R. J., (1999). Transcellular uptake mechanisms of the intestinal epithelial barrier part one. *Pharmaceutical Science & Technology Today, 2*(4), 144–151.
87. Lerner, A., & Matthias, T., (2015). Changes in intestinal tight junction permeability associated with industrial food additives explain the rising incidence of autoimmune disease. *Autoimmunity Reviews, 14*(6), 479–489.
88. Bischoff, S. C., et al., (2014). Intestinal permeability-a new target for disease prevention and therapy. *BMC Gastroenterology, 14*(1), 189.

89. Maher, S., Mrsny, R. J., & Brayden, D. J., (2016). Intestinal permeation enhancers for oral peptide delivery. *Advanced Drug Delivery Reviews, 106*, 277–319.
90. Bohn, T., et al., (2015). Mind the gap-deficits in our knowledge of aspects impacting the bioavailability of phytochemicals and their metabolites—a position paper focusing on carotenoids and polyphenols. *Molecular Nutrition & Food Research, 59*(7), 1307–1323.
91. Rastogi, H., & Jana, S., (2016). Evaluation of physicochemical properties and intestinal permeability of six dietary polyphenols in human intestinal colon adenocarcinoma Caco-2 cells. *European Journal of Drug Metabolism and Pharmacokinetics, 41*(1), 33–43.
92. Yu, M., et al., (2016). Advances in the transepithelial transport of nanoparticles. *Drug Discovery Today, 21*(7), 1155–1161.
93. Tang, D. W., et al., (2013). Characterization of tea catechins-loaded nanoparticles prepared from chitosan and an edible polypeptide. *Food Hydrocolloids, 30*(1), 33–41.
94. Renukuntla, J., et al., (2013). Approaches for enhancing oral bioavailability of peptides and proteins. *International Journal of Pharmaceutics, 447*(1, 2), 75–93.
95. Guri, A., Gülseren, I., & Corredig, M., (2013). Utilization of solid lipid nanoparticles for enhanced delivery of curcumin in cocultures of HT29-MTX and Caco-2 cells. *Food & Function, 4*(9), 1410–1419.
96. Yun, Y., Cho, Y. W., & Park, K., (2013). Nanoparticles for oral delivery: Targeted nanoparticles with peptidic ligands for oral protein delivery. *Advanced Drug Delivery Reviews, 65*(6), 822–832.
97. Li, X., (2011). *Oral Bioavailability: Basic Principles, Advanced Concepts, and Applications* (Vol. 16). John Wiley & Sons.
98. Kettiger, H., et al., (2013). Engineered nanomaterial uptake and tissue distribution: From cell to organism. *International Journal of Nanomedicine, 8*, 3255.
99. Des, R. A., et al., (2007). An improved *in vitro* model of human intestinal follicle-associated epithelium to study nanoparticle transport by M cells. *European Journal of Pharmaceutical Sciences, 30*(5), 380–391.
100. Acosta, E., (2009). Bioavailability of nanoparticles in nutrient and nutraceutical delivery. *Current Opinion in Colloid & Interface Science, 14*(1), 3–15.
101. Plapied, L., et al., (2011). Fate of polymeric nanocarriers for oral drug delivery. *Current Opinion in Colloid & Interface Science, 16*(3), 228–237.
102. Zhang, Z. H., et al., (2013). Studies on lactoferrin nanoparticles of gambogic acid for oral delivery. *Drug Delivery, 20*(2), 86–93.
103. Misaka, S., Müller, F., & Fromm, M. F., (2013). Clinical relevance of drug efflux pumps in the gut. *Current Opinion in Pharmacology, 13*(6), 847–852.
104. Wise, D. L., (2000). *Handbook of Pharmaceutical Controlled Release Technology*. CRC Press.
105. Zhang, M., et al., (2003). Simulation of drug release from biodegradable polymeric microspheres with bulk and surface erosions. *Journal of Pharmaceutical Sciences, 92*(10), 2040–2056.
106. Chirico, S., et al., (2007). Analysis and modeling of swelling and erosion behavior for pure HPMC tablet. *Journal of Controlled Release, 122*(2), 181–188.
107. Arifin, D. Y., Lee, L. Y., & Wang, C. H., (2006). Mathematical modeling and simulation of drug release from microspheres: Implications to drug delivery systems. *Advanced Drug Delivery Reviews, 58*(12, 13), 1274–1325.

108. Bealer, E. J., et al., (2020). Protein-polysaccharide composite materials: Fabrication and applications. *Polymers, 12*(2), 464.
109. Fathi, M., Martin, A., & McClements, D. J., (2014). Nanoencapsulation of food ingredients using carbohydrate-based delivery systems. *Trends in Food Science & Technology, 39*(1), 18–39.
110. Fathi, M., et al., (2013). Cellular automata modeling of hesperetin release phenomenon from lipid nanocarriers. *Food and Bioprocess Technology, 6*(11), 3134–3142.
111. Fathi, M., et al., (2013). Hesperetin-loaded solid lipid nanoparticles and nanostructure lipid carriers for food fortification: Preparation, characterization, and modeling. *Food and Bioprocess Technology, 6*(6), 1464–1475.
112. Pandey, M., Verma, R. K., & Saraf, S. A., (2010). Nutraceuticals: New era of medicine and health. *Asian J. Pharm. Clin. Res., 3*(1), 11–15.
113. Pradhan, N., et al., (2015). Facets of nanotechnology as seen in food processing, packaging, and preservation industry. *BioMed Research International, 2015.*
114. Qi, L., et al., (2004). Preparation and antibacterial activity of chitosan nanoparticles. *Carbohydrate Research, 339*(16), 2693–2700.
115. Javeri, I., (2016). Application of "nano" nutraceuticals in medicine. In: *Nutraceuticals* (pp. 189–192). Elsevier.
116. Scampicchio, M., et al., (2008). Amperometric electronic tongue for food analysis. *Microchimica Acta, 163*(1, 2), 11–21.
117. Sondi, I., & Salopek-Sondi, B., (2004). Silver nanoparticles as antimicrobial agent: A case study on *E. coli* as a model for gram-negative bacteria. *Journal of Colloid and Interface Science, 275*(1), 177–182.

CHAPTER 3

Nanoparticulate Approaches for Improved Nutrient Bioavailability

ABDUL QADIR, MOHD. AQIL, and DIPAK KUMAR GUPTA

Department of Pharmaceutics, School of Pharmaceutical Education and Research, Jamia Hamdard (Deemed University), M. B. Road, New Delhi – 110062, India

ABSTRACT

Nanoparticles (NPs) are described as minute dispersions particles or solid particles with a size varying 10–1000 nm. The NPs are made either by dissolving, entrapping, encapsulating, or attaching to a nanoparticle matrix. Currently, biodegradable polymeric NPs, especially those coated with a hydrophilic polymer like poly ethylene glycol are quite a hit as potential drug delivery devices due to their advantages. Oral delivery of NPs is a desirable route to dispense therapeutics or bioactive compounds in long-term treatments. In order to reduce the carrier-induced undesirable cytotoxicity, food polymers are best to employ in designing such delivery systems. Different methods of preparation of NPs include: Nanoprecipitation, nano-emulsion technique and reverse-phase evaporation etc. The parameters used to evaluate the NPs are: particle Size, surface charge, surface morphology, encapsulation efficiency, differential thermal analysis (DTA) and differential scanning calorimetry (DSC) etc. Methods to enhance oral bioavailability of nutraceuticals include: safety of labile compounds, delay of gastric retention time, lymphatic uptake etc., In the food industry, the application of nano-technology is at the infant stage due to insufficient knowledge regarding the safety of NPs. Food proteins have exhibited potential for development and incorporation in nutraceuticals and provide controlled release via the oral route. To make NPs widely amenable, it is important to explore and develop methods for assuring their safety and characterization. Combinatorial techniques can be applied for characterization of NPs in food matrices.

Future research should focus on the development and validation of methods for analysis of NPs in food and other samples.

3.1 INTRODUCTION TO NANOPARTICLES (NPS)

Nanoparticles (NPs) are described as minute dispersions particles or solid particles with a size varying 10–1000 nm. The NPs are made either by dissolving, entrapping, encapsulating, or attaching to a nanoparticle matrix. NPs are defined as per the method of preparation used, such as NPs, nanospheres, or nanocapsules. In nanocapsules systems, API is entrapped in a cavity enclosed by a unique polymer membrane, while in nanospheres system API is uniformly suspended physically in a matrix system. Currently, biodegradable polymeric NPs, especially those coated with a hydrophilic polymer like poly (ethylene glycol) (PEG) identified as long-circulating particles are quite hit as potential drug delivery devices due to their numerous advantages such as the capability to disseminate for a prolonged period of time, targeted drug delivery, as transporters of DNA in gene therapy, ability to deliver proteins, peptides, and genes [1–5].

Numerous kinds of nanoscale materials are produced by nanotechnology [6]. Researchers discovered the versatility of NPs when they found the nanosize alter the physiochemical characterizes of a substance, e.g., the optical properties. The physical properties of 20 nm gold (Au), platinum (Pt), palladium (Pd), and silver (Ag) NPs are wine red color, yellowish gray, black, and dark black colors, respectively. These unique properties of NPs such as colors, size, and shape, are utilized in bio-imaging applications [7]. Any changes in these properties alter the absorption properties of the NPs, and therefore different absorption colors are detected. NPs are not simple molecules but complex one and contain three layers:

1. **Upper Layers:** i.e., the surface layer, which are activated by type of small molecules, metal ions, surfactants, and polymers.
2. **The Shell Layer:** Completely different from the core layer in chemically material.
3. **The Core:** Fundamentally central portion of the NP and typically refers to the NP itself [8, 9].

NPs possess numerous benefits such as: reduce the toxicity, improve bioactivity, advance targeting, and offer multipurpose means to control the release profile of the encapsulated moiety [10].

Oral delivery of NPs is a desirable route to dispense therapeutics or bioactive compounds in long-term treatments due to various advantages, which include: patient compliance and ease of administration. Drugs that are made on polymer-based delivery systems have been explored extensively for the biomedical and pharmaceutical sectors in order to enhance the targeted delivery of bioactive compounds and to protect them as well. The key mode of actions entailed for modification of bioactive molecule absorption by polymeric NPs are:

- Safeguarding the API or bioactive molecule from the abrasive environment of the GI tract;
- Extending residence time in the gut via mucoadhesion;
- Endocytosis of the particles and Permeabilizing effect of the polymer.

In order to reduce the carrier-induced undesirable cytotoxicity, food polymers are best to employ in evolving such delivery systems in oral consumption. Food origin polymers are best due to their property similar to soft condensed matter with which we interact daily. Another added advantage is: natural, soft materials, biodegradable, biocompatible, and biofunctional. These food-based polymers are optimum choices to deliver API in therapeutics and functional foods. These include nanostructured vehicles like association colloids, lipid-based nano-encapsulator, bio-polymeric NPs, nanotubes, nanoemulsions (NEs), and nano-fibers made from food-grade ingredients such as food biopolymers (proteins, carbohydrates), fats, low molecular weight surfactants and co-polymers (protein-carbohydrate conjugates) [11–13].

3.2 PREPARATION OF NANOPARTICLES (NPS)

Pharmaceutical industries have vested their interest in the effective delivery of bioactive agents, peptides, and APIs to the systemic circulation and eventually to the targeted organ or cells due to current progress and development in biotechnology.

3.2.1 NANOPRECIPITATION

This method is applicable to lipophilic drugs because of the miscibility of the solvent with the aqueous phase [14]. In short, organic solvents such as acetone are used to dissolve the lipids and the drug. Then this organic mixture is mixed with water containing surfactant. Promptly after this step, the organic solvent is separated from the colloidal suspension under reduced pressure by Rota evaporation. The ensuing particle suspension is filtered via a 1.0-mm cellulose nitrate membrane filter; tailored in size by mechanical extrusion to obtain a nanoparticle formulation [15].

3.2.2 NANOEMULSION TECHNIQUES

Among many methods for preparing NPs, nanoemulsion is one. It is defined as the heterogeneous mixture of different oils with minute-diameter oil droplets in water (20–500 nm). They have potential application in numerous chemicals, pharmaceutical, and cosmetic industries due to their safe transdermal applications worldwide. There are numerous benefits of NEs like the possibility to solubilize hydrophobic compounds in the oil phase, the ability to customize the surface of the oil droplets with polymers to prolong circulation times, and passive targeting of tumors and/or actively targeting ligands [14]. NEs composed of oils are formulated by coarse homogenization trailed by high-energy ultrasonication method [17, 18]. Briefly, the aqueous phase is prepared by adding soya lecithin into the deionized water, and stirred at high speed. Organic solvents are used to dissolve the candidate drug and then dispersed in oil. Subsequently, evaporation of the aqueous phase is done by heating at 70–75°C. The remainder of the oil phase which comprises the entrapped drug is slowly added to the aqueous phase to make a uniform solution which eventually makes the coarse oil-in-water (O/W) emulsion [19]. The obtained coarse emulsion is ultrasonicated to get the desired nano-sized oil droplets.

3.2.3 REVERSE-PHASE EVAPORATION

This is a widely used technique to make NPs of various types of drugs. Lipids such as selective phospholipids, in pure form or mixed with other lipids like cholesterol or long-chain alcohols are used. This lipid combination is further mixed with organic solvent, afterwards the solvent is isolated under reduced pressure via Rota evaporator. The resultant system is then purged with nitrogen to get reverse-phase vesicles which are formed after re-dissolving the lipids in the organic phase. To enhance the solubility of lipids in ether, chloroform or methanol can be added. The system is preserved under nitrogen and the water phase and the resulting two-phase system is sonicated for 2–5 min, until the mixture becomes either clear or a homogeneous opalescent dispersion that does not separate for at least 30 min after sonication. The organic solvent is then separated by Rota vapor under reduced pressure. After removal of bulk of the solvent, viscous gel appears; later an aqueous suspension is formed after 5–10 minutes. Finally, the obtained product is either dialyzed or centrifuged to eliminate non-encapsulated material and residual organic solvent [15].

3.3 EVALUATION OF NANOPARTICLES (NPS)

Evaluation of NPs can be performed for some parameters, which are discussed in subsections.

3.3.1 PARTICLE SIZE AND SURFACE CHARGE

Determination of particle size and zeta potential of the NPs was done by photon correlation spectroscopy and laser Doppler Anemometry, using a Mastersizer 2000 and Zetasizer 2000, respectively (Malvern Instruments, South borough, MA). Before analysis, samples were diluted with suitable media and filtered (0.22 mm pore size) to obtain an appropriate range. The size analysis was performed at 25°C. It was recorded for 180s for each measurement. The polydispersity values of nanoparticle dispersions after homogenization varied between 0.2 and 0.5. The mean hydrodynamic diameter was generated by cumulative analysis. The zeta potential measurement was analyzed using an aqueous dip cell in the automatic mode. Particles with zeta potentials > +30 mV or < −30 mV generally marks the stability of the formulation [20, 21].

3.3.2 SURFACE AND INTERIOR MORPHOLOGY

Nanoparticle morphology was analyzed under a transmission electron microscopy (TEM): The freshly-prepared NPs suspension diluted with suitable media and put on a copper grid sealed with nitrocellulose and allowed to get dry then stained with phosphotungstic acid (1% w/v). It is further analyzed by a transmission electron microscope [20].

3.3.3 ENCAPSULATION EFFICIENCY (EE)

The preparation of nanoparticles is a blend of coated and uncoated (free drug) medicament portions [22]. The separation between the coated and uncoated medicament is the initial step of the technique that can be determined by using a dialysis membrane in which the nanoparticle sample is immersed in a phosphate buffer solution (PBS) for 120 minutes [23].

3.3.4 DIFFERENTIAL THERMAL ANALYSIS (DTA) AND DIFFERENTIAL SCANNING CALORIMETRY (DSC)

Measure the temperature and heat flow difference between a sample and a reference material. They can be used to measure phase changes, melting point, purity, evaporation, sublimation, crystallization, pyrolysis, heat capacity, polymerization, aggregation, compatibility, etc. The methods can be used to track the degradation process of NPs by identifying the formed by-product, and simultaneously to inspect the food quality change along with the addition of NPs [24].

3.3.5 STORAGE STABILITY

Nanoparticle dispersions were stored at 4°C for 20 days. Particle size and turbidity were analyzed immediately after preparation and after storage for 1, 3, 10, and 20 days [20].

3.3.6 IN VITRO RELEASE STUDIES

The *in-vitro* drug release study was performed by the dialysis tube diffusion method. Some milliliters of the formulation should be placed in the dialysis bag that should be tied in such a way that air could not pass through it. The dialysis bag placed in the cell containing the suitable aqueous medium and should be maintained at 37°C with continuous agitation. The cell should be closed to avoid vaporization of the aqueous medium. Samples of the dialysate are then taken out at different time intervals, and at the same time, the same amount of same fresh sample should be added to keep the volume of the cell constant. Withdrawing of samples should be performed in triplicate and then analyzed for the estimation of drugs by using any chromatographic technique [25, 26].

3.4 BIOAVAILABILITY OF NUTRACEUTICALS

According to the FDA, the definition of bioavailability is the rate and amount of API absorbed from the dosage form and becomes available at the site of action. Bioavailability in its definition explains two key points: (i) the absorption rate-how quickly the bioactive agent which goes into the systemic circulation and (ii) the absorption extent-the amount of bioactive available in the systematic circulation. Poor bioavailability indicates the inability of the API or bioactive agents to reach the targeted site hence reducing the therapeutic efficacy, which leads to failed biological results. Typically, some different steps determine the biological fate of bioactive agents after ingestion:

- release of the bioactive agent from the dietary matrix;
- Digestion by enzymes within the intestine;
- Adherence and uptake by the mucosal layer of the intestine;
- Transfer across the gut wall (passing through and/or between the epithelium cells) to the lymphatic system or portal vein;
- Systemic distribution and deposition (storage);
- Metabolic and functional use;
- Excretion (via urine or feces).

The efficacy of nutraceutical products in preventing diseases depends on protective the bioavailability of the active ingredients. This represents a difficult challenge, given that only a small proportion of molecules remain available following oral administration, due to insufficient gastric residence time, low permeability and/or solubility within the gut, as well as instability under conditions encountered in food processing or in the gastrointestinal (GI) tract, all of which limit the activity and potential health benefits of nutraceutical molecules [18].

A number of external and internal factors affect the overall bioavailability rate of consumed drug. The external factors consist of: nature of the bioactive agent, composition, and structure of the food matrix; whereas internal factors comprise gender, age, health, nutrient status, and life phase. Several definitions of bioavailability are only confined it to nutrients which is defined as the amount of the nutrient used, stored, absorbed, or excreted. Macronutrient's nutrients such as: carbohydrates, proteins, and fats typically have very high bioavailability, which is more than 90% of the consumed amount in the gut. While the bioavailability of micronutrients (vitamins and minerals) and nutraceuticals (flavonoids and carotenoids) differ depending on their molecular and physicochemical properties. For instance, lipophilic bio-actives have limited or poor bioavailability due to their poor solubility, high melting point, chemical instability, and ingredient interactions. On the whole, numerous components regulate the bioavailability of nutrients and nutraceuticals as mentioned in Figure 3.1.

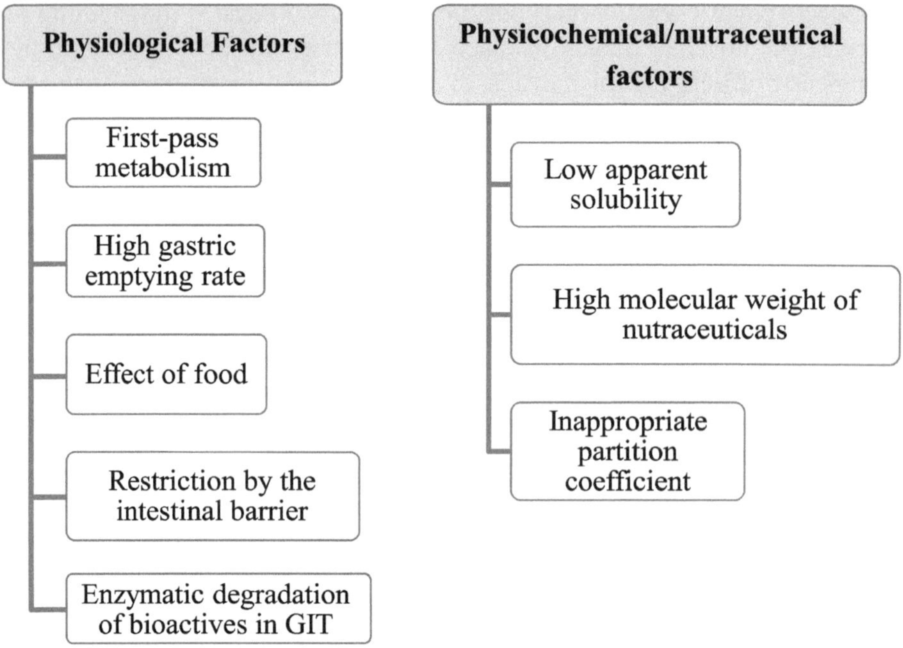

FIGURE 3.1 Factors determining the bioavailability of nutraceuticals and other bioactive components.

In this chapter, we focus on the effective nanoparticle delivery systems for nutraceuticals and related active ingredients; it is necessary to understand the biological processes that regulate uptake and bioavailability [27].

3.5 EXTERNAL FACTORS AFFECTING THE NUTRACEUTICAL BIOAVAILABILITY

Bioaccessibility is an important phase in the bioavailability of nutraceuticals in foods. There are several factors that influence the bioaccessibility of nutraceuticals, such as those factors that act before food ingestion (external

GIT factors) and factors that manifest during the food digestion (internal GIT factors). The previous studies provide more focus about the effect of GIT mechanisms and its environment, which affects the bioavailability of bioactive agents and their transport to tissues and organs. Additionally, external GIT factors should also be considered as they have a major influence on nutraceuticals bioavailability that act on the food matrix prior to oral consumption. Study and evaluation of such factors play an important role in the development of medical and functional foods designing via reverse engineering methods. With the advancement in technology, the food products during the process can be customized to get the best nutritional values, attractive sensory attributes along and health benefits. Major external GIT factors altering the bioavailability of nutraceuticals are physicochemical properties of nutraceuticals, characteristics of food matrices, properties of nutraceutical delivery systems, level of processing, and conditions of food storage. For instance, nutraceuticals solubility is a key factor that influences food processing and preparation of food nano-carriers as well as the behavior of nutraceuticals within the GIT [28].

3.6 METHODS TO ENHANCE ORAL BIOAVAILABILITY OF NUTRACEUTICALS

Nutraceuticals go through various physiological and physicochemical barriers after consumption that which decrease the dose reaching the systemic circulation. To increase the bioavailability of bioactive; many researchers have designed numerous products with diverse delivery systems to minimize or overcome these limiting factors. In this method, we will explore different techniques that investigators applied while designing an optimum delivery system for nutraceuticals.

3.6.1 SAFETY OF LABILE COMPOUNDS

Nutraceuticals on oral intake pass through complicated digestion processes which include physiological or physiochemical environmental changes. From the mouth to the colon, the vast GIT tract environment may cause instability to the chemical structures of active ingredients. Enumerate factors including pH variations, ionic strength, enzyme degradations, mechanistic motilities, etc., are possibly responsible for the degradation of nutraceuticals.

Hence, dosage forms which can avoid or protect these gastric instabilities and promote effective oral dosing are encouraged [29].

3.6.2 DELAY OF GASTRIC RETENTION TIME

The oral digestion process involves various complex steps that make the material pass through various sites in the GIT. Gastric retention time is not always sufficient for proper absorption to allow desired results of nutraceuticals which cause incomplete absorption of nutrients, excessive compound excretion, and a decrease in the dose-responsive efficiency of therapeutic purposes. Developing and designing such delivery systems which offer the following advantages is highly acceptable: reduce the gastric movement with higher viscosity or ability to slow down the gastric movement of bioactive compounds, enhances the residence time in the GI tract and make a greater percentage of bioactives available at the targeted site for absorption prior to gastric emptying [29].

3.6.3 IMPROVEMENT OF AQUEOUS SOLUBILITY

The bioactive compounds need to be solubilized, suspended, or dispersed in the aqueous environment of the GIT. Hydrophilic nutrients are easy to solubilize in aqueous environment compared to lipophilic compounds, which show poor solubility and often get precipitate as clusters after adding to the aqueous environment. Formation of these big clusters, which lack the desired particle size requirement, prevents intestinal absorption of lipophilic nutrients, and the latter get eliminated quickly via excretion mechanisms. Hence, low aqueous solubility is a key aspect which precincts the absorption of lipophilic compounds. Dosage forms which can overcome this hindrance and increase solubility or dispersion of such ingredients will invariably increase the concentration of the bioactive at the required site in the body; thus, will yield the desired biological results [29].

3.6.4 CONTROLLED/DELAYED RELEASE

A controlled release dosage form is required to maintain the desired concentration of the bioactive in the systemic circulation in order to obtain optimum therapeutic results. The novel controlled/delayed drug delivery systems are capable of providing this uniform and continuous release of bioactive substances in the system for prolonged time, simultaneously preventing the GIT complex environment and ensuring proper absorption. In this kind of delivery system, the vehicle carrying the active moiety gets disintegrated by the enzymes present in the GI milieu. The rate and time of release of bioactive compounds can be regulated by choosing the correct material with more tolerance to the digestive system and applying more protective layers on vehicle surfaces [29].

3.6.5 LYMPHATIC UPTAKE

Typically, the majority of the compounds with good aqueous solubility get absorbed in the blood via portal vein in the small intestine; then they are metabolized in the liver. For lipophilic substances; the lymphatic uptake system is a better option as it skips the first-pass metabolism and increases the bioavailability of parenteral drugs. The extent of lymphatic uptake depends on the capability of bioactive compounds to couple with lipoprotein within enterocyte. The lipid-based delivery systems with nanoscale particle size have been reported as an effective way to enhance direct intestinal lymphatic uptake of lipophilic compounds [29].

3.6.6 IMPROVEMENT OF INTESTINAL PERMEABILITY

Diverse form of materials has exhibited the ability to alter the physical barrier function of the intestinal wall. Consumption of dietary lipids can influence the intestinal membrane fluidity also by interaction with mucoadhesive polymers. While designing delivery vehicles, the focus should be kept on all components making up vehicle which can provide maximum support to intestinal membrane fluidity. For example, chitosan is a positively-charged mucoadhesive polymer which mitigates intestinal membrane integrity and tight junction widening that permits the paracellular absorption of lipophilic compounds [29].

3.6.7 MODULATION OF METABOLIC ACTIVITIES

The first barrier which diminishes the bioavailability is the limited absorption; the second one being the first-pass metabolism that decreases the systemic dosage level of nutraceuticals. Including materials which can prevent physical or chemical activity of metabolic enzymes on the delivery vehicle may reasonably improve the bioavailability of bioactive in the systemic circulation. While working on it, the safety should be kept on priority as these enzyme inhibitors may sometimes produce toxicity due to impaired detoxification activity [29].

3.7 RELEASE MECHANISMS OF NANOPARTICLES (NPS) FOR NUTRACEUTICALS

One of the main objectives of controlling drug release is to retain the drug concentration within the therapeutic range in the blood. Therefore, it is ideal to make drug carriers that have low dosing rate and provide controlled drug release. Zero-order drug release profile is aimed to get the controlled release in which the drug is uniformly released. Drug release from a nanocarrier is affected by various factors drug, polymer, and excipient, the ratio of ingredients, physical or chemical interaction among components, and manufacturing methods, etc. Drug release can be categorized into four segments: diffusion, solvent, chemical interaction, and stimulated release determined by the mechanism of drug discharge from the vehicle as shown in Figure 3.2 [30].

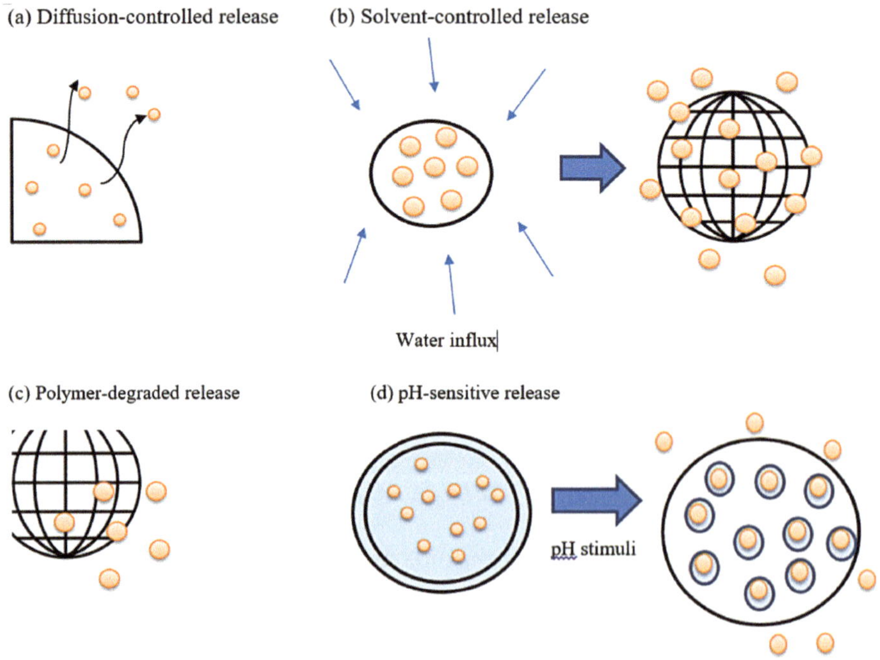

FIGURE 3.2 Various mechanisms of drug release from nano-carriers.

3.7.1 DIFFUSION-CONTROLLED RELEASE

In this mechanism, in a capsule-like systems, the drug is released where the drug is either melted or dispersed in a core. The movement of the drug happened due to difference in concentration gradient across the membrane. Initially, the drug gets dissolved in the central part and afterwards diffuses via membrane. The matrix type nanospheres do not have membrane barriers but have a diffusion-controlled release profile, in which the drug molecules are continuously released in the polymer matrix. Consequently, these systems usually have high initial release, but then over time, the release rate decreases due to an increase in the drug molecule diffusion distance inside the carrier [30].

3.7.2 SOLVENT-CONTROLLED RELEASE

The solvent-controlled release depends on osmosis-controlled release and swelling controlled release. The former happens in a vehicle covered with a semi-permeable polymeric membrane, in which water moves from outside of the carrier to the inside core loaded with drug, i.e., from low drug concentration to high drug concentration. This mechanism causes a zero-order release profile until the constant concentration gradient is maintained across the membrane. The hydrophilic polymeric systems get easily diffused into the system in an aqueous solution, including body fluids. The water diffused across the system causes swelling of the polymeric particles, which leads to drug release, and this kind of release is known as a swelling-controlled release system. Diffusion rate of water and the chain relaxation rate of polymers determine the drug release rate. The swelling-controlled systems consist of polymeric materials with three-dimensionally crosslinked network like hydrogels. In this, mesh size plays a key role in regulating the drug release. Semi-empirical Peppas model can be employed to calculate drug release from hydrogels. This model helps in defining the release mechanism (e.g., Fickian or non-Fickian diffusion) [31, 32]. Zero-order drug release can be achieved by swelling-controlled systems subjected to the initial drug distribution or polymer composition in the system [33].

3.7.3 DEGRADATION-CONTROLLED RELEASE

Biodegradable polymers are used for drug carrier constitution, for example, polyesters, polyamides, and polysaccharides which release the drug via enzymatic decomposition, which in turn cause ester or amide bond degradation or hydrolysis. A matrix composed of polymers such as poly-lacticcoglycolic acid (PLGA), poly-lactic acid (PLA), or polycaprolactone (PCL)consequently goes through the process of degradation, and altogether the matrix is deteriorated simultaneously. Contrarily, the matrix consisting of polymeric anhydrides or ortho-esters ordinarily erodes from the surface towards the center and results in faster degradation of the polymer than water diffuses into the matrix. However, in the NPs, there is a small-sized matrix and hence shows a low diffusion length for water and a restricted zone of crystallization. Overall polymer degradation continues to stimulate the release process rather than only surface erosion. Polymer systems that are biodegradable are usually preferred because they can degrade in the body [30].

3.7.4 STIMULI-CONTROLLED RELEASE

Internal or external stimuli controlled the release of drug from nano-carriers that are stimuli-responsive, like ionic strength, temperature, pH, sound, and electric or magnetic fields.

As it is feasible to confine the stimuli, these carriers for target-specific delivery of drugs have been explored. For example, for tumors site specific delivery of drugs, nanocarriers having pH sensitive linkers have been developed as well as to take advantage of weakly acidic pH of various solid tumors. To increase the difference between drug release, i.e., extracellular drug release and intracellular drug release pH-sensitive carriers are established. Temperature induced phase transition of the polymer results in the drug release from carriers that are thermo-sensitive [33].

3.8 ROLE OF NUTRACEUTICALS LOADED NANOPARTICLES (NPS) IN VARIOUS DISEASE CONDITIONS

General principles of nanotechnology are usually followed for nanoformulation of nutraceuticals. Consequently, the nanotechnology platforms are being exceedingly used to develop the delivery systems for nutraceuticals and natural bioactive products having poor solubility in water. The market prediction for these technologies proposes a multiplex increase in their marketing potential over the next 5 years. A list of nutraceuticals, materials used in the development of NPs, NPs size and their targets like specific tissue/tumor, are summarized in Table 3.1.

TABLE 3.1 The Phytochemicals and the Materials Used for Preparing Nanoparticles for Various Diseases

Phytochemicals	Targets Site	Materials Used
Curcumin	Leukemia, colon, breast, prostate cancer cells	PLGA
	Cervical cancer cells	Alginate-chitosan
	Breast cancer cells	Silk
	Cervical cancer cells	Casein
Ellagic Acid	Kidney	PLGA-polycaprolactone
Dibenzoylmethane	Cervical cancer cells	Polylactic acid
Eugenol	Bacteria	Chitosan
Ferulic acid	Liver	Bovine serum albumin
Naringenin	Liver	Polyvinyl acetate
Quercetin	Brain	Polylactide
	Brain, liver	PLGA
	Stomach, intestine	Glyceryl monoste
Resveratrol	Neuronal cell line	Poly-caprolactone-PEG
Simvastatin	Plasma	Glyceryl monooleate/poloxamer 407
Thymoquinone	Leukemic cells	PLGA
Ursolic acid	Liver	Soybean phospholipid-poloxamer 188

For example, low systemic bioavailability is shown by curcumin; by developing several nanoparticle formulations; its biologic activity and bioavailability can be greatly increased. Various studies have proposed curcumin for chemoprevention and as a safe alternative for cancer therapy with its anticancer properties and anti-inflammatory properties, yet the compound has not been accepted unequivocally by the cancer community. A clinical trial of colorectal cancer patients showed that the postoperatively administered curcumin's systemic bioavailability is less in humans [34].

Nanoparticulate curcumin has been developed by Bisht et al. by utilizing the cross-linked polymeric NPs that are composed of N-vinyl-2-pyrrolidone, PEG acrylate, and N-isopropyl acrylamide which has been subjected to pancreatic cancer cell lines test [35].

A purified compound viz. triptolide found in Chinese traditional origin medicine has immunosuppressive, antineoplastic, anti-inflammatory, and antifertility properties [36]. A study by Mei et al. reported penetration of triptolide into the skin and its anti-inflammatory efficacy got increased by the preparation of SLN (solid lipid nanoparticles) for transdermal delivery [37, 38]. It is predicted that the above strategy enhances the bioavailability of drug at the site of action, as well as reduces the dose required, and also reduces side effects that are dose-dependent such as stinging and irritation.

Another group of nutraceuticals, i.e., polyphenols have established anti-inflammatory properties and thus have high capability for cancer therapy. The low bioavailability and short half-life of polyphenols is a challenge for the treatment of cancer. Polyphenol-loaded NPs is one of the substitutes to free compounds [16].

3.9 CONCLUSION

In the food industry, the application of nanotechnology is at the infant stage due to insufficient knowledge regarding the safety of NPs in food and food-related products like: (a) which hinders its further development; (b) consumers are also cautious to consume NPs food products due to its uncertain safety profile and potential health risk; (c) this constrains authorities in developing proper legislation. Food proteins have exhibit potential to get developed and incorporate in nutraceuticals and provide controlled release via the oral route. Food proteins are also suggested to be safe. The well-defined advantages of food protein matrices are high nutritional value, abundant renewable sources, consumer acceptability due to their natural and easy digestion by the digestive enzyme.

3.10 FUTURE PERSPECTIVE

To make NPs widely available, it is important to explore and develop methods for assuring their safety and characterization of NPs. As highlighted in this chapter, techniques are not available to singly detect and characterize all the vital attributes of NPs used in food, nutraceuticals, and food additives. Combinatorial techniques can be applied for characterization of NPs in food matrices. Future research should focus on the development and validation of methods for analysis of NPs in food and other samples. In most cases, pre-treatment of solid food samples is inevitable, but caution is recommended to select appropriate methods and develop applicable protocols that result in minimal disturbance of the NPs within a sample.

KEYWORDS

- **characterization of nanoparticles**
- **method of preparation**
- **nanoparticles**
- **nutrient bioavailability**

REFERENCES

1. Langer, R., (2000). Biomaterials in drug delivery and tissue engineering: One laboratory's experience. *Acc. Chem. Res., 33*, 94–101.
2. Bhadra, D., Bhadra, S., Jain, P., & Jain, N. K., (2002). Pegnology: A review of PEGylated systems. *Pharmazie, 57*, 5–29.
3. Kommareddy, S., Tiwari, S. B., & Amiji, M. M., (2005). long-circulating polymeric nanovectors for tumor-selective gene delivery. *Technol. Cancer Res. Treat, 4*, 61525.
4. Lee, M., & Kim, S. W., (2005). Polyethylene glycol-conjugated copolymers for plasmid DNA delivery. *Pharm. Res., 22*, 1–10.
5. Laurent, S., Forge, D., Port, M., Roch, A., Robic, C., Vander, E. L., & Muller, R. N., (2010). Magnetic iron oxide nanoparticles: Synthesis, stabilization, vectorization, physicochemical characterizations, and biological applications. *Chem. Rev., 110*. http://dx.doi.org/10.1021/cr900197g, 2574–2574.
6. Dreaden, E. C., Alkilany, A. M., Huang, X., Murphy, C. J., & El-Sayed, M. A., (2012). The golden age: Gold nanoparticles for biomedicine. *Chem. Soc. Rev., 41*, 2740–2779. http://dx.doi.org/10.1039/C1CS15237H.
7. Shin, W. K., Cho, J., Kannan, A. G., Lee, Y. S., & Kim, D. W., (2016). Cross-linked composite gel polymer electrolyte using mesoporous methacrylate-functionalized SiO_2 nanoparticles for lithium-ion polymer batteries. *Sci. Rep., 6*, 26332. http://dx.doi.org/10.1038/srep26332.
8. Khan, I., Saeed, K., & Khan, I., (2019). Nanoparticles: Properties, applications, and toxicities. *Arabian Journal of Chemistry, 12*(7), 908–931.
9. Khurana, A., Tekula, S., Saifi, M. A., Venkatesh, P., & Godugu, C., (2019). Therapeutic applications of selenium nanoparticles. *Biomedicine & Pharmacotherapy, 1, 111*, 802–812.
10. Hu, B., & Huang, Q. R., (2013). Biopolymer based nano-delivery systems for enhancing bioavailability of nutraceuticals. *Chinese Journal of Polymer Science, 31*(9), 1190–1203.
11. desRieux, A., Fievez, V., Garinot, M., Schneider, Y. J., & Preat, V., (2006). *J. Control. Release, 116*, 1.
12. Donald, A., (2004). *Nat. Mater., 3*, 579.
13. Chen, L., Remondetto, G. E., & Subirade, M., (2006). Food protein-based materials as nutraceutical delivery systems. *Trends in Food Science & Technology, 17*(5), 272–283.
14. Rawat, M. K., Jain, A., Mishra, A., Muthu, M. S., & Singh, S., (2010). Development of repaglinide loaded solid lipid nanocarrier: Selection of fabrication method. *Curr. Drug Deliv., 7*, 44–50.

15. Nair, H. B., Sung, B., Yadav, V. R., Kannappan, R., Chaturvedi, M. M., & Aggarwal, B. B., (2010). Delivery of antiinflammatory nutraceuticals by nanoparticles for the prevention and treatment of cancer. *Biochemical Pharmacology., 80*(12), 1833–1843.
16. Barras, A., Mezzetti, A., Richard, A., Lazzaroni, S., Roux, S., & Melnyk, P., (2009). Formulation and characterization of polyphenol-loaded lipid nanocapsules. *Int. J. Pharm., 379*, 270–277.
17. Ganta, S., & Amiji, M., (2009). Coadministration of paclitaxel and curcumin in nanoemulsion formulations to overcome multidrug resistance in tumor cells. *Mol. Pharm., 6*, 928–939.
18. Ganta, S., Sharma, P., Paxton, J. W., Baguley, B. C., & Garg, S., (2009). A pharmacokinetics and pharmacodynamics of chlorambucil delivered in long-circulating nanoemulsion. *J. Drug Target.*
19. Anton, N., Benoit, J. P., & Saulnier, P., (2008). Design and production of nanoparticles formulated from nano-emulsion templates: A review. *J. Control Release, 128*, 185–199.
20. Giroux, H. J., Houde, J., & Britten, M., (2010). Preparation of nanoparticles from denatured whey protein by pH-cycling treatment. *Food Hydrocolloids., 24*(4), 341–346.
21. Hunter, R., & Midmore, H., (2001). *J. Colloid Interf. Sci., 237*, 147.
22. Maddan, T. D., Harrigan, P. R., Tai, L. C. L., Bally, M. B., Mayer, L. D., Redelmeier, T. E., Loughrey, H. C., et al., (1990). The accumulation of drugs within large unilamellar vesicles exhibiting a proton gradient: A survey. *Chem. Phys. Lipids., 53*, 37.
23. Padamwar, M. N., & Pokharkar, V. B., (2006). Development of vitamin loaded topical liposomal formulation using factorial design approach: Drug deposition and stability. *International Journal of Pharmaceutics, 320*(1, 2), 37–44.
24. Dudkiewicz, A., Luo, P., Tiede, K., & Boxall, A., (2012). Detecting and characterizing nanoparticles in food, beverages, and nutraceuticals. In: *Nanotechnology in the Food, Beverage and Nutraceutical Industries* (pp. 53–81). Woodhead publishing.
25. Harivardhan, R. L., Vivek, K., Bakshi, N., & Murthy, R. S., (2006). Tamoxifen citrate loaded solid lipid nanoparticles (SLN™): Preparation, characterization, *in vitro* drug release, and pharmacokinetic evaluation. *Pharmaceutical Development and Technology., 11*(2), 167–177.
26. Laouini, A., Jaafar-Maalej, C., Limayem-Blouza, I., Sfar, S., Charcosset, C., & Fessi, H., (2012). Preparation, characterization, and applications of liposomes: State of the art. *Journal of Colloid Science and Biotechnology., 1*(2), 147–168.
27. Acosta, E., (2009). Bioavailability of nanoparticles in nutrient and nutraceutical delivery. *Current Opinion in Colloid & Interface Science, 14*(1), 3–15.
28. Dima, C., Assadpour, E., Dima, S., & Jafari, S. M., (2020). Bioavailability of nutraceuticals: Role of the food matrix, processing conditions, the gastrointestinal tract, and nano delivery systems. *Comprehensive Reviews in Food Science and Food Safety.*
29. Ting, Y., Jiang, Y., Ho, C. T., & Huang, Q., (2014). Common delivery systems for enhancing *in vivo* bioavailability and biological efficacy of nutraceuticals. *Journal of Functional Foods, 7*, 112–128.
30. Son, G. H., Lee, B. J., & Cho, C. W., (2017). Mechanisms of drug release from advanced drug formulations such as polymeric-based drug-delivery systems and lipid nanoparticles. *Journal of Pharmaceutical Investigation, 47*(4), 287–296.
31. Korsmeyer, R. W., Gurny, R., Doelker, E., Buri, P., & Peppas, N. A., (1983). Mechanisms of potassium chloride release from compressed, hydrophilic, polymeric matrices: Effect of entrapped air. *J. Pharm. Sci., 72*, 1189–1191.

32. Peppas, N. A., Bures, P., Leobandung, W., & Ichikawa, H., (2000). Hydrogels in pharmaceutical formulations. *Eur. J. Pharm. Biopharm., 50*, 27–46.
33. Lee, J. H., & Yeo, Y., (2015). Controlled drug release from pharmaceutical nanocarriers. *Chemical Engineering Science, 125*, 75–84.
34. Dhillon, N., Aggarwal, B. B., Newman, R. A., Wolff, R. A., Kunnumakkara, A. B., Abbruzzese, J. L., et al., (2008). Phase II trial of curcumin in patients with advanced pancreatic cancer. *Clin. Cancer Res., 14*, 4491–4499.
35. Bisht, S., Feldmann, G., Soni, S., Ravi, R., Karikar, C., Maitra, A., & Maitra, A., (2007). Polymeric nanoparticle-encapsulated curcumin ("nanocurcumin"): A novel strategy for human cancer therapy. *Journal of Nanobiotechnology, 5*(1), 1–18.
36. Chen, B. J., (2001). Triptolide. a novel immunosuppressive and anti-inflammatory agent purified from a Chinese herb *Tripterygium wilfordii* hook F. *Leuk Lymphoma, 42*, 253–265.
37. Mei, Z., Chen, H., Weng, T., Yang, Y., & Yang, X., (2003). Solid lipid nanoparticle and microemulsion for topical delivery of triptolide. *Eur. J. Pharm. Biopharm., 56*, 189–196.
38. Barras, A., Mezzetti, A., Richard, A., Lazzaroni, S., Roux, S., & Melnyk, P. (2009). Formulation and characterization of polyphenol-loaded lipid nanocapsules. *Int J Pharm, 379*, 270–277.

CHAPTER 4

Adulteration and Safety Issues in Nutraceuticals and Functional Foods

SHUJAT ALI,[1,7] SYED WADOOD ALI SHAH,[2] MUHAMMAD AJMAL SHAH,[3] MUHAMMAD ZAREEF,[1] MUHAMMAD ARSLAN,[1] MD. MEHEDI HASSAN,[1] SHUJAAT AHMAD,[4] IMDAD ALI,[5] MUMTAZ ALI,[6] and SHAFI ULLAH[2,5]

[1]*School of Food and Biological Engineering, Jiangsu University, Zhenjiang – 212013, P. R. China*

[2]*Department of Pharmacy, University of Malakand, Khyber Pakhtunkhwa – 18800, Pakistan*

[3]*Department of Pharmacognosy, Faculty of Pharmaceutical Sciences, Government College University, Faisalabad, Pakistan*

[4]*Department of Pharmacy, Shaheed Benazir Bhutto University Sheringal, Dir (Upper), Khyber Pakhtunkhwa, Pakistan*

[5]*H.E.J. Research Institute of Chemistry, International Center for Chemical and Biological Sciences, University of Karachi, Karachi – 75270, Pakistan*

[6]*Department of Chemistry, University of Malakand, Khyber Pakhtunkhwa – 18800, Pakistan*

[7]*College of Electrical and Electronic Engineering, Wenzhou University, Wenzhou 325035, PR China*

ABSTRACT

The consumption of nutraceuticals and functional foods, especially those that originated from plants, has been increasing owing to the communal concept that they are natural substances and are free from hazards. However, adulteration and safety issues in the production and selling of these substances are a universal concern for consumers, health professionals, regulators, and stakeholders. Particularly, adulteration by the unlawful addition of other ingredients is of main concern since dishonest manufacturers can misrepresent these substances to offer speedy effects and to promote sales. This illegal practice extremely endangers human health with several chronic and acute diseases and disregards the public rights for safer food. The intent of this chapter is to offer a base reference document for understanding adulteration and safety issues in nutraceuticals and functional foods, their evaluation, impacts on human health, and ways to prevent these issues. This will offer a background for future quantitative and innovative research. The adulteration and safety issues are described in terms of economically and criminally motivated adulteration, unintentional adulteration, undeclared labeling, and regulatory issues. The study provides major causes and evaluation of adulteration and safety issues. In the later part of the chapter, their impacts on public health and ways to prevent these issues are discussed. This study provides a foundation for future research regarding food safety, food adulteration, and food defense.

4.1 INTRODUCTION

Besides, food is something having general nutrition, aroma, and taste; the additional categories of food have been recognized, such as "nutraceuticals" and "functional foods." These are the substances that have more advantages than simple foods and are possibly equivalence with formally recognized "vitamins" [1].

A worldwide debate concerning nutraceuticals, functional food, and dietary supplements is whether they should be regarded as medicine or food. Likewise, it is difficult to distinguish between functional foods, nutraceuticals, herbal medicines, food additives, or nutrients owing to their drug-like health-related properties [2]. Other terms such as "medicinal foods and dietary supplements" are also used to refer to these substances. Health Canada defines a nutraceutical as a substance that is obtained from food and supposed to have advantages to health and/or prevent chronic diseases [3]. In other words, any safe food extract additive that has logically established health advantages for the prevention and treatment of illnesses [4]. Zeisel defined the term nutraceuticals as the diet supplements that bring a concentrated form or isolated form of a reputed bioactive ingredient from a food, obtainable in a nonfood matrix and could be applied to encourage health [5]. American Dietetic Association (ADA) described the nutraceuticals as any substance having a food-constituent and give health or medical benefits, such as treatment and prevention of diseases, for example, minerals (selenium), vitamins, and animal (carnitine, carnosine, chitosan), and plants (ginger, garlic, Ginkgo biloba) extracts [6].

Similarly, the term functional food is defined as food that should have an appropriate effect on health or well-being and minimize the risk for disease [7]. In another study, the term is defined as foods like conventional substances that are used up as a constituent of a usual diet and beyond basic nutritional functions have established biological advantages and/or minimize the hazard of long-lasting disease [8]. Food for special dietary use and foods for specified health use (FOSHU) have highlighted those functional foods are regarded as foods, determine their properties in amounts that can usually be projected to be used in the diet, and are eaten as a constituent of a conventional food form [9]. Natural and traditional foods can be sold or advertised as functional foods, provided they are attended by the somewhat new representation of their health advantages [10]. Functional food constituents offer health-promoting properties other than usual nutrition and when compared to dietary supplement constituents, have unique regulatory requirements, and need diverse safety measures. Functional food constituents are not the same as a dietary supplement, while they may have the same chemical functional groups and may have same health benefits [11]. In a simple way, functional food ranges from any improved food or food products that may offer health advantages, while in another aspect, foods that have possibly disease-preventing and health-promoting properties [12]. Furthermore, the term functional food sometimes is confusing, as nearly all foods, irrespective of whether they have additional constituents, somehow affect health by offering nutrients and calories and can be regarded as "functional." Nevertheless, effective, or not, the terminology of functional food has emerged as the main one and must be elucidated in order to educate its scope and its spot, as well as to ease the improvement of a generally recognized regulatory outline [13].

Nutraceuticals and functional foods are natural constituents that may be consumed in combination, individually, or added to beverage for health benefits or technologic purposes and essentially have a suitable safety outline that determines safe for eating by humans [14]. Medical drugs have the risk of adverse effects or toxicity, hence, the search for harmless functional food and nutraceutical-based tactics are of great interest for maintaining the good health of human beings. This led to a worldwide revolution in functional food and nutraceuticals. The option of disease management and health regulation by natural methods has been assumed by a substantial percentage of the global population [15]. Previously, researchers have tried to standardize the definitions of functional foods and nutraceuticals. The Nutraceutical Research and Education Act (NREA) was proposed by Stephen De Felice in 1999 [1]. However, the proposal was for the time being laid to rest, and no considerable contribution was established regarding this. Since then, the terms functional foods and nutraceuticals have been inflated. Generally, the dietary supplements are presented to be safer and natural, hence most of the worldwide population favor these substances for health care advancement over pharmacological medicines [16]. Quick growth in study on nutraceuticals and functional foods is a vital and integral component of the revolution. For the successful use of functional foods and nutraceuticals in the management of human health the safety and efficacy are two essential key sets for the purpose [17]. The safety and efficacy are rapidly improving due to the sophisticated and modern technologies.

The problem of adulteration and safety in the nutraceuticals and functional food occurs from manufacturing level to consumption. Some food processors, restaurant owners, manufacturers, and transporters are responsible for this wrong act of adulteration. Nutraceuticals and functional foods may be adulterated by means of different inexpensive substances, toxic artificial colors, and harmful chemicals [18]. Uses of harmful chemicals to get quick effects and to attract consumers is among the commonly used practices [19]. The consumption of such unsafe substances negatively affects public health with frequent chronic and acute infections. Adulteration of nutraceutical and functional food has been observed several years ago, and this unethical act is growing day by day, especially in developing countries [20]. Major causes of nutraceutical and functional food adulteration include dishonest importers, traders, cultivators, manufacturers, and processing agencies [21]. Particularly, in developing countries, these unethical practices are involved in the adulteration of functional foods and nutraceuticals, and there is no proper and strict laws and principles to control the adulteration and safety issues [22]. The rules may comprehend the offenses like lack of hygiene, fake licenses, poor quality of food, substandard infrastructure food impurity, food adulteration, selling products with expired dates and incorrect information on food packages. However, the issue is the appropriate and sustained implementation of the rules and regulations by the dependable establishments. Also, shortage of test instruments, reagents, and skill-persons is much noticeable.

This chapter provides a base reference document to understand adulteration and safety issues in nutraceuticals and functional foods, their evaluation, impacts on human health, and ways to prevent these issues. The adulteration and safety issues are described in terms of economically and criminally motivated adulteration, unintentional adulteration, undeclared labeling, and regulatory issues. Various approaches to the evaluation of adulteration and safety issues in nutraceuticals and functional foods are critically discussed. In the last part of the chapter, the impacts of adulteration and safety issues on public health and ways to prevent these issues are discussed (Figure 4.1). We believe that the study provides a background for future innovative and quantitative research. Furthermore, it may assist researchers to understand the foundation for future research regarding food safety, food adulteration, and food defense.

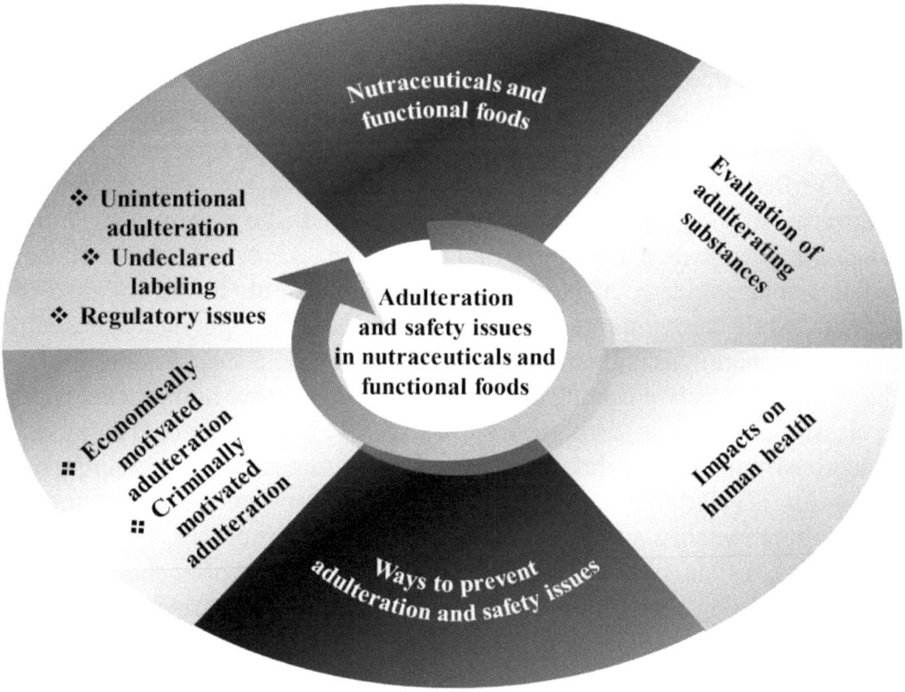

FIGURE 4.1 Understanding the terms nutraceuticals and functional food, adulteration, and safety issues, evaluation of adulterating substances, impact on human health, and ways to prevent adulteration and safety issues.

4.2 WHAT ARE NUTRACEUTICALS AND FUNCTIONAL FOODS?

Nutraceutical is a pragmatic term, and its meaning is not uniform globally, the term has different descriptions worldwide [1]. This terminology was presented by Stephen DeFelice in 1989, Chairman and Founder of the Foundation for Innovation in Medicine (FIM) [23]. Basically, the word nutraceutical originated from "nutrition" and "pharmaceutical" and followed the way of the cosmeceuticals term, which was introduced by Albert Kligman in 1980 and Raymond Reed in 1961 [24]. According to DeFelice, "nutraceuticals are food or food constituent that offers health advantages and medicinal values, counting the prevention and/or treatment of a disease" [1]. Regardless of the substantial use of the term nutraceutical in common practice and selling, it has no clear regulatory definition or absolute legal standing [25]. Nutraceuticals are health-indorsing substances, sold with the health-promoting claims of improving different mental and physical actions of the body, generally without pharmaceutical constituents [26]. A nutraceutical is almost molding drug and food into one formulation, which acts neither as a pharmaceutical product nor a simple food [23]. Nevertheless, it is a wide-ranging word that includes minerals, amino acids, vitamins, botanicals, and herbs [27]. Thus, both fortified food products and dietary supplements can be considered as nutraceuticals [28]. Nutraceuticals can be divided into two classes, established nutraceuticals and potential nutraceuticals. An established nutraceutical is one that holds enough clinical data to demonstrate health and medical benefits, while a potential nutraceutical is one that has a potential for health benefits. Generally, majority of nutraceuticals exist in the 'potential' class and waiting to declare established [29]. In broader sense, nutraceuticals can be classified into these three groups [30], i.e., (1) Nutrients, substances with recognized nutritional roles, such as vitamins, minerals, amino acids, and fatty acids; (2) Herbals, concentrated botanical extracts and products; and (3) Dietary supplements, substances obtained from other sources, for example, chondroitin sulfate, pyruvate, and steroid hormone. They offer functions and advantages such as weight-loss supplements, meal replacements and sports nutrition.

The word functional food was familiarized for the first time in the mid-1980s in Japan, referred to as treated food comprising ingredients that can affect body functions [31, 32]. The definition of functional food is not the same all over the world, and the term is occasionally used miscellaneous with the other terms related to food [1]. Certainly, a wide variety of food products are categorized as functional foods, with various constituents, both classified and not classified as nutrients, regulate body activities relevant to the reduction of the risk of disease. Hence, functional food is to be understood as a concept and no universally accepted and simple definition of the term exists until now. Some elaborate definitions of the term are: (a) daily diet foodstuffs, resulted from naturally available substances and when ingested possess certain biological advantages [33]; (b) daily diet foodstuffs, obtained from naturally occurring substances and when ingested can help to regulate body process [34]; (c) daily diet foodstuffs, and when ingested establish physiological advantages and minimize the risk of chronic illness [35]; and (d) daily diet foodstuffs, that may offer health advantages beyond that of the traditional nutrients it contains [35]. Based on the definitions, functional food appears as a sole concept that justifies a category of its own, different from designer food, vita-food, pharma-food, nutraceutical, and medi-food. Functional food is also a concept that is related to nutrition and not pharmacology. It is not drugs and must be food. Furthermore, their role concerning disease is reducing the risk rather than treatment. However, it should be underlined that a functional food will not essentially be equally efficient for all members of a population, and that corresponding individual biological needs with selected food constituent consumptions may develop a key task on their body response [36]. Functional food can be classified into different categories; based on the bioactive ingredients, it can be divided into fibers, probiotics, phytochemicals, minerals, vitamins, herbs, and proteins, etc. Functional foods offer physiological advantages that distinguish them from normal foods. Its effectiveness is resulting from bioactive constituents and depends on numerous technical features. The bioactive constituents in functional foods assist in the stoppage of infections and improve body performance of the individual beyond their recognized nutritional role. They directly involved in adjusting body systems, such as the endocrine, circulatory, nervous, digestive, and immune systems [37].

All the aforementioned definitions for nutraceuticals and functional food pointed out that there are no consent and a certain definition of nutraceuticals and functional food. Consistently, there is a need to emphasize the terms nutraceuticals and functional foods as drug or food; in fact, it blurs the demarcation between food and drug (Figure 4.2). For example, cholesterin, obtained from red yeast rice is a cholesterol-lowering ingredient and is essentially a supplement identical to lovastatin [38]. Similarly, tryptophan, an amino acid derivative, is necessary for metabolism in small amounts, while it, at higher doses, in the form of 5-hydroxy-L-tryptophan behave like a drug for the treatment of insomnia via enhancing brain serotonin production [39]. But it was legally banned from the market since tryptophan administration resulted in the eosinophilia-myalgia syndrome (EMS). Nutraceuticals are considered to have at least one constituent of essential macro or micro-nutrients that are the active part of foods. Consequently, many nutraceuticals having food phytochemicals such as sulfur compounds (from garlic), carotenoids (lycopene from tomato), curcumin, glycosinolates, isoflavonoids, phytosterols, essential fatty acids, proanthocyanins, proteins, vitamins, amino acids (e.g., arginine), peptides (e.g., carnosine) antioxidants and polysaccharides, etc., are now existing in the market [40, 41]. The exact value that any consumer would place on nutraceuticals and functional foods directly depends on the consumer's self-image. For example, a hypothetical consumer might realize himself as normally existing within a range of 75–85% efficiency [42]. Below 75% efficiency, there is no equilibrium and no longer feeling good himself and even in low range one may feel sick, motivating a need for some sort of medication and treatment. However, within this range (75–85%) one feels in equilibrium with his surroundings. The high level of this range (85%) is the best one and give healthy lifestyle habits, and good feeling. The goal is to uphold oneself in the 75–85% range, but environment consideration of limited food choice or limits on physical action disturbs to stay within this optimal range. Hence, nutraceuticals, and functional foods may put that goal within easier reach to be accomplished.

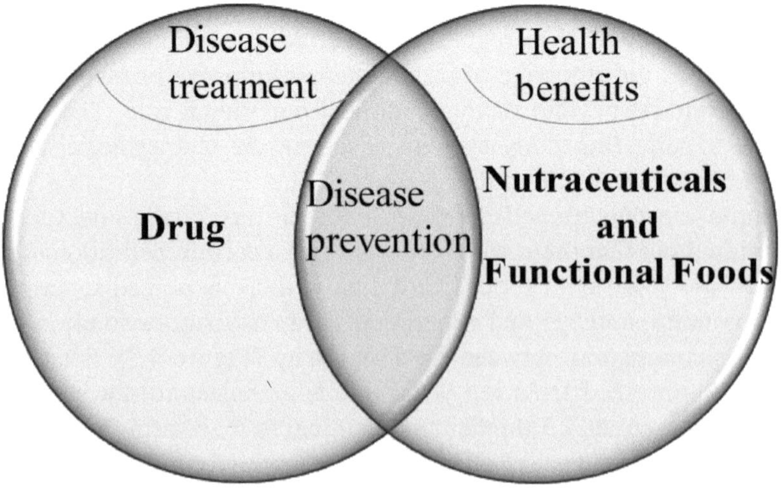

FIGURE 4.2 Demarcation between nutraceuticals and functional food and drug.

4.3 ADULTERATION AND SAFETY ISSUES

Nutraceuticals and functional foods are marketed as natural products deprived of side effects and are sold with therapeutic claims. Unfortunately, any unacknowledged ingredients may cause dangerous hazards to consumer's health. Furthermore, according to some reports, they can be substandard, adulterated, falsified, counterfeit, and unregistered [43]. Regarding the definitions, there is an enduring debate concerning the exact descriptions for adulterated nutraceuticals and functional food. According to the World Health Organization (WHO), adulterated products are impure formulations, debased, or corrupted [44]. Adulteration may occur due to the addition of an external or inferior element or substance. Adulteration in nutraceuticals and functional food can also be determined as the existence of an undeclared material, or a component is changed from its standard limits, and that a profile is improbable to happen [45]. Adulteration may be unintentionally or purposefully. The motivation behind purposeful adulteration in nutraceuticals and functional foods is ultimately for economic income [46]. Natural products have conventionally used in drug research, especially in the treatment of metabolic syndrome disorders, immunosuppression, and malignant diseases [47]. Worldwide request for plant phytochemicals for use in functional foods and nutraceuticals has been rapidly increasing. The sources of many phytochemicals are restricted, and the preparation of such substances with desired molecules are cost-consuming and involves a lengthy process, so this can lead to intentional or purposefully adulteration [18]. Plants are among the most common source of nutraceuticals and functional food, these sources may be contaminated during plantation and manufacturing, and by heavy metals, fertilizers, microbial agents, pesticides, etc. All these incidences may result in food-borne illnesses such as liver injury, gastric complaints, and other life-threatening infections [48]. Hence, the safety and security of fresh and processed nutraceuticals and functional food are essential by defining specifications in appropriate detail. Furthermore, issues related to the stability of active ingredients and pathogenic control have also been observed. It should also be highlighted that safety issues do not occur only with pesticides, synthetic drugs, and additional species; but also, pollens, insects, dust, parasites, microbes, rodents, molds, fungi, heavy metals, and toxins pose a serious problem with herbal formulations related nutraceuticals and functional foods [49]. Other serious hazards may result from synthetic chemicals, which are unsafe and hence not permissible in order to intensify a claimed biological effect or to change an immediate physiological action

[50]. The consumption of nutraceuticals and functional foods, particularly those having plants phytochemicals as constituents, has been rising owing to the public concept that they are natural substances and pretense no hazards to human health. Adulteration by the unlawful addition of medicinal ingredients or their equivalents is of main concern since dishonest manufacturers can misrepresent these substances to provide speedy effects and to upsurge sales. The adulteration of functional foods and nutraceuticals is an emerging issue and that an operative check by food regulatory establishments is required to protect consumers. Some important issues regarding adulteration and safety of nutraceuticals and functional foods are given below.

4.3.1 ECONOMICALLY MOTIVATED ADULTERATION

This kind of adulteration is the origin of public health hazards, and the term was demarcated in the Open Meeting on Economically Motivated Adulteration in May 2009. According to this definition, economically motivated adulteration is "the fake, intended addition or substitution of material in a product for reducing the cost of its production or to enhance the seeming value of the product" [51]. Economically motivated adulteration may include dilution of foodstuffs with amplified amounts of cheap substances as well as the substitution or addition of an ingredient to hide dilution, and as a result, this may pose a possible or known health hazard to consumers [52]. The result or impact of economically motivated adulteration is an actual public health hazard and the misbranding or adulteration creates the potential for harm. Similarly, it may threaten more risk than conventional adulteration and safety because the contaminants are not traditional. Fraudsters may apply additives which are not registered among those predictable food safety adulterants. For instance, fraudsters applied melamine since it inventively mimicked high-quality protein in common protein content quality control tests, it was an unpredicted food contaminant; meanwhile it is a plasticizer and used in making plastic products [53]. The concept of safety covers all threats associated with consumers' health, irrespective of the source, production, processing, and traditional efforts. This is the commonsense development for an impression as well as for well-known, the adulteration response now starts at the interference stage (that is, to know about the risks) then passes to the reply stage (that is, public-private mutual coordination). When the response stage develops well, fraud attention will logically change to the preclusion stage [54].

4.3.2 CRIMINALLY MOTIVATED ADULTERATION

This kind of adulteration is carried by professional people for their own benefits. Criminals form a system to perform a food-adulteration crime, and when some action take place against them, they disperse, however they return to their usual and re-organize into a new criminal system for perpetrating a new fraud [55]. Since the chance is for a minor fraud to be circulated across consumers, less erudite criminals who are planned should not be neglected. As compare to old-style organized crimes, these are frequently networks or groups, interrupting any single connection in the network will not certainly break the chain or the capability for a new fraud [54, 56]. It is significant to highlight that there could be proficient hurdles and protectors in place, but the nature of a developing and evolving danger is that new slits always happen. People associated with adulteration are not all a civil law violators or criminals and may not be measured wrong in many cultures [57]. An infinite number of producers may relate to fraud, increased brand recognition and brand growth of a product, and hence essentially rises the fraud opportunity [58]. Then, the guardian led to a big fraud occasion. These are objects that protect or monitor the foodstuffs and may include individual companies, customs, local or federal law enforcement, non-governmental organizations, and trade associations, and so on to minimize such incidences. Furthermore, the risk of detection should be increase, the necessary technology that commit the fraud should be make expensive and strict laws and regulations should be implemented. Steps have been proposed to decrease the chances of fraud, but narrowing of focus in detection and modification to a procedure could unintentionally create new gaps that may be used by food criminals. Food adulteration by criminals is resourceful in nature and denotes an important challenge to both government and industry [59]. The detection and exploring process is further complicated because the fraud network generally is intelligent, clandestine, resilient, and sophisticated at stealthily sidestepping detection [60].

4.3.3 UNINTENTIONAL ADULTERATION

Unintentional adulteration of nutraceuticals and functional food may occur from naturally occurring substandard, poor storage conditions, drought, lack of rainfall, etc. Furthermore, unintentional adulteration may be owing to lack of knowledge about the authentic source, the similarity of sources in aroma or morphology, non-availability of the reliable source, careless processing, confusion in language names between indigenous systems and local dialects and other unidentified reasons [61]. Not all adulterations are intentional, it is noted that the nutraceuticals and functional foods are also adulterated unintentionally, sometimes, suppliers are uneducated and not aware of their counterfeit supply [62].

In other words, unintentional adulteration is the addition of unsolicited substances due to carelessness, ignorance, lack of proper hygiene, and lack of facilities during the processing of nutraceuticals and functional foods. This can be developed from contamination by the entry of harmful residues from packing material, dust, and stones, spoilage of food by rodents, fungi, and bacteria [63]. Similarly, inherent adulteration such as the presence of organic compounds, certain chemicals, or radicals naturally happening in foods like poisonous varieties of mushrooms, pulses, fish, and seafoods may also occur [64]. Other possible sources for unintentional adulteration are operations carried out for veterinary medicine and animal husbandry, crop husbandry, treatment, manufacture, processing, preparation, packaging, packing, and transport, etc., [64].

Nowadays, in developed countries, modern instruments and techniques are using to maintain high-quality standards. WHO rejects any raw material regarding medicinal plants which has more than 5% of any other part of the same authentic plant [65]. According to these standards, adulteration, whether unintentional or intentional, should be excluded. Also, traders and suppliers must be educated about the control and management of unintentional adulterations.

4.3.4 UNDECLARED LABELING

Most of the problems associated with the use of functional foods and nutraceuticals arise mainly from the organization of many of these products as dietary supplements or foods in some countries. Furthermore, the presence of unidentified allergens is ubiquitous, and several such substances have been found in packaged foodstuffs without preventive labeling [66]. Consequently, food products without precautionary or mandatory labeling are not safe for allergic consumers. Hence, safety quality, and efficacy of these ingredients should be considered before marketing [67]. Similarly, production standards, quality tests, and appropriate labeling are less precise and, in some cases, manufacturers, and producers may not be licensed and certified. Therefore, the undeclared labeling has become a foremost concern to both the general public and national health authorities [68]. It is necessary that nutraceuticals and functional foods be accompanied by comprehensive information for safe use, such as how to store the product, how to use the product, regulatory information, and side effects. The information needs to be labeled on the packaging or leaflet put into the product package [67].

Although producers may use health benefits to sponsor their foodstuffs, the ultimate determination of health claims is to help consumers by offering detail on healthy eating outlines that may assist in minimizing the risk of cancer, heart sickness, high blood pressure, osteoporosis, dental cavities, and birth defects. Various kinds of health-related claims are permissible in food labeling. This information represents about ingredients or other nutrients in food and its health-related effects [69]. Indirect health claims may also be specified, which indirectly declares a disease-diet relationship. Indirect claims may seem in symbols, vignettes, and brand names, when used with detailed nutrient information. Though, all labels having indirect claims must also stand for the full health claim [12]. Foods nominated as "functional food" and credited definite health claims should obtain logical scrutiny prior to specific health benefits are allowed. In countries, where rich cultural tradition belief occurs, the health-promoting features of various food components are related to specific health claims, but these claims may not be scientifically proved to establish experience with these practices [70]. Some health claims have been legally documented, and future research and studies will fully document and approve or disprove the supposed health benefits [71]. Nutraceuticals and functional foods-related professionals should continue to recommend validified and hazard-free items and follow the rules of food components and specific foods as both preventative and treatment therapy for health problems. With the new substitute methods for defining the technical basis for health benefits, attention in health claims is expected to remain high, and innovative claims seem to influence food labels within the conceivable future [12].

4.4 REGULATORY ISSUES

Several studies have been dedicated to nutraceuticals and functional foods, research publications in reputed journals regarding nutraceuticals and functional foods show auspicious projections for the applications of these constituents in foodstuffs, hence create worth for producers and advantages for consumers health [72]. This scenario results in a crucial demand for regulation, which would make safe this new collection of foods [67]. There were no specific guidelines or was registered by any health establishments in Europe to monitor nutraceuticals and functional foods until 1997 when "Green Paper on Food Law" started a new provocation to the foundation of European Food Law [18]. Afterward, in 2000 this law was favored by the "White Paper on Food Safety." Since then, most of the "White Paper" proposals have been applied. Though there is no uniform regulatory framework for nutraceuticals and functional foods, the guidelines to be applied are abundant and related to the kind and origin of the foodstuff. Furthermore, regulations on food supplements, novel foods, and nutritional foods may also be appropriate to nutraceuticals and functional foods depending on the kind and origin of the product as well as on their use. The foundation of European Union regulation on food products, such as nutraceuticals and functional foods is 'safety.' Conclusions on the safety basis of regulations are based on risk investigation, in which logical risk analysis is achieved by the European Food Safety Authority (EFSA) and risk management was made by the European Commission (EC). In the risk management stage, both the safety value and other authentic factors were measured in selecting a suitable way to deal with adulteration and safety issues [67]. Now, the EU implemented an instruction to harmonize the regulation of food products across the EU and initiated a basic licensing system to assist the public make knowledgeable choices about the consumption of nutraceuticals and functional foods [73]. However, in the developing countries where poorly regulated food products and many unregistered foodstuffs are sold freely on the market with no or slight limitation. Furthermore, the public misconception that natural products are not contaminated and are free of adverse effects often result in the unrestrained intake and inappropriate use, and this has caused hazards and health difficulties. This misconception is not only in developing countries, but it also occurs in developed countries, where the people frequently resort to "natural" products without any suitable information or awareness on the related risks [74].

4.5 EVALUATION OF ADULTERATING SUBSTANCES

The analytical technique chosen for the evaluation of adulterating substances and detection of the adulterants depends on several factors such as required sensitivity, number of targeted substances, nature of the substances, the complexity of the formulation and physical behaviors (i.e., liquids, solids, gas, etc.), [75]. Generally, sample preparation approaches include the extraction of adulterating substance by organic solvent. The suitable solvent system is to be chosen, and the suspension or solution is to be sonicated, agitated, centrifuged, filtered, and further diluted. Despite the fast and simple procedure for sample preparation, complexity in the co-extraction of several different compounds should be considered, as this can influence the determination of the targeted analyte [56]. Hence, in addition to simple extraction methods, pre-concentration steps and clean-up measures are required based on the detection procedure applied for evaluation. For example, if mass spectrometry is applied, co-extracted matrix-compounds can possibly affect the evaluation of target substances and result in matrix effects through the ionization process and cause enhancement or suppression of signal, and hence make inaccurate analytical results [76]. This effect can be minimized by applying sophisticated techniques such as solid-phase extraction [77], liquid-liquid extraction [78], and surface-enhanced Raman spectroscopy (SERS) procedure [79]. SERS technology is based on definite vibrational spectroscopy with very high sensitivity at the molecular level constructed on Raman peaks for the detection of targeted substances. This technique offers a rapid, accurate, simple way for real samples examination [80, 81].

Chemometrics, a multivariate data scrutiny tool generally used in combination with data-rich instrumental approaches such as nuclear magnetic resonance, infrared spectroscopy, and mass spectrometry [82]. This is a prevailing data minimizing tool regarding food fraud, used qualitatively for classifying or grouping unknown samples with analogous features and quantitatively for analyzing adulterants in food samples [83]. Reports have demonstrated the application of partial least squares multivariate models of the infrared spectra and main component analysis to detect contaminants in various samples [84–86].

In addition to the technologies such as nuclear magnetic resonance spectroscopy, mass spectrometry, near-infrared spectroscopy, Raman spectroscopy, Fourier transform infrared spectroscopy, etc., many others approach also exist including electronic noses and tongues [87, 88], electrochemical detection [89], nanosensors, and nanoparticle-based detection systems [90], apt sensor-based detection [89] and quantification via ELISA [91]. In the current age of systems-level thinking and the resultant interdisciplinary collaborations between the physical sciences, biology, and engineering (Figure 4.3), it is clear that novel developments and research will establish an easy and cost-effective technology for determination of adulteration and safety issues in nutraceuticals and functional food [92, 93].

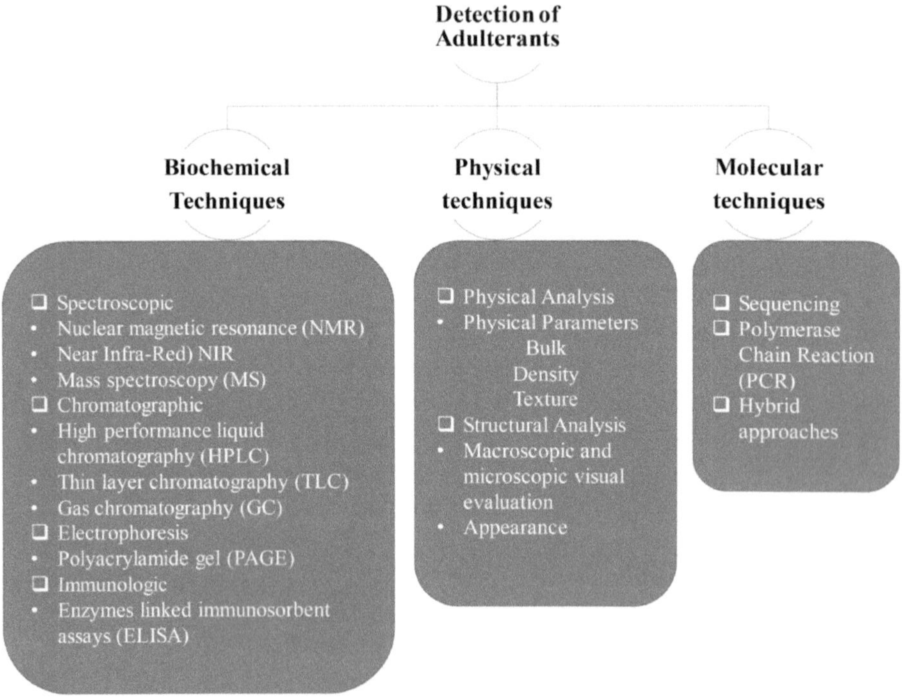

FIGURE 4.3 Various methods used for the detection of adulterants.

4.6 IMPACTS ON HUMAN HEALTH

Adulteration and safety issues in nutraceuticals and functional foods are food hazards that gaining concern and recognition. Irrespective of the reason of the risk, adulteration, and safety issues, evaluation is the responsibility of both the government and industries. Food fraud and adulteration is an intentional practice for financial gain, while safety issues may be unintentional acts with unintentional hazards. Food-associated public health hazards are more dangerous than conventional food safety issues due to unconventional contaminants. Up-to-date intervention systems are not well-planned to mark a huge number of potential contaminants. Consumption of adulterated nutraceuticals and functional foods may cause serious complications [18, 94].

The majority of the nations have considered this problem as a priority and initiated strict efforts to control food-related issues, specifically in the developing countries. Safety issues in developing countries are of interest to many international organizations and are broadly recognized [95]. Efforts to cope with the demand of nutraceuticals and functional foods in the developing regions will apparently not work in the absence of a deliberate approach to monitor safety issues. Significant causes of illnesses such as diarrhea, kidney failure, digestive problems, have been raised in different regions of the world. These adulterants are reported to be responsible for the increasing incidents of diseases.

4.7 WAYS TO PREVENT ADULTERATION AND SAFETY ISSUES

There are challenges to proactively notice and completely eradicate adulterants and safety issues from nutraceuticals and functional foods. Ethical manufacturers ensure and precisely determine the chemicals and ingredients present in the products, but they do not always account for the hazard of unexpected or unknown substances. Consumers, scientists, and regulators are at a disadvantage when they are experienced by un-wanted substances that can lead to shocking health issues and subsequently undermine public trust. Reports have shown that many such foodstuffs claiming to be safe and beneficial have been exposed with a compelling indication of adulteration and safety issues. In other words, the phenomena of adulteration and safety issues becoming dangerous in different parts of the world, especially in developing nations. Hence, all countries are suggested to impose more harsh rules and regulations and licensing measures to advance regulatory ethics to create suitable pre- and post-marketing checkups for analysis and to protect their public. The whole supply chain from the manufacturers and importers via wholesalers to sellers will have to be monitored properly because inspection at the selling level only will not cause enough positive influences. Regular inspection by authoritative agencies should do it in a designed way for monitoring adulteration and safety issues. Similarly, a consumer awareness campaign is to be initiated to alert people about the risks associated with adulterated nutraceuticals and functional foods. Adequate actions by the related civil societies, agencies, social organizations, electronic, and print media, and consumers can make changes to ensure the safety and security of such food items. In this way, a combined effort is to be set to obtain safe food for a healthy life. For example, the Food and Drug Administration (FDA) of the United States (US), contained trained personnel with the competence of implementation of laws to screen the quality of drugs and foods existing in the US markets [96].

Testing nutraceuticals and functional food against publicly available standards are effective measures in responding to contaminants and adulterants arising from new processing methods or new sources of raw materials [97]. Sensitive evaluation procedures can be developed or modified in a community way and rigorous standards to notice new impurities. To date, no legal requirements mandate food ingredients analysis before product use [97]. Both public health and public confidence are threatened by the lack of standards. Hence, nutraceuticals and functional food manufacturers should provide appropriate specifications by using quality monographs. Stakeholders can contribute to set standards, and this will ensure that manufacturers, distributors, importers, exporters, and consumers know that the nutraceuticals and functional food possess appropriate quality features.

Adulteration and safety issues that result in public health risks are often unknown till it is too late and may only be known by chance rather than from a proper risk-based study; hence it is required to develop analytical models for the future. Several approaches are in use to detect the presence of adulterants and safety issues in nutraceuticals and functional foods; however, these tactics rely on the adulterant or their sources, and on this basis, the substances could not be declared totally free of adulteration. Currently, technologies available for the detection of adulteration and safety issues include the vibrational spectroscopies: mid-infrared, near-infrared, Raman; mass spectrometry, as well as NMR spectroscopy. More sophisticated techniques are now developing to verify the provenance claims made about nutraceuticals and functional foods [43]. Previously, analytical screening approaches have been applied to recognize adulteration and safety issues, these techniques worked well when the nature of the adulterant was known. The evolving forensics methods such as spectroscopy or isotope analysis do not require the adulterants to be known and are frequently applied for adulteration analysis [98, 99]. However, these assessments are costly and could not be used as a tool for simple verification and not as a form of screening for routine batch release. Hence, for common practice, these approaches are not used as online, real-time monitoring, either a preventative control within an established quality plan [56]. Concurrent risk analysis studies on social and economic factors such as animal disease outbreaks, pressure on food prices and nature actions causing crop harm, etc., together with related predictive modeling can be used to forecast the potential for adulteration and safety issues. Policy measures have announced the need for the implementation of both predictive procedures, detection, and reaction approaches. Prediction of adulteration depends on the suitable investigation of intelligence through the application of expert knowledge and predictive tools [100].

4.8 CONCLUDING REMARKS AND OUTLOOK

Nutraceutical and functional foods are natural substances and may be consumed in combination, individually, and may be added to beverage or food for specific health benefits or technologic purposes, must have an acceptable safety profile signifying the safety for consumption by the public. The risk of adverse effects and toxicity of general medicines leads us to the attention of harmless functional food and nutraceutical-based tactics for health management. Nutraceuticals and functional foods are in high demand by consumers throughout the world. These products are usually active to maintain a healthy life by preventing diseases. However, adulteration, and safety issues are growing problems globally, and reports have exposed numerous ambiguities with respect to their definition, registration, claim, sales tactics, safety, and efficacy. Consumption of adulterated nutraceuticals and functional foods badly affects human health by causing many acute illnesses. The benefits and risks of these substances are not as recognized as for conventional drugs, and there is a lack of described possible adverse effects regarded the misuse of such items.

Although functional foods and nutraceuticals have significant potential in the health management and prevention of diseases. However, nutritionists, regulatory toxicologists and health professionals should strategically work together to design proper regulation to provide the desired health and therapeutic benefits to the public. The effects of various processing approaches on the effectiveness and biological availability of nutraceuticals and functional foods remain to be determined. Governmental authorities and lawmakers are more must reinforce the current regulation for health-promoting beverages and other food items, specifically with food labeling. Laws and regulations regarding functional foods and nutraceuticals may be modified in the better interest of public health. For safety, proper manufacturing methods, patenting focus, formula, and formulations and specified applications should be evaluated and modified. An adequate and enabling documented explanation is essential for getting a license. Inventors should periodically revise their findings and adopt whether to pursue a patent for maintaining as trade secrets or to use for new discoveries. Comprehensive analysis and searches on related technology would support inventors to pinpoint the market position and assess the patentability of their products. Furthermore, frequent evaluations and revisits of developments in patenting policy and regulatory should be endorsed for setting up reasonable research and trading strategy for the encouragement of safer business. The future is fertile with opportunities to adopt and measure established frameworks and systems from the more sophisticated regulatory environments of the developed markets such as in Europe and North America. The time is right for businesses and entrepreneurs from around the globe to take advantage of the situation and launch services and protocols with the potential to become existent standards in the developing countries' regulatory networks.

KEYWORDS

- **adulteration**
- **Eosinophilia-Myalgia-syndrome**
- **European Commission**
- **functional foods**
- **nutraceuticals**
- **safety issues**
- **surface-enhanced Raman spectroscopy**

REFERENCES

1. Kalra, E. K., (2003). Nutraceutical-definition and introduction. *AAPS Pharmsci., 5*(3), 27, 28.
2. Aronson, J. K., (2017). Defining 'nutraceuticals': neither nutritious nor pharmaceutical. *British Journal of Clinical Pharmacology, 83*(1), 8–19.
3. Bishop, K. S., et al., (2015). From (2000). years of Ganoderma lucidum to recent developments in nutraceuticals. *Phytochemistry, 114*, 56–65.
4. Dillard, C. J., & German, J. B., (2000). Phytochemicals: Nutraceuticals and human health. *Journal of the Science of Food and Agriculture, 80*(12), 1744–1756.
5. Zeisel, S. H., (1999). *Regulation of" Nutraceuticals.*" American Association for the Advancement of Science.
6. Bloch, A., & Thomson, C., (1995). Position of the American dietetic association. phytochemicals and functional foods. *Journal of the American Dietetic Association, 95*(4), 493–496.

7. Roberfroid, M. B., (1999). Concepts in functional foods: The case of inulin and oligofructose. *The Journal of Nutrition, 129*(7), 1398S–1401S.
8. Clydesdale, F. M., (1997). A proposal for the establishment of scientific criteria for health claims for functional foods. *Nutrition Reviews, 55*(12), 413–422.
9. Bailey, R., (1999). Foods for specified health use (FOSHU) as functional foods in Japan. *Canadian Chemical News, 51*, 18–19.
10. Kwak, N. S., & Jukes, D. J., (2001). Functional foods. Part 1: The development of a regulatory concept. *Food Control, 12*(2), 99–107.
11. Milner, J. A., (2000). Functional foods: The US perspective. *The American Journal of Clinical Nutrition, 71*(6), 1654S–1659S.
12. Arvanitoyannis, I. S., & Houwelingen-Koukaliaroglou, M. V., (2005). Functional foods: A survey of health claims, pros and cons, and current legislation. *Critical Reviews in Food Science and Nutrition, 45*(5), 385–404.
13. Roberfroid, M., (2002). Functional food concept and its application to prebiotics. *Digestive and Liver Disease, 34,* S105–S110.
14. Katan, M. B., & ROOS, N. M., (2004). Promises and problems of functional foods. *Critical Reviews in Food Science and Nutrition, 44*(5), 369–377.
15. Putnam, S. E., et al., (2007). Natural products as alternative treatments for metabolic bone disorders and for maintenance of bone health. *Phytotherapy Research: An International Journal Devoted to Pharmacological and Toxicological Evaluation of Natural Product Derivatives, 21*(2), 99–112.
16. Ekor, M., (2014). The growing use of herbal medicines: Issues relating to adverse reactions and challenges in monitoring safety. *Frontiers in Pharmacology, 4*, 177.
17. Govindaraghavan, S., & Sucher, N. J., (2015). Quality assessment of medicinal herbs and their extracts: Criteria and prerequisites for consistent safety and efficacy of herbal medicines. *Epilepsy & Behavior, 52,* 363–371.
18. Orhan, I. E., et al., (2016). Adulteration and safety issues in nutraceuticals and dietary supplements: Innocent or risky. In: Grumezescu, A. M., (ed.), *Nutraceuticals, Nanotechnology in the Agri-Food Industry* (pp. 153–182).
19. López, M. I., et al., (2014). Multivariate screening in food adulteration: Untargeted versus targeted modelling. *Food Chemistry, 147*, 177–181.
20. Handford, C. E., Campbell, K., & Elliott, C. T., (2016). Impacts of milk fraud on food safety and nutrition with special emphasis on developing countries. *Comprehensive Reviews in Food Science and Food Safety, 15*(1), 130–142.
21. MacMahon, S., et al., (2012). A liquid chromatography-tandem mass spectrometry method for the detection of economically motivated adulteration in protein-containing foods. *Journal of Chromatography A, 1220,* 101–107.
22. Rahman, M. A., et al., (2015). Food Adulteration: A serious public health concern in Bangladesh. *Bangladesh Pharmaceutical Journal, 18*(1), 1–7.
23. Andlauer, W., & Fürst, P., (2002). Nutraceuticals: A piece of history, present status, and outlook. *Food Research International, 35*(2, 3), 171–176.
24. Millikan, L. E., (2001). Cosmetology, cosmetics, cosmeceuticals: Definitions and regulations. *Clinics in Dermatology, 19*(4), 371–374.
25. Hardy, G., Hardy, I., & Ball, P. A., (2003). Nutraceuticals-a pharmaceutical viewpoint: Part II. *Current Opinion in Clinical Nutrition & Metabolic Care, 6*(6), 661–671.

26. Gulati, O. P., & Ottaway, P. B., (2006). Legislation relating to nutraceuticals in the European Union with a particular focus on botanical-sourced products. *Toxicology, 221*(1), 75–87.
27. Dickinson, A., (2011). History and overview of DSHEA. *Fitoterapia, 82*(1), 5–10.
28. Espín, J. C., García-Conesa, M. T., & Tomás-Barberán, F. A., (2007). Nutraceuticals: Facts and fiction. *Phytochemistry, 68*(22–24), 2986–3008.
29. DeFelice, S. L., (1995). The nutraceutical revolution: Its impact on food industry R&D. *Trends in Food Science & Technology, 6*(2), 59–61.
30. Hathcock, J., (2001). Dietary supplements: How they are used and regulated? *The Journal of Nutrition, 131*(3), 1114S–1117S.
31. Arai, S., et al., (2001). A mainstay of functional food science in Japan: History, present status, and future outlook. *Bioscience, Biotechnology, and Biochemistry, 65*(1), 1–13.
32. Arai, S., (1996). Studies on functional foods in Japan: State of the art. *Bioscience, Biotechnology, and Biochemistry, 60*(1), 9–15.
33. Roberfroid, M. B., (2002). Global view on functional foods: European perspectives. *British Journal of Nutrition, 88*(S2), S133–S138.
34. Smith, B. L., Marcotte, M., & Harrison, G., (1997). A comparative analysis of the regulatory framework affecting functional food development and commercialization in Canada, Japan, the European Union, and the United States of America. *Journal of Nutraceuticals, Functional & Medical Foods, 1*(2), 45–87.
35. Bigliardi, B., & Galati, F., (2013). Innovation trends in the food industry: The case of functional foods. *Trends in Food Science & Technology, 31*(2), 118–129.
36. Roberfroid, M., (2011). Defining functional foods and associated claims. In: *Functional Foods* (pp. 3–24). Elsevier.
37. Ferrari, C. K., (2007). Functional foods and physical activities in health promotion of aging people. *Maturitas, 58*(4), 327–339.
38. Man, R. Y., et al., (2002). Cholesterin inhibits cholesterol synthesis and secretion in hepatic cells (HepG2). *Molecular and Cellular Biochemistry, 233*(1, 2), 153–158.
39. Belongia, E. A., Mayeno, A. N., & Osterholm, M. T., (1992). The eosinophilia-myalgia syndrome and tryptophan. *Annual Review of Nutrition, 12*(1), 235–254.
40. Ferrari, C. K., (2004). Functional foods, herbs, and nutraceuticals: Towards biochemical mechanisms of healthy aging. *Biogerontology, 5*(5), 275–289.
41. Nicoletti, M., (2012). Nutraceuticals and botanicals: Overview and perspectives. *International Journal of Food Sciences and Nutrition, 63*(sup1), 2–6.
42. Burdock, G. A., Carabin, I. G., & Griffiths, J. C., (2006). The importance of GRAS to the functional food and nutraceutical industries. *Toxicology, 221*(1), 17–27.
43. Otles, S., & Cagindi, O., (2012). Safety considerations of nutraceuticals and functional foods. In: *Novel Technologies in Food Science* (pp. 121–136). Springer.
44. Attaran, A., et al., (2012). How to achieve international action on falsified and substandard medicines. *BMJ, 345*, e7381.
45. Dhanya, K., & Sasikumar, B., (2010). Molecular marker-based adulteration detection in traded food and agricultural commodities of plant origin with special reference to spices. *Current Trends in Biotechnology and Pharmacy, 4*(1), 454–489.
46. Villani, T. S., et al., (2015). Chemical investigation of commercial grape seed derived products to assess quality and detect adulteration. *Food Chemistry, 170*, 271–280.
47. Butler, M. S., Robertson, A. A., & Cooper, M. A., (2014). Natural product and natural product derived drugs in clinical trials. *Natural Product Reports, 31*(11), 1612–1661.

48. Linscott, A. J., (2011). Food-borne illnesses. *Clinical Microbiology Newsletter, 33*(6), 41–45.
49. Posadzki, P., Watson, L., & Ernst, E., (2013). Contamination and adulteration of herbal medicinal products (HMPs): An overview of systematic reviews. *European Journal of Clinical Pharmacology, 69*(3), 295–307.
50. Wang, J., Chen, B., & Yao, S., (2008). Analysis of six synthetic adulterants in herbal weight-reducing dietary supplements by LC electrospray ionization-MS. *Food Additives and Contaminants, 25*(7), 822–830.
51. Spink, J., & Moyer, D. C. (2011). Defining the public health threat of food fraud. *Journal of Food Science, 76*(9), R157–R163.
52. Spink, J., (2009). Defining food fraud and the chemistry of the crime. In: *Proceedings of the FDA Open Meeting, Economically Motivated Adulteration*, College Park, MD, USA. Wiley Online Library.
53. Roth, A. V., et al., (2008). Unraveling the food supply chain: Strategic insights from China and the 2007 recalls. *Journal of Supply Chain Management, 44*(1), 22–39.
54. Spink, J., & Moyer, D. C., (2011). Defining the public health threat of food fraud. *Journal of Food Science, 76*(9), R157–R163.
55. Spink, J., Moyer, D. C., & Speier-Pero, C., (2016). Introducing the food fraud initial screening model (FFIS). *Food Control, 69*, 306–314.
56. Manning, L., & Soon, J. M., (2014). Developing systems to control food adulteration. *Food Policy, 49*, 23–32.
57. Northcutt, J. K., & Parisi, M. A., (2013). *Major Food Laws and Regulations* (pp. 73–96). Guide to US food laws and regulations. Chichester: Wiley Blackwell.
58. Manning, L., (2016). Food fraud: Policy and food chain. *Current Opinion in Food Science, 10*, 16–21.
59. Esteki, M., Regueiro, J., & Simal-Gándara, J., (2019). Tackling fraudsters with global strategies to expose fraud in the food chain. *Comprehensive Reviews in Food Science and Food Safety, 18*(2), 425–440.
60. Van, R. S. M., Huisman, W., & Luning, P. A., (2017). Food fraud vulnerability and its key factors. *Trends in Food Science & Technology, 67*, 70–75.
61. Marvin, H. J., et al., (2016). A holistic approach to food safety risks: Food fraud as an example. *Food Research International, 89*, 463–470.
62. Mitra, S., & Kannan, R., (2007). A note on unintentional adulterations in Ayurvedic herbs. *Ethnobotanical Leaflets, 2007*(1), 3.
63. Shaheen, S., et al., (2019). Types and causes of adulteration: Global perspectives. In: *Adulteration in Herbal Drugs: A Burning Issue* (pp. 9–16). Springer.
64. Bansal, S., et al., (2017). Food adulteration: Sources, health risks, and detection methods. *Critical Reviews in Food Science and Nutrition, 57*(6), 1174–1189.
65. Unit, P., & Organization, W. H., (1992). *Quality Control Methods for Medicinal Plant Materials*. Geneva: World Health Organization.
66. Ford, L. S., et al., (2010). Food allergen advisory labeling and product contamination with egg, milk, and peanut. *Journal of Allergy and Clinical Immunology, 126*(2), 384–385.
67. Coppens, P., Da Silva, M. F., & Pettman, S., (2006). European regulations on nutraceuticals, dietary supplements, and functional foods: A framework based on safety. *Toxicology, 221*(1), 59–74.

68. Almada, A. L., (2014). Nutraceuticals and functional foods: Aligning with the norm or pioneering through a storm. In: *Nutraceutical and Functional Food Regulations in the United States and Around the World* (pp. 3–11). Elsevier.
69. Cecchini, M., & Warin, L., (2016). Impact of food labelling systems on food choices and eating behaviors: A systematic review and meta-analysis of randomized studies. *Obesity Reviews, 17*(3), 201–210.
70. Sattigere, V. D., Ramesh, K. P., & Prakash, V., (2018). Science-based regulatory approach for safe nutraceuticals. *Journal of the Science of Food and Agriculture.*
71. Gul, K., Singh, A., & Jabeen, R., (2016). Nutraceuticals and functional foods: The foods for the future world. *Critical Reviews in Food Science and Nutrition, 56*(16), 2617–2627.
72. Shahidi, F., (2009). Nutraceuticals and functional foods: Whole versus processed foods. *Trends in Food Science & Technology, 20*(9), 376–387.
73. Ruckman, S. A., (2008). Regulations for nutraceuticals and functional foods in Europe and the United Kingdom. In: *Nutraceutical and Functional Food Regulations in the United States and Around the World* (pp. 221–238). Elsevier.
74. Rishton, G. M., (2008). Natural products as a robust source of new drugs and drug leads: Past successes and present-day issues. *The American Journal of Cardiology, 101*(10), S43–S49.
75. Cordella, C., et al., (2002). Recent developments in food characterization and adulteration detection: Technique-oriented perspectives. *Journal of Agricultural and Food Chemistry, 50*(7), 1751–1764.
76. Vaclavik, L., Krynitsky, A. J., & Rader, J. I., (2014). Mass spectrometric analysis of pharmaceutical adulterants in products labeled as botanical dietary supplements or herbal remedies: A review. *Analytical and Bioanalytical Chemistry, 406*(27), 6767–6790.
77. Becue, I., Poucke, C. V., & Peteghem, C. V., (2011). An LC-MS screening method with library identification for the detection of steroids in dietary supplements. *Journal of Mass Spectrometry, 46*(3), 327–335.
78. Strano-Rossi, S., et al., (2015). Liquid chromatography-high resolution mass spectrometry (LC-HRMS) determination of stimulants, anorectic drugs and phosphodiesterase 5 inhibitors (PDE5I) in food supplements. *Journal of Pharmaceutical and Biomedical Analysis, 106*, 144–152.
79. Li, H., et al., (2019). Rapid quantitative analysis of Hg^{2+} residue in dairy products using SERS coupled with ACO-BP-AdaBoost algorithm. *Spectrochimica Acta Part A: Molecular and Biomolecular Spectroscopy, 223*, 117281.
80. Craig, A. P., Franca, A. S., & Irudayaraj, J., (2013). Surface-enhanced Raman spectroscopy applied to food safety. *Annual Review of Food Science and Technology, 4*, 369–380.
81. Zheng, J., & He, L., (2014). Surface-enhanced Raman spectroscopy for the chemical analysis of food. *Comprehensive Reviews in Food Science and Food Safety, 13*(3), 317–328.
82. Esteki, M., et al., (2018). A review on the application of chromatographic methods, coupled to chemometrics, for food authentication. *Food Control, 93*, 165–182.
83. Iwaniak, A., et al., (2015). Chemometrics and cheminformatics in the analysis of biologically active peptides from food sources. *Journal of Functional Foods, 16*, 334–351.
84. Huang, L., et al., (2014). Nondestructive measurement of total volatile basic nitrogen (TVB-N) in pork meat by integrating near-infrared spectroscopy, computer vision and electronic nose techniques. *Food Chemistry, 145*, 228–236.

85. Chen, Q., et al., (2008). Determination of total polyphenols content in green tea using FT-NIR spectroscopy and different PLS algorithms. *Journal of Pharmaceutical and Biomedical Analysis, 46*(3), 568–573.
86. Sheng, R., et al., (2019). Model development for soluble solids and lycopene contents of cherry tomato at different temperatures using near-infrared spectroscopy. *Postharvest Biology and Technology, 156*, 110952.
87. Ouyang, Q., Zhao, J., & Chen, Q., (2013). Classification of rice wine according to different marked ages using a portable multi-electrode electronic tongue coupled with multivariate analysis. *Food Research International, 51*(2), 633–640.
88. Chen, Q., et al., (2011). Discrimination of green tea quality using the electronic nose technique and the human panel test, comparison of linear and nonlinear classification tools. *Sensors and Actuators B: Chemical, 159*(1), 294–300.
89. Ouyang, Q., et al., (2017). Rapid and specific sensing of tetracycline in food using a novel upconversion aptasensor. *Food Control, 81*, 156–163.
90. Liu, Y., et al., (2018). Turn-on fluorescence sensor for Hg^{2+} in food based on FRET between aptamers-functionalized upconversion nanoparticles and gold nanoparticles. *Journal of Agricultural and Food Chemistry, 66*(24), 6188–6195.
91. Perestam, A. T., et al., (2017). Comparison of real-time PCR and ELISA-based methods for the detection of beef and pork in processed meat products. *Food Control, 71*, 346–352.
92. Ellis, D. I., et al., (2012). Fingerprinting food: Current technologies for the detection of food adulteration and contamination. *Chemical Society Reviews, 41*(17), 5706–5727.
93. Chen, Q., et al., (2013). Recent advances in emerging imaging techniques for non-destructive detection of food quality and safety. *TrAC Trends in Analytical Chemistry, 52*, 261–274.
94. ElAmrawy, F., et al., (2016). Adulterated and counterfeit male enhancement nutraceuticals and dietary supplements pose a real threat to the management of erectile dysfunction: A global perspective. *Journal of Dietary Supplements, 13*(6), 660–693.
95. Akhtar, S., (2015). Food safety challenges: A Pakistan's perspective. *Critical Reviews in Food Science and Nutrition, 55*(2), 219–226.
96. Coté, T. R., et al., (2005). Botulinum toxin type A injections: Adverse events reported to the US Food and Drug Administration in therapeutic and cosmetic cases. *Journal of the American Academy of Dermatology, 53*(3), 407–415.
97. Griffiths, J., et al., (2009). Functional food ingredient quality: Opportunities to improve public health by compendial standardization. *Journal of Functional Foods, 1*(1), 128–130.
98. CABANero, A. I., Recio, J. L., & Ruperez, M., (2006). Liquid chromatography coupled to isotope ratio mass spectrometry: A new perspective on honey adulteration detection. *Journal of Agricultural and Food Chemistry, 54*(26), 9719–9727.
99. Woodbury, S. E., et al., (1995). Detection of vegetable oil adulteration using gas chromatography combustion/isotope ratio mass spectrometry. *Analytical Chemistry, 67*(15), 2685–2690.
100. Rausch, E., Cassidy, M. F., & Buede, D., (2009). *Does the Accuracy of Expert Judgment Comply with Common Sense: Caveat Emptor*. Management Decision.

CHAPTER 5

Nutraceuticals-Loaded Nano-Sized Delivery Systems: Potential Use in the Prevention and Treatment of Cancer

MOHAMMED JAFAR,[1] SYED SARIM IMAM,[2] SULTAN ALSHEHRI,[2] CHANDRA KALA,[3] and AMEEDUZZAFAR ZAFAR[4]

[1]*Department of Pharmaceutics, College of Clinical Pharmacy, Imam Abdulrahman Bin Faisal University, Dammam, Saudi Arabia*

[2]*Department of Pharmaceutics, College of Pharmacy, King Saud University, Riyadh, Saudi Arabia*

[3]*Faculty of Pharmacy, Maulana Azad University, Jodhpur – 342802, Rajasthan, India*

[4]*Department of Pharmaceutics, College of Pharmacy, Jouf University, Sakaka, Aljouf, Saudi Arabia*

ABSTRACT

Cancer is considered as one of the most life-threatening diseases, wherein uncontrolled growth of abnormal cells takes place. Nutraceutical is any compound which is a nutritious food or a fraction of nutritious food which gives health or clinical boons, together with the prevention and treatment of disease. This chapter aims to provide to its readers a key scientific knowledge about how important are these nutraceuticals in terms of effectively preventing and treating various types of cancer via a novel nanocarrier strategy. Various types of lipid type, polymeric type, and inorganic nanocarriers have been investigated to improve the bioaccessibility and therapeutic success of nutraceuticals. It is also briefly explained in this chapter in its subsections, the importance of the combination approach of nutraceuticals and chemo-therapeutic agents utilizing nanocarriers, over simple nutraceutical loaded nanocarriers. Regardless of efficient manufacturing procedures of nano-nutraceutical delivery systems, the caliber, strength, potency, and untoward effects should work out and discourse on top preference. To make use of the full prospective of nanocarriers, additional preclinical and clinical investigations are required for nanoformulation nutraceuticals. It is apprehended that the continued attempts in the field of nutraceutical delivery using the variety of novel nanocarriers would yield many rewarding outcomes.

5.1 INTRODUCTION

Cancer is considered as one of the most life-threatening diseases, wherein unrestrained growth of abnormal cells takes place [1]. According to an International cancer research institution, GLOBOCAN, it is estimated that about 18.1 million new cancer cases were identified throughout the world in 2018 and approximately 9.6 million cancer deaths took place. As per one of the important reports of the American Cancer Society, the second most general cause of mortality in the US is cancer, which is surpassed only by heart disease [2]. It is also reported that behavioral factors, mainly substandard nutrition, high alcohol intake, physical inertia, and high weight gain, are responsible for 25% of incident cancers that takes place in the US, and therefore, these can be prevented [3, 4]. It is not only sufficient to modify the lifestyle to prevent cancer, but also it is required to minimize the spread of cancer in individuals. Therefore, the search for newer strategies for the effective treatment of cancer is in progress. However, nutraceuticals are emerging as a new approach in decreasing the progression of cancer.

Nutraceutical, a composite expression from 'nutrition' and 'pharmaceutical,' was coined by Stephen L. DeFelice, chairman of the Foundation for Innovation in Medicine (FIM), Cranford, in 1989. According to Stephen nutraceutical is any compound which is a nutritious food or a fraction of nutritious food which gives health or clinical boons, together with the prevention and treatment of disease [5–7]. In the recent past, nutraceuticals have acquired ample recognition in the field of cancer investigation due to their pleiotropic sequel and reasonably innocuous nature [8]. These compounds include vitamins, carotenoids, prebiotics, probiotics, dietary fiber, phenolics, and fatty acids [9]. Nutraceuticals prevent cancer through many different mechanisms such as inhibiting efflux transporters such as Breast Cancer Resistance Protein (BCRP), P-glycoprotein (P-gp), multidrug resistance protein (MRP), inhibiting cell proliferation and differentiation, or by reducing the harmful effects of anticancer drugs [10, 11].

The vast majority of nutraceuticals have been explored for the prevention of cancer throughout the world, but most of them suffer from poor bioavailability in humans due to their poor aqueous solubility and poor permeability [12, 13]. Some of the other factors responsible for poor rates and extents of absorption of nutraceuticals are: (i) abrupt delivery of active constituents from the nutritious foodstuff [14]; (ii) Insoluble complex formation with the different constituents of GIT; and/or (iii) Intestinal first-pass metabolism [15–17].

To conquer these obstacles nanoformulations have transpired as competent vehicles because of their nano size and other promising attributes. Nano formulations improve aqueous solubility and stability, provide moisture protection to foodstuff, prolongs drug release, manipulate texture and flavor. Moreover, nanocarriers can influence the pharmacodynamic and pharmacokinetic profile of nutraceuticals [18]. Different types of nanocarriers viz liposomes, micelles, polymeric NPs, etc., have been used to improve the biological performance of nutraceuticals.

5.2 NUTRACEUTICALS IN CANCER PREVENTION AND TREATMENT

It is mainly the non-toxic nature of nutritional substances, which makes them gain plenty of engrossment for their capability in cancer prevention and treatment. It is reported that nutraceuticals by the modulation of miRNAs, cellular signaling, and epigenome, could prevent cancer progression in individuals [19]. Moreover, another interesting property of nutraceuticals is that they are pleiotropic, i.e., they can down-regulate several signaling pathways. All of these good qualities make nutraceuticals excellent suitor for accomplishing greater treatment outcomes in cancer patients, as solely targeted moiety usually malfunction in clinical trials [20, 21]. The major signaling pathways influenced by nutraceuticals are Pi3K/Akt/mTOR pathway, insulin-like growth factor receptor (IGFR), MAPK/ERK pathway, Ras/Raf signaling pathway, epidermal growth factor receptor (EGFR) family receptors, B-catenin signaling pathway sonic hedgehog signaling pathway, etc., [7, 22]. By acting against these molecular targets, nutraceuticals inhibit the rapid increase of cancer cells, elicit cell cycle arrest, and conquer angiogenesis/invasion/metastasis. Thus, the cytotoxic outcomes of nutraceuticals are arbitrated through the action against different factors, viz survivin, vascular endothelial growth factor (VEGF), matrix metalloproteinases (MMPs), etc. Many different nutraceuticals such as curcumin, 3,30-diindolylmethane (DIM), resveratrol, indole-3-carbinol, epigallocatechin-3-gallate, lycopene, and curcumin are known to down-regulate the signal transductions like Pi3K, NF-kB, Akt and other signal transduction pathways which are required for the spread of cancer [23].

Besides restraining these traditional targets, nutraceuticals are also been showed to harmonize the pharmacokinetics of anticancer drugs by regulating ATP-binding cassette (ABC) such as MRP, P-gp, BCRP, etc. [83]. The simultaneous oral administration of curcumin increases the AUC and C_{max} of etoposide by 1.50 and 1.36-fold, respectively [24]. It is not only that the nutraceuticals only enhance the bioavailability of simultaneously administered anticancer drugs, but they also extenuate the toxic effects of administered drugs. It is reported that a simultaneous intake of co-enzyme Q10 prevents anthracycline-induced cardiotoxicity [25]. Nevertheless, the majority of nutraceuticals, because of their poor aqueous solubility and poor permeability, exhibit poor bioavailability and thus low therapeutic response. To overcome these hurdles, nanocarriers strategies have appeared to be more effective approaches and showed their capability to improve the clinical performance of nutraceuticals, and the same is explained in the following section.

5.3 NANOCARRIERS-BASED NUTRACEUTICALS FOR CANCER PREVENTION AND TREATMENT

Among the novel anticancer drugs, 40% are lipophilic, and this nature is a big hindrance for a new medicine invention project [26]. The traditional formulation methods to handle these drugs involve pH adjustment, use of cosolvents, particle size reduction, and use of surfactants which are generally harmful. In one of the commercial paclitaxel i.v. formulation (Taxol®), they used Cremophor EL, a surfactant which is linked with acute hypersensitivity reactions, and neurotoxicity [27]. Furthermore, the major hurdle with these conventional formulations is that they showed a huge amount of "collateral damage" to some normal cells, and this is due to their abnormal distribution in few compartments of the body [28].

The abnormal pharmacokinetic pattern showed by the nutraceuticals after their oral administration could be greatly minimized by encapsulation them into an emerging novel nano-drug carrier systems-based approach [29]. In recent years different types of nano-drug carrier systems, for instance, liposomes, solid lipid nanoparticles (SLNs), micelles, polymeric nanoparticles (NPs), polymeric conjugates, carbon nanotubes, quantum dots, etc., have been investigated to regulate both pharmacokinetics and pharmacodynamics of the drug.

Many different mechanisms have been suggested for improving the oral bioavailability of these nanocarriers. One important example is chitosan-based nanocarriers can cross the tight-junctions via paracellular route and ability to modulate P-gp present on epithelial cells. Another interesting example is polymeric NPs prepared using PLGA are absorbed through a distinctive Payer's patches. Liposomes have the potential to regulate P-gp and/or CYP450 to enhance intracellular concentration and stimulate lipoprotein/chylomicron production. Moreover, entrapment of nutraceuticals into novel nanocarriers could safeguard them from the harsh gastrointestinal (GI) environment. In the latest report, it is explained that incorporation of epigallocatechin gallate (EGCG) in liposomes greatly decreases its degradation in an artificial intestinal fluid by approximately 10-folds [30].

Few nutraceuticals like fatty acids, Vitamin E, etc., are also used as additives in the formulation of nanocarriers, and in these systems besides their additive role, the nutraceuticals play a vital role in improving the oral bioavailability of several chemotherapeutic agents. For example, the use of d-a-Tocopheryl polyethylene glycol 1000 succinate (Vitamin E TPGS), a non-ionic surfactant, inhibits P-glycoprotein via ATPase inhibition and thus regulates the pharmacokinetics of the administered P-gp substrate [31]. These novel nanocarrier systems also exhibit improved permeation and retention effect through active and also passive targeting [32]. Because of the leaky vasculature of cancer endothelial cells, these nanocarriers could efficiently gain entry into cancer cells using passive targeting. Nevertheless, these novel nano-drug delivery carriers have the potential for the site-specific delivery of chemotherapeutic agents by binding a suitable directing moiety (ligands) on the exterior of nanocarriers utilizing active targeting strategy [10, 33]. Hyaluronic acid, RGD peptides, and folic acids are few among the various ligands widely used for the targeted delivery of chemotherapeutic agents because of their overexpression on cancer cells [34–36].

5.3.1 LIPID BASED NANOCARRIERS

The different types of lipid-based nanocarriers like liposomes, SLNs, nanostructured lipid carriers (NLCs), self-emulsifying systems have been used for improving the bioavailability and therapeutic effects of nutraceuticals (Table 5.1).

The lipid-based nanocarriers usually contain a solid lipid core, which has potential in accumulating medications with hydrophilic and lipophilic nature into its lipid fabric. SLN is accurately embraced with more than a single solid lipid, which shows a melting point of 40°C and even higher. Subsequently, at the beginning of the 1990s, the benefits of control release property of SLNs have been emerged [48], including cellular toxicity, augmented compatibility, and high *in vivo* tolerance [49, 50]. Compared to SLN, NLCs, which contain suitable blends of both liquid and solid lipids seems to possess the benefits of elevated medication carrying potential, improved storage steadiness, and efficient drug discharging characteristics [51, 52]. These newer lipid-based NPs have the potential to circumvent P-gp through paracellular penetration and capability to uptake by microfold cell [53]. One of the recent studies conducted on SLN reports that the *in-vitro* cytotoxicity of a nutraceutical aloe-emodin against human breast cancer MCF-7 cells and human hepatoma HepG2 cells was drastically increased compared to its pure form, after encapsulation of it into SLNs [41], on the other hand, there was a great reduction in its toxicity on normal human mammary epithelial MCF-10A cells was recorded. This could be due to high cellular ingestion of SLNs as compared to the pure aloe-emodin formulation. In another similar study conducted by Ramalingam and Ko, it is reported about the enhanced bioavailability of resveratrol from SLNs formulated using N-trimethyl chitosan (TMC)-g-palmitic acid (PA) [54]. This investigation showed that the relative bioavailability of resveratrol from TMC-g-PA SLNs was 3.8-fold higher than that from its conventional suspension dosage form.

Liposomes are colloidal vesicular transporters, which are produced by the hydration of phospholipids. The nanosized liposomes are made up of phospholipids composed of the polar head as well as non-polar fatty acid chains, which aids them accommodated in individual minor structural phospholipid units both the hydrophilic and hydrophobic drug molecules accessing their delivery to the targeted sites [55]. Liposomes are amenable to surface modification with various targeting ligands such as sialic acid, aptamers, folic acid, etc., [56]. However, to make liposomes long circulatory *in-vivo* their surface has been modified by incorporating in its formulation polyethylene glycol. In an investigation, it is showed that PEG-modified liposome of ursolic acid has improved *in vitro* cytotoxicity in EC-304 cells as compare to pure ursolic acid [57].

TABLE 5.1 Outline of Few Current Investigations on Various Types of Nanocarriers Utilized in the Delivery of Nutraceuticals in Cancer Management

Nutraceuticals	Chief Excipients	*In-Vitro*/*In-Vivo* Model	Developments	References
Apigenin	2-Distearoyl-sn-glycero-3-phosphocholine (DSPC)	Human colon cancer cell lines HCT-15, and HT-29 *In-vivo* nude mice xenograft model	Enhanced *in-vitro* cytotoxic activity of apigenin and 5-fluorouracil combination was attributed to maximal reversal of Warburg effect. The increased *in-vivo* chemotherapeutic potential of apigenin was due to the passive targeting achieved by the liposomal drug-loaded nanocarrier. Synergistic effect of apigenin with 5-fluorouracil.	[37]
Quercetin	Compritol	Human MCF-7 and MCF-10 A cell lines	Quercetin-Solid lipid nanoparticles (QT-SLNs) inhibited MCF-7 cells growth with a low IC_{50} (50% inhibitory concentration) value, compared to the free QT. QT-SLNs induced a significant decrease in the viability and proliferation of MCF-7 cells, compared to the free QT.	[38]
Curcumin	Krill lipids	A549 lung cancer cells Human umbilical vein endothelial cells	Sustained *in-vitro* drug release Powerful antioxidant activity. Improved *in-vitro* cytotoxic activity against specific cancer cells	[39]
Resveratrol	1,2-dipalmitoyl-sn-glycero-3-phosphocholine (DPPC)	Human colon cancer cells HT-29	Improved *in-vitro* cytotoxicity against human colon cancer cells	[40]
Aloe-emodin	Lecithin, Ploxomer 188, Poloxomer 407	MCF-7, MCF-10A, and HepG2 cell lines	Improved *in-vitro* cytotoxicity of Aloe-emodin solid lipid nanoparticles as compare to free Aloe-emodin	[41]

TABLE 5.1 *(Continued)*

Nutraceuticals	Chief Excipients	*In-Vitro*/*In-Vivo* Model	Developments	References
Fisetin	Polylactic acid (PLA)	Xenograft mouse model of breast cancer cells	Improved bioavailability Reduced toxicity Inhibited tumor volume and weight without altering body weight	[42]
Quercetin	Chitosan	Tumor xenograft mice with A549 and MDA MB 468 cells.	Decreased tumor weight and volume	[85]
Epigallocatechin-3-gallate	Polyethylene glycol (PEG), Polylactic acid (PLA)	22Rv1 cells implanted tumor xenograft in athymic nude mice	Decreased tumor size and volume Reduced prostate-specific antigen levels in serum	[43]
Genistein	Mannitol, PLGA, TPGS	HepG2 cells	Surface modified nanoparticles showed superior *in-vitro* cytotoxicity, and *in-vivo* antitumor activity than plain or linear nanoparticles of genistein	[44]
Thymoquinone	Polyethylene glycol (PEG), Polyvinyl pyrrolidone (PVP)	MCF-7, and HBL-100	Improved *in-vitro* cytotoxicity of Thymoquinone nanoparticles than pure thymoquinone	[45]
Curcumin	AUNPs Folic acid Polyvinyl pyrrolidone (PVP)	Human breast epithelial and mouse fibroblast cell lines. Breast cancer orthotopic mouse model	Folate-coated cancer cell targeting using CurAu-PVP NCs is a promising approach for tumor-specific therapy of breast cancer without harming normal cells. Improved *in-vivo* activity	[46]
Resveratrol	AUNPs	Human breast (MDAMB-231), pancreatic (PANC-1), and prostate (PC-3) cancers cell lines	Improved anticancer effect	[47]

The isotropic mixture of surfactant, co-surfactant, and oil constitutes a novel self-emulsifying drug delivery system SNEDDS [58]. These novel nanocarriers were proved to be highly effective in the delivery of nutraceuticals of various potential including chemoneutraceuticals. In some recently reported investigations, it was found that well-known nutraceuticals like curcumin, piperine, and naringenin, when given in the form of SNEDDS their bioavailability was increased many folds as compared to their traditional formulations. Moreover, the bioavailability of these nutraceuticals was doubled when they are administered in combination in SNEDDS [59, 60]. This could be attributed to the synergistic effect of neutraceuticals. One more investigation showed that self-micro emulsifying drug delivery system (SMEDDS) of *Brucea javanica* oil significantly inhibited the growth of tumor cells and drastically decreased S180 cancer [61].

Polymers show different characteristics in their composition. For attaining favorable drug delivery to the targeted area, the best suitable option is the polymeric NPs of colloidal nanosized systems (1 nm$<d<$1000 nm) [62]. Based on the structural differences, the polymeric NPs are classified as nanospheres and nanocapsules. Nanospheres generally consist of a polymeric matrix with three drug-loading patterns: (1) to encapsulate drugs into the spheres; (2) to absorb drugs onto the surface; (3) to disperse drugs within the polymeric network.

In contrast to nanospheres, nanocapsule score-shell possesses the ability to dissolve drugs in the core or to absorb drugs on the shell when present in the drug-loading form [63, 64].

A variety of polymers such as biodegradable polymers, polysaccharides polymers have been extensively used to deliver nutraceuticals to their specific site. Some of the reported biodegradable polymers exploited for the delivery of nutraceuticals are polylactic acid co-glycolic acid (PLGA), polycaprolactone (PCL), polylactic acid (PLA) and their copolymers are poly(ethylene glycol) (PEG), d-a-tocopheryl polyethylene glycol 1000 succinate (TPGS), etc., [65]. Similarly, polysaccharides-based polymers viz alginate, chitosan, pectin, etc., have also been used in the encapsulation and site-specific delivery of nutraceuticals [66, 67]. Polymeric NPs possess many benefits over conventional nanocarriers such as pH-dependent prolonged drug release, amenable to surface modification because of the existence of functional groups for site-specific drug delivery systems. Moreover, few polymers such as chondroitin sulfate, and hyaluronic acid have the potential to target CD44 overexpressing cancer cells [68, 69]. Jiang et al. [86] developed bovine serum albumin PCL NPs of curcumin and showed enhancement in inhibition of the growth of three-dimensional LNCaP multicellular tumors as compared to native curcumin. This improvement in the cytotoxicity could be attributed to the efficient cellular uptake of the NPs via caveolin endocytosis. Some of the recently developed polymeric NPs for delivery of nutraceuticals are summarized in Table 5.1.

Nano-micelles consisting of polymeric and surfactant nano-micelles are known to be as rising novel carrier systems for nutraceuticals delivery. Apart from their smaller size, improved drug solubility and stability [70], and lower adverse effects and high biocompatibility [71] aids them to become potential candidates for poorly aqueous soluble drug delivery. Fewer amphiphilic molecules, when added to special solvents, adapt to self-assemble and results in core-shell monomers called nano-micelles [72]. Liu et al. [52] described novel curcumin containing a nano micelle formulation using a polyvinyl caprolactam-polyvinyl acetate-polyethene glycol (PVCL-PVA-PEG) graft copolymer. Nano micelle curcumin was formulated and optimized and then further evaluated for *in vitro* cytotoxicity, *in vitro* cellular uptake, *in vitro* antioxidant activity, and also various *in-vivo* activities. The solubility, chemical stability, and antioxidant activity of curcumin were greatly improved after the encapsulation of it into the PVCL-PVA-PEG nanomicelles. Moreover, the formulated curcumin nanomicelles are stable at storage conditions, they had good cellular tolerance, and also their *in-vitro* and *in-vivo* activities were significantly improved when compared with a free curcumin solution.

Polymeric hydrogels are cross-linked water-loving polymer meshes, and they possess the potential to pledge localized, prolonged release of nutraceuticals. These novel polymeric hydrogels possess a great attraction to water, but they are protected from solubilizing because of either by physical entanglements, covalent cross-linking, or by non-covalent attractions [73, 74]. These polymeric hydrogels could also undergo phase transition while exposed to atmospheric temperature [67]. A most useful characteristic of these polymeric hydrogels is that they are amenable for surface modification by active targeting ligands. It is reported that folate functionalized PEG cross-linked acrylic polymer (FA-CLAP) hydrogel were successfully utilized for the targeted delivery of curcumin [75]. Moreover, it is proved in the study that folate-functionalized hydrogel demonstrated higher uptake in HeLa cell lines than non-functionalized hydrogels. One more investigation carried out on the same nutraceutical curcumin, curcumin hydrogels were formulated using gelatin, hyaluronan, and showed improved *in vitro* cytotoxicity against A549 lung adenocarcinoma cells than pure curcumin

[76]. Moreover, these hydrogels showed comparatively greater apoptosis rates than pure curcumin demonstrated via Annexin V-FITC/PI analysis.

5.3.2 INORGANIC NANOCARRIERS

Inorganic type of nanocarriers which could deliver nutraceuticals include quantum dots, nanosilica, carbon nanotubes, magnetic nanomaterials, silver, and gold nanoparticles (AuNPs) [77, 78]. These nanocarriers possess some unique properties like smaller size, different shapes, varying content, high surface volume ratio, and most importantly, their potential for surface modification makes them excellent candidates to be extensively used in nutraceutical delivery. Carbon nanotubes, are hydrophobic tubular meshworks of carbon atoms showing length and diameter of around 1 to 100 nm and 1 to 4 nm respectively, turned out to be utilized for the delivery of nutraceuticals [79, 80]. The main problems with these carbon nanotubes are, they are highly insoluble in almost all solvents and they are held by several toxicity issues, but these problems could be overcome by modifying them chemically and thus ameliorating their biocompatibility, alleviating their toxicity and make them water-soluble carriers [1].

Among the various inorganic nanocarriers AuNPs have been voluminously investigated for nutraceuticals delivery because of their well-defined surface chemistry, simplicity in synthesis, and magnificent biocompatibility [57]. In one of the reported investigations, apigenin AuNPs were formulated and biocompatible property of prepared NPs was shown by non-toxicity on normal epidermal cells (HaCat). Moreover, these NPs showed *in vitro* compatibility with squamous epidermal cancer (A431) and also with human cervical squamous cell cancer (SiHa) cells. Epigallocatechin-3-gallate and green tea consolidated AuNPs have also been designed and adjudged particularly harmful towards MCF-7 cells, and Ehrlich's Ascites Carcinoma, while showing no toxicity in normal primary mouse hepatocytes and the reason attributed to it is that they possess greatest antioxidant characteristics [81]. Magnetic NPs, besides their small size, possess excellent magnetic properties, because of which they were successfully utilized in the delivery of nutraceuticals. In one of the reported research study curcumin magnetic NPs were designed and expanded with the aim to enhance its bioavailability and thus efficacy [82]. The investigation showed a 2.5-fold increase in oral bioavailability of curcumin as compare to pure curcumin. Moreover, this nanoparticle formulation significantly reduced pancreatic tumor growth in an HPAF-II xenograft mouse model and increases the life span of mice by slowing down the cancerous cell growth.

5.4 CONCLUSION AND FUTURE PROSPECTS

Scientific investigations done during the last few decapods have given substantial affirmation of the appreciable flexibility of nutraceuticals and the various targets that make them highly potential candidates for cancer prevention and treatment. While, preclinical and cell culture studies showed that nutraceuticals have potential antitumor activity and other fitness boons, therapeutic application of the aforesaid compounds is lean. An application of novel nanocarrier built delivery systems has facilitated scientists to conquer the physicochemical and biological hindrances of nutraceuticals. Nearly all of polymers and other additives used in the design and development of nanocarriers have been approved as safe by FDA. Moreover, these novel carrier systems result in site-specific delivery of nutraceuticals in cancer tissue because of improved tissue permeation and retention effect. The most interesting and important characteristics of nanocarriers are that their surfaces could be modified using specific ligands in order to attain cancerous cell-specific delivery of nutraceuticals. During the past few years, a combined approach of nutraceuticals and anticancer drugs utilizing nanocarriers has gained much attention. It is apprehended that the continued attempts in the field of nutraceutical delivery using the variety of novel nanocarriers and also combination-based strategies would yield many rewarding outcomes.

KEYWORDS

- **cancer**
- **chemotherapy**
- **insulin-like growth factor receptor**
- **nanoparticles**
- **nutraceuticals**
- **targeted delivery systems**
- **vascular endothelial growth factor**

REFERENCES

1. Perez-Herrero, E., & Fernandez-Medarde, A., (2015). Advanced targeted therapies in cancer: Drug nanocarriers, the future of chemotherapy. *European Journal of Pharmaceutics and Biopharmaceutics, 93*, 52–79.
2. Siegel, R. L., Miller, K. D., & Jemal, A., (2016). Cancer statistics-2016. *CA: A Cancer Journal for Clinicians, 66*, 7–30.
3. Gonzalez-Vallinas, M., Gonzalez-Castejon, M., Rodr-Aguez-Casado, A., & De Molina, A. R., (2013). Dietary phytochemicals in cancer prevention and therapy: A complementary approach with promising perspectives. *Nutrition Reviews, 71*, 585–599.
4. Makarem, N., Lin, Y., Bandera, E. V., Jacques, P. F., & Parekh, N., (2015). Concordance with world Cancer Research Fund/American Institute for Cancer Research (WCRF/AICR) guidelines for cancer prevention and obesity-related cancer risk in the Framingham Offspring cohort (1991–2008). *Cancer Causes & Control, 26*, 277–286.
5. Brower, V., (1998). Nutraceuticals: Poised for a healthy slice of the healthcare market? *Nature Biotechnology, 16*, 728–732.
6. DeFelice, S. L., (1995). The nutraceutical revolution: Its impact on food industry R&D. *Trends in Food Science & Technology, 6*, 59–61.
7. Li, Y., Ahmad, A., Kong, D., Bao, B., & Sarkar, F. H., (2014). Recent progress on nutraceutical research in prostate cancer. *Cancer and Metastasis Reviews, 33*, 629–640.
8. Nair, H. B., Sung, B., Yadav, V. R., Kannappan, R., Chaturvedi, M. M., & Aggarwal, B. B., (2010). Delivery of antiinflammatory nutraceuticals by nanoparticles for the prevention and treatment of cancer. *Biochemical Pharmacology, 80*, 1833–1843.
9. Wang, J., Guleria, S., Koffas, M. A., & Yan, Y., (2016). Microbial production of value-added nutraceuticals. *Current Opinion in Biotechnology, 37*, 97–104.
10. Saneja, A., Khare, V., Alam, N., Dubey, R. D., & Gupta, P. N., (2014). Advances in P-glycoprotein- based approaches for delivering anticancer drugs: Pharmacokinetic perspective and clinical relevance. *Expert Opinion on Drug Delivery, 11*, 121–138.
11. Trottier, G., Bostr€om, P. J., Lawrentschuk, N., & Fleshner, N. E., (2010). Nutraceuticals and prostate cancer prevention: A current review. *Nature Reviews Urology, 7*, 21–30.
12. Bethune, S. J., Schultheiss, N., & Henck, J. O., (2011). Improving the poor aqueous solubility of nutraceutical compound pterostilbene through cocrystal formation. *Crystal Growth & Design, 11*, 2817–2823.
13. McClements, D. J., Li, F., & Xiao, H., (2015). The nutraceutical bioavailability classification scheme: Classifying nutraceuticals according to factors limiting their oral bioavailability. *Annual Review of Food Science and Technology, 6*, 299–327.
14. Moelants, K. R., Lemmens, L., Vandebroeck, M., Van, B. S., Van, L. A. M., & Hendrickx, M. E., (2012). Relation between particle size and carotenoid bioaccessibility in carrot and tomato-derived suspensions. *Journal of Agricultural and Food Chemistry, 60*, 11995–12003.
15. D'Ambrosio, D. N., Clugston, R. D., & Blaner, W. S., (2011). Vitamin a metabolism: An update. *Nutrients, 3*, 63–103.
16. Fernandez-García, E., Carvajal-Lerida, I., Jaren-Galan, M., Garrido-Fern_andez, J., Perez-Galvez, A., & Hornero-Mendez, D., (2012). Carotenoids bioavailability from foods: From plant pigments to efficient biological activities. *Food Research International, 46*, 438–450.

17. Hurst, S., Loi, C. M., Brodfuehrer, J., & El-Kattan, A., (2007). Impact of physiological, physicochemical and biopharmaceutical factors in absorption and metabolism mechanisms on the drug oral bioavailability of rats and humans. *Expert Opinion on Drug Metabolism & Toxicology, 3*, 469–489.
18. Díaz, M. R., & Vivas-Mejia, P. E., (2013). Nanoparticles as drug delivery systems in cancer medicine: Emphasis on RNAi-containing nanoliposomes. *Pharmaceuticals, 6*, 1361–1380.
19. Li, Y., Go, V., & Sarkar, F. H., (2015). The role of nutraceuticals in pancreatic cancer prevention and therapy: Targeting cellular signaling, micro RNAs, and epigenome. *Pancreas, 44*, 1–10.
20. Ahmad, A., Ginnebaugh, K. R., Li, Y., Padhye, S. B., & Sarkar, F. H., (2015). Molecular targets of naturopathy in cancer research: Bridge to modern medicine. *Nutrients, 7*, 321–334.
21. Piermartiri, T., Pan, H., Figueiredo, T. H., & Marini, A. M., (2015). linolenic acid, a nutraceutical with pleiotropic properties that targets endogenous neuroprotective pathways to protect against organophosphate nerve agent-induced neuropathology. *Molecules, 20*, 20355–20380.
22. Khare, V., Alam, N., Saneja, A., Dubey, R. D., & Gupta, P. N., (2014). Targeted drug delivery systems for pancreatic cancer. *Journal of Biomedical Nanotechnology, 10*, 3462–3482.
23. Sarkar, F. H., Li, Y., Wang, Z., & Kong, D., (2010). The role of nutraceuticals in the regulation of WNT and hedgehog signaling in cancer. *Cancer and Metastasis Reviews, 29*, 383–394.
24. Lee, C. K., Ki, S. H., & Choi, J. S., (2011). Effects of oral curcumin on the pharmacokinetics of intravenous and oral etoposide in rats: Possible role of intestinal CYP3A and P-gp inhibition by curcumin. *Biopharmaceutics & Drug Disposition, 32*, 245–251.
25. Conklin, K. A., (2005). Coenzyme q10 for prevention of anthracycline-induced cardiotoxicity. *Integrative Cancer Therapies, 4*, 110–130.
26. Yadollahi, R., Vasilev, K., & Simovic, S., (2015). Nanosuspension technologies for delivery of poorly soluble drugs. *Journal of Nanomaterials, 2015*, 1–13.
27. Nehate, C., Jain, S., Saneja, A., Khare, V., Alam, N., Dhar, D. R., et al., (2014). Paclitaxel formulations: Challenges and novel delivery options. *Current Drug Delivery, 11*, 666–686.
28. Kim, B. Y., Rutka, J. T., & Chan, W. C., (2010). Nanomedicine. *New England Journal of Medicine, 363*, 2434–2443.
29. Siddiqui, I. A., Adhami, V. M., Bharali, D. J., Hafeez, B. B., Asim, M., Khwaja, S. I., et al., (2009). Introducing nanochemoprevention as a novel approach for cancer control: Proof of principle with green tea polyphenol epigallocatechin-3-gallate. *Cancer Res., 69*, 1712–1716.
30. Zou, L. Q., Peng, S. F., Liu, W., Gan, L., Liu, W. L., Liang, R. H., et al., (2014). Improved *in vitro* digestion stability of 3-epigallocatechin gallate through nanoliposome encapsulation. *Food Research International, 64*, 492–499.
31. Collnot, E. M., Baldes, C., Schaefer, U. F., Edgar, K. J., Wempe, M. F., & Lehr, C. M., (2010). Vitamin E TPGS P-glycoprotein inhibition mechanism: Influence on conformational flexibility, intracellular ATP levels, and role of time and site of access. *Molecular Pharmaceutics, 7*, 642–651.

32. Blanco, E., Shen, H., & Ferrari, M., (2015). Principles of nanoparticle design for overcoming biological barriers to drug delivery. *Nat. Biotechnol., 33*, 941–951.
33. Thanki, K., Gangwal, R. P., Sangamwar, A. T., & Jain, S., (2013). Oral delivery of anticancer drugs: Challenges and opportunities. *Journal of Controlled Release, 170*, 15–40.
34. Allen, T. M., (2002). Ligand-targeted therapeutics in anticancer therapy. *Nature Reviews Cancer, 2*, 750–763.
35. Sutradhar, K. B., & Amin, M. L., (2014). Nanotechnology in cancer drug delivery and selective targeting. *ISRN Nanotechnology, 2014*, 1–12.
36. Zhong, Y., Meng, F., Deng, C., & Zhong, Z., (2014). Ligand-directed active tumor targeting polymeric nanoparticles for cancer chemotherapy. *Biomacromolecules, 15*, 1955–1969.
37. Sen, K., Banerjee, S., & Mandal, M., (2019). Dual drug loaded liposome bearing apigenin and 5-fluorouracil for synergistic therapeutic efficacy in colorectal cancer. *Colloids Surf B Biointerfaces, 180*, 9–22. doi: 10.1016/j.colsurfb.2019.04.035.
38. Niazvand, F., Orazizadeh, M., Khorsandi, L., Abbaspour, M., Mansouri, E., & Khodadadi, A., (2019). Effects of quercetin-loaded nanoparticles on MCF-7 human breast cancer cells. *Medicina (Kaunas, Lithuania), 55*(4), 114. https://doi.org/10.3390/medicina55040114.
39. Ibrahim, S., Tagami, T., Kishi, T., & Ozeki, T., (2018). Curcumin marinosomes as promising nano-drug delivery system for lung cancer. *Int J Pharm., 540*(1, 2), 40–49. doi: 10.1016/j.ijpharm.2018.01.051.
40. Soo, E., Thakur, S., Qu, Z., Jambhrunkar, S., Parekh, H. S., & Popat, A., (2016). Enhancing delivery and cytotoxicity of resveratrol through a dual nanoencapsulation approach. *J. Colloids and Interface Science, 462*, 368–374.
41. Chen, R., Wang, S., Zhang, J., Chen, M., & Wang, Y., (2015). Aloe-emodin loaded solid lipid nanoparticles: Formulation design and *in vitro* anticancer study. *Drug Delivery, 22*, 666–674.
42. Feng, C., Yuan, X., Chu, K., Zhang, H., Ji, W., & Rui, M., (2019). Preparation and optimization of poly (lactic acid) nanoparticles loaded with fisetin to improve anticancer therapy. *Int. J. Biol. Macromol. 125*, 700–710.
43. Sanna, V., Singh, C. K., Jashari, R., Adhami, V. M., Chamcheu, J. C., Rady, I., Sechi, M., Mukhtar, H., & Siddiqui, I. A., (2017). Targeted nanoparticles encapsulating (-)-epigallocatechin- 3-gallate for prostate cancer prevention and therapy. *Sci. Rep., 7*, 41573.
44. Wu, B., Liang, Y., Tan, Y., Xie, C., Shen, J., Zhang, M., et al., (2016). Genistein-loaded nanoparticles of star-shaped diblock copolymer mannitol-core PLGA-TPGS for the treatment of liver cancer. *Materials Science and Engineering C: Materials for Biological Applications, 59*, 792–800.
45. Bhattacharya, S., Ahir, M., Patra, P., Mukherjee, S., Ghosh, S., Mazumdar, M., et al., (2015). PEGylated-thymoquinone-nanoparticle mediated retardation of breast cancer cell migration by deregulation of cytoskeletal actin polymerization through miR-34a. *Biomaterials, 51*, 91–107.
46. Mahalunkar, S., Yadav, A. S., Gorain, M., Pawar, V., Braathen, R., Weiss, S., Bogen, B., Gosavi, S. W., & Kundu, G. C., (2019). Functional design of pH-responsive folate-targeted polymer-coated gold nanoparticles for drug delivery and *in vivo* therapy in breast cancer. *International Journal of Nanomedicine, 14*, 8285–8302. https://doi.org/10.2147/IJN.S215142.

47. Thipe, V. C., Panjtan, A. K., Bloebaum, P., Raphael, K. A., Khoobchandani, M., Katti, K. K., Jurisson, S. S., & Katti, K. V., (2019). Development of resveratrol-conjugated gold nanoparticles: Interrelationship of increased resveratrol corona on antitumor efficacy against breast, pancreatic and prostate cancers. *International Journal of Nanomedicine, 14*, 4413–4428. https://doi.org/10.2147/IJN.S204443.
48. Souto, E. B., & Doktorovová, S., (2009). Chapter 6-solid lipid nanoparticle formulations: Pharmacokinetic and biopharmaceutical aspects in drug delivery. In: Düzgünes, N., (ed.), *Methods in Enzymology* (pp. 105–129). Amsterdam (AMS): Academic Press.
49. Doktorovova, S., Souto, E. B., & Silva, A. M., (2014). Nanotoxicology applied to solid lipid nanoparticles and nanostructured lipid carriers: A systematic review of *in vitro* data. *Eur. J. Pharm. Biopharm., 87*(1), 1–18.
50. Doktorovová, S., Kovačević, A. B., Garcia, M. L., et al., (2016). Preclinical safety of solid lipid nanoparticles and nanostructured lipid carriers: Current evidence from *in vitro* and *in vivo* evaluation. *Eur. J. Pharm. Biopharm., 108*(Supplement-C), 235–252.
51. Das, S., Ng, W. K., & Tan, R. B., (2012). Are nano structured lipid carriers (NLCs) better than solid lipid nanoparticles (SLN): Development, characterizations, and comparative evaluations of clotrimazole-loaded SLN and NLCs? *Eur. J. Pharm. Sci., 47*(1), 139–151.
52. Liu, D., Li, J., Cheng, B., et al., (2017). Ex vivo and *in vivo* evaluation of the effect of coating a coumarin-6-labeled nanostructured lipid carrier with chitosan-n-acetylcysteine on rabbit ocular distribution. *Mol Pharm., 14*(8), 2639–2648.
53. Weber, S., Zimmer, A., & Pardeike, J., (2014). Solid lipid nanoparticles (SLN) and nanostructured lipid carriers (NLC) for pulmonary application: A review of the state of the art. *European Journal of Pharmaceutics and Biopharmaceutics, 86,* 7–22.
54. Ramalingam, P., & Ko, Y. T., (2016). Improved oral delivery of resveratrol from N-trimethyl chitosan-g-palmitic acid surface-modified solid lipid nanoparticles. *Colloids and Surfaces B Biointerfaces, 139*, 52–61.
55. Peptu, C. A., Popa, M., Savin, C., et al., (2015). Modern drug delivery systems for targeting the posterior segment of the eye. *Curr. Pharm. Des., 21*(42), 6055–6069.
56. Sercombe, L., Veerati, T., Moheimani, F., Wu, S. Y., Sood, A. K., & Hua, S., (2015). Advances and challenges of liposome assisted drug delivery. *Frontiers in Pharmacology, 6*, 286. https://doi.org/10.3389/fphar.2015.00286.
57. Zhao, T., Liu, Y., Gao, Z., Gao, D., Li, N., Bian, Y., et al., (2015). Self-assembly and cytotoxicity study of PEG-modified ursolic acid liposomes. *Materials Science and Engineering C: Materials for Biological Applications, 53*, 196–203.
58. Gursoy, R. N., & Benita, S., (2004). Self-emulsifying drug delivery systems (SEDDS) for improved oral delivery of lipophilic drugs. *Biomedicine & Pharmacotherapy, 58*, 173–182.
59. Vecchione, R., Quagliariello, V., Calabria, D., Calcagno, V., De Luca, E., Iaffaioli, R. V., et al., (2016). Curcumin bioavailability from oil in water nano-emulsions: *In vitro* and *in vivo* study on the dimensional, compositional, and interactional dependence. *Journal of Controlled Release*. http://dx.doi.org/10.1016/j.jconrel. 2016.05.004.
60. Khan, A. W., Kotta, S., Ansari, S. H., Sharma, R. K., & Ali, J., (2015). Self-nanoemulsifying drug delivery system (SNEDDS) of the poorly water-soluble grapefruit flavonoid naringenin: Design, characterization, *in vitro* and *in vivo* evaluation. *Drug Delivery, 22*, 552–561.

61. Shao, A., Chen, G., Jiang, N., Li, Y., Zhang, X., Wen, L., et al., (2013). Development and evaluation of self-micro emulsifying liquid and granule formulations of *Brucea javanica* oil. *Archives of Pharmacal Research, 36,* 993–1003.
62. Ghanghoria, R., Tekade, R. K., Mishra, A. K., et al., (2016). Luteinizing hormone-releasing hormone peptide tethered nanoparticulate system for enhanced antitumoral efficacy of paclitaxel. *Nanomedicine, 11*(7), 797–816.
63. Meyer, H., Stöver, T., Fouchet, F., et al., (2012). Lipidic nanocapsule drug delivery: Neuronal protection for cochlear implant optimization. *Int. J. Nanomedicine, 7,* 2449–2464.
64. Tekade, R. K., Youngren-Ortiz, S. R., Yang, H., et al., (2014). Designing hybrid onconase nanocarriers for mesothelioma therapy: A Taguchi orthogonal array and multivariate component-driven analysis. *Mol. Pharm., 11*(10), 3671–3683.
65. Kumari, A., Yadav, S. K., & Yadav, S. C., (2010). Biodegradable polymeric nanoparticles-based drug delivery systems. *Colloids and Surfaces B: Biointerfaces, 75,* 1–18.
66. Arora, D., Sharma, N., Sharma, V., Abrol, V., Shankar, R., & Jaglan, S., (2016). An update on polysaccharide-based nanomaterials for antimicrobial applications. *Applied Microbiology and Biotechnology, 100,* 2603–2615.
67. Saneja, A., Nehate, C., Alam, N., & Gupta, P. N., (2016). Recent advances in chitosan-based nanomedicines for cancer chemotherapy. In: *Chitin and Chitosan for Regenerative Medicine* (pp. 229–259). Springer.
68. Lo, Y. L., Sung, K. H., Chiu, C. C., & Wang, L. F., (2013). Chemically conjugating polyethylenimine with chondroitin sulfate to promote CD44-mediated endocytosis for gene delivery. *Molecular Pharmaceutics, 10,* 664–676.
69. Platt, V. M., & Szoka, F. C. Jr., (2008). Anticancer therapeutics: Targeting macromolecules and nanocarriers to hyaluronan or CD44, a hyaluronan receptor. *Molecular Pharmaceutics, 5,* 474–486.
70. Alvarez-Rivera, F., Fernández-Villanueva, D., Concheiro, A., et al., (2016). α -lipoic acid in soluplus® polymeric nanomicelles for ocular treatment of diabetes-associated corneal diseases. *J. Pharm. Sci., 105*(9), 2855–2863.
71. Vadlapudi, A. D., Cholkar, K., Vadlapatla, R. K., et al., (2014). Aqueous nanomicellar formulation for topical delivery of biotinylated lipid prodrug of acyclovir: Formulation development and ocular biocompatibility. *J. Ocul. Pharmacol. Ther., 30*(1), 49–58.
72. Cholkar, K., Patel, A., Vadlapudi, A. D., et al., (2012). Novel nanomicellar formulation approaches for anterior and posterior segment ocular drug delivery. *Recent Pat. Nanomed.,* 2(2), 82–95.
73. Hoare, T. R., & Kohane, D. S., (2008). Hydrogels in drug delivery: Progress and challenges. *Polymer, 49,* 1993–2007.
74. Ladet, S., David, L., & Domard, A., (2008). Multi-membrane hydrogels. *Nature, 452,* 76–79.
75. Pillai, J. J., Thulasidasan, A., Anto, R. J., Chithralekha, D. N., Narayanan, A., & Kumar, G., (2014). Folic acid conjugated cross-linked acrylic polymer (FA-CLAP) hydrogel for site specific delivery of hydrophobic drugs to cancer cells. *Journal of Nanobiotechnology, 12,* 25.
76. Teong, B., Lin, C. Y., Chang, S. J., Niu, G. C. C., Yao, C. H., Chen, I. F., et al., (2015). Enhanced anticancer activity by curcumin-loaded hydrogel nanoparticle derived aggregates on A549 lung adenocarcinoma cells. *Journal of Materials Science: Materials in Medicine, 26,* 1–15.

77. Anselmo, A. C., & Mitragotri, S., (2015). A review of clinical translation of inorganic nanoparticles. *The AAPS Journal, 17*, 1041–1054.
78. Santos, H. A., Bimbo, L. M., Peltonen, L., & Hirvonen, J., (2015). Inorganic nanoparticles in targeted drug delivery and imaging. In: *Targeted drug delivery: Concepts and Design* (pp. 571–613). Springer.
79. Bianco, A., Kostarelos, K., & Prato, M., (2011). Making carbon nanotubes biocompatible and biodegradable. *Chemical Communications, 47*, 10182–10188.
80. Nagai, H., Okazaki, Y., Chew, S. H., Misawa, N., Yamashita, Y., Akatsuka, S., et al., (2011). Diameter and rigidity of multiwalled carbon nanotubes are critical factors in mesothelial injury and carcinogenesis. *Proceedings of the National Academy of Sciences, 108*, 1330–1338.
81. Mukherjee, S., Ghosh, S., Das, D. K., Chakraborty, P., Choudhury, S., Gupta, P., et al., (2015). Gold-conjugated green tea nanoparticles for enhanced antitumor activities and hepatoprotection synthesis, characterization, and *in vitro* evaluation. *Journal of Nutritional Biochemistry, 26,* 1283–1297.
82. Yallapu, M. M., Ebeling, M. C., Khan, S., Sundram, V., Chauhan, N., Gupta, B. K., et al., (2013). Novel curcumin-loaded magnetic nanoparticles for pancreatic cancer treatment. *Molecular Cancer Therapeutics, 12*, 1471–1480.
83. Cui, H., Zhang, A. J., Chen, M., & Liu, J. J., (2015). ABC transporter inhibitors in reversing multidrug resistance to chemotherapy. *Current Drug Targets, 16*(12), 1356–1371. https://doi.org/10.2174/1389450116666150330113506 (accessed 10 August 2021).
84. Zhao, J., Lee, P., Wallace, M. J., & Melancon, M. P., (2015). Gold nanoparticles in cancer therapy: Efficacy, biodistribution, and toxicity. *Current Pharmaceutical Design, 21*, 4240–4251.
85. Baksi, R., Singh, D. P., Borse, S. P., Rana, R., Sharma, V., & Nivsarkar, M. (2018). In vitro and in vivo anticancer efficacy potential of Quercetin loaded polymeric nanoparticles. *Biomed Pharmacother 106*, 1513–1526
86. Jiang, Y., Lu, H., Dag, A., Hart-Smith, G., & Stenzel, M. H., (2016). Albumin-polymer conjugate nanoparticles and their interactions with prostate cancer cells in 2D and 3D culture: comparison between PMMA and PCL. *Journal of Materials Chemistry. B, 4*(11), 2017–2027. https://doi.org/10.1039/c5tb02576a (accessed 10 August 2021).

CHAPTER 6

Nutrition Nutraceuticals: A Proactive Approach for Healthcare

CONOR P. AKINTOLA,[1] DEARBHLA FINNEGAN,[1] NIAMH HUNT,[1] RICHARD LALOR,[2] SANDRA O'NEILL,[2] and CHRISTINE LOSCHER[1]

[1]*Immune Modulation Group, School of Biotechnology, Dublin City University, Dublin, Ireland*

[2]*Fundamental and Translational Immunology Group, School of Biotechnology, Dublin City University, Dublin, Ireland*

ABSTRACT

The important link between nutrition and health has led to the discovery of nutraceuticals which are food derived products that exhibit health boosting properties beyond their nutritional value. This chapter analyses the current body of data describing the benefits of the most commonly studied nutraceuticals including the new generation of protein-derived nutraceuticals that are currently under development. Many of the nutraceuticals are found in abundance in superfoods such as milk, fish oils, tomatoes, berries and dark chocolate, which are promoted as health boosting foods. This chapter provides evidence of their health protecting aspects including anti-inflammatory, anti-oxidant, pro-anabolic, liporegulatory, and glucoregulatory that have the potential to promote metabolic, cardiovascular, and tissue health when consumed in the right amounts. Furthermore, nutraceuticals such as curcumin, have shown synergistic effects with established chemotherapeutic strategies to treat cancer. Despite the strong evidence of their beneficial properties, this chapter identifies several important questions which remain unanswered such as whether there is a need for regulation of nutraceuticals to the same extent as drug products and the role of these nutraceuticals in the management of chronic diseases. There is no clinical data to examine whether or not, supplementing a diet, with a given nutraceutical, has any long-term adverse effects while many over the counter food supplements have health claims with no robust evidence, The European Commission is working to address this issue and in the context of health promotion or disease management strong clinical evidence with medical oversight or clear communication and education from the manufacturer with regard to its use will be required. Age is a major factor when it comes to the prevalence of chronic disease where the use of nutraceuticals to promote healthy aging is an area where it can have the greatest impact, particularly since the World Health Organization estimates that by 2050, there will be over two billion people worldwide aged over 60. Future efforts at mining food sources for health boosting bioactives should focus on the hunt for anti-inflammatory and antioxidant nutraceuticals to support healthy aging in this ever-growing population.

6.1 INTRODUCTION

Two commonly quoted phrases, whose meanings are still highly relevant today are, "you are, what you eat," which has its roots in the 1826 publication "The Physiology of Taste" by Anthelme Brillat-Savarin and "Let food be thy medicine, and medicine be thy food" a phrase allegedly spoken by Hippocrates in 400 BC [1]. An individual's health status and the development of human diseases are influenced by genetics, in addition to environmental factors, where diet is a major contributing factor [2]. Nutritional status, notably, malnutrition plays a critical role in the development of human disease [3]. The lack of essential micronutrients such as vitamins has long been associated with the development of disease. For example, vitamin C deficiency causes a breakdown of epithelial junctions in the body resulting in the onset of scurvy [4], while deficits in vitamin D causes weak bone structure and increased incidences of rickets and osteopathologies [5]. There are also detrimental effects noted with protein malnutrition, including dysfunction in skeletal muscle physical and metabolic functions, and complications for critical organ health [6]. Often overlooked, is the phenomenon of "overnutrition," which is a problem associated with middle-to high-income populations. Overnutrition is a consequence of direct, easy access to more food than is nutritionally required and similar to malnutrition, it plays an integral role in the incidence of human disease. In developed countries, diets high in refined sugar, salt, and "bad fats" such as saturated fats or trans-fatty acids correlate directly to increased rates of dietary-related morbidity such as hypertension, diabetes, cancer, and cardiovascular vascular disease (CVD) [7]. Excess intake of dietary protein has also been linked with the occurrence of renal disease [8].

In 2020, the World Health Organization (WHO) estimates that the global population will approach 7.8 billion, with an estimated 1.9 billion adults' overweight, of which 650 million are classed as obese. In the United States of America (USA), 27.6% of adult men and 33.2% of adult women in the USA are obese [9, 10]. An increased incidence of obesity amongst a population will place additional pressure on global resources, most notably the health care sector. Perhaps more worryingly, is that obesity rates in children and adolescents are on the rise. In the USA, approximately 11.6% of children and adolescents are overweight [9, 10], and similar percentages are observed in studies on children and adolescents in other developed countries [11]. However, recent studies have shown that the rates of childhood obesity have plateaued in developed countries [12]. Lack of exercise and unhealthy eating habits both directly influence national and global rates of obesity. In 2018, a survey by Sport Ireland and the health service executive (HSE), found that only 17% of primary school and 10% of post-primary school children engage in at least 60 minutes of exercise per day. Eating habit surveys conducted by the HSE also note an increased prevalence of unhealthy eating habits [13].

Higher rates of obesity are correlated with increased incidences of chronic, non-communicable diseases (NCDs) such as CVD, metabolic syndromes (Met-S), cancers, and immune-mediated disorders [14]. According to the WHO, 71% of all global deaths are attributed to NCDs, accounting for 41 million deaths per year. NCDs can manifest in several forms, such as type 1 diabetes mellitus and asthma, which are conditions that arise from birth [15]. However, there is a continuing and concerning rise in dietary-related NCDs globally [16, 17] that have a massive economic burden to global healthcare systems and have significant impacts on disability-adjusted life years (DALYs) [18]. Moreover, pharmaceutical, therapeutic strategies for treating dietary-related NCDs present several challenges in terms of efficacy and side effects. For example, Infliximab, a tumor necrosis factor-alpha (TNF-α) agonist commonly employed to treat rheumatoid arthritis (RA) and Crohn's disease (CD), displays immunogenic properties in humans, which results in resistance to the treatment over time. While combination therapies with other immunosuppressive drugs improve Infliximab efficacy, these therapies are associated with increased susceptibility to viral and bacterial infections [19]. Another example is the use of nicotinic acid (NA), a drug to treat hyperlipidemia that exhibits common side effects including skin itch and flushing [20]. Therefore, for many NCDs, newer therapies with fewer side effects are required.

Nutraceutical is a term derived from "nutrition" and "pharmaceutical" that applies to products that are isolated from herbal, dietary supplements, and functional foods such as dairy, cereals, and beverages. These products exhibit properties beyond nutritional value such as health-boosting properties [21]. The supplementation of diets with nutraceuticals derived from functional foods or other sources have been shown to be useful in the treatment of chronic human NCDs such as CVD, Met-S, and chronic inflammatory diseases [22]. In today's society, there is a trend in consumers becoming increasingly health-conscious with a desire to acquire products that are derived from natural sources. The popularity of functional foods in the nutraceutical industries are therefore increasing as consumers seek viable alternatives to conventional therapies that are often more expensive, high-tech treatments that have unforeseen or undesirable side effects. Nutraceuticals have received considerable interest due to their safety and therapeutic effects as unlike their pharmaceutical counterparts, no studies have observed the accumulation of nutraceuticals in the body's tissues during prolonged usage. As a consequence, it is unlikely that there will be no long-term negative side effects, however this remains to be investigated [23–25]. Consequently, these industries are rapidly expanding with a net value of $230.9 billion globally in 2018, which is projected to reach $336.1 billion by 2023 [26].

There has been major investment in the research and development of nutraceuticals for the prevention and treatment of NCDs as a proactive approach to healthcare. In this chapter, we will discuss nutraceutical strategies that may be used to prevent or ameliorate the symptoms and treat the major NCDs that are highly prevalent in today's society, such as CVD, Met-S, cancers, and immune disorders. Specifically, the current nutraceuticals that will be discussed are, curcumin, omega 3 polyunsaturated fatty acids (PUFAs), carotenoids, flavonoids, and specific amino acids, due to the fact that these are functional molecules that are naturally occurring in foods deemed "superfoods" and also available as individual purified supplements in health food stores and pharmacies. Technological strategies used to mine food for bioactive nutraceuticals will be discussed, specifically highlighting the work ongoing into uncovering and screening of bioactive peptides from both plant-based and animal sources. We will also highlight clinical trials that have investigated the potential benefits of nutraceuticals in the prevention and treatment of NCDs.

6.2 NATURAL NUTRACEUTICALS

6.2.1 CURCUMIN

Curcumin is a natural bioactive that has received a lot of attention as a nutraceutical strategy for several human diseases. It has been described to possess several health boosting properties including immunomodulatory, anticancer, anti-cardiovascular disease, anti-diabetic, antioxidant, and anti-ageing [26, 27]. Curcumin is a polyphenolic compound found in the commonly available spice, turmeric [26] that was shown to target multiple cell signaling pathways as well as intracellular and extracellular molecules. In particular, it is thought to target nuclear factor kappa-light-chain-enhancer of activated B cells (NFκB), a pathway critical to driving a pro-inflammatory immune response [28]. Chronic activation of NFκB and consequent chronic inflammation is linked with several human diseases including cancer, inflammatory disease, CVD, and metabolic disease [28–30].

Curcumin modulates several immune cell types, including dendritic cells, cells critical to the activation of the adaptive immune response [31]. The activation of dendritic cells with bacterial lipopolysaccharide induces a pro-inflammatory state where the activation of NFκB is a critical part of this process [32]. These activated cells secrete a panel of inflammatory mediators, including cytokines such as interleukin (IL)-12 and TNF, and costimulatory cell surface markers required for the activation of adaptive immune responses [33]. Curcumin inhibits the secretion of these cytokines and the expression of the co-stimulatory cell surface markers MHCI/II, CD80, and CD86 on LPS activated dendritic cells. Curcumin impairs the translocation of the NFκB p65 subunit which is an important process in the activation of inflammatory responses [31]. Curcumin similarly inhibits the activity of macrophages, cells important in the induction and maintenance of adaptive immune responses. Macrophages can be differentiated into different cell phenotypes and studies demonstrate that curcumin can skew macrophages from a pro-inflammatory M1 phenotype towards an anti-inflammatory M2 phenotype [34].

Immune-mediated inflammatory disorders, such as RA and inflammatory bowel disease (IBD), share common underlying mechanism that involve pro-inflammatory cytokines such as TNF and the activation of NFκB. Given the anti-inflammatory potential of curcumin, several studies were designed to investigate the efficacy of curcumin in treating inflammatory disorders. In a rat model of RA, curcumin supplementation of 200 mg/kg of body weight for three weeks resulted in reduced infiltration of inflammatory cells into the synovium of the knee joint. Furthermore, curcumin reduced mTOR signaling within the synovium, which is often overexpressed in RA lesions, thought to propagate the disease [35]. In a pilot study in humans, curcumin reduced RA severity compared to diclofenac sodium, a NSAID commonly used to treat RA [36].

Curcumin has also been documented as improving the clinical symptoms of IBD, through ant-inflammatory activity and by directly and beneficially influencing the gut microbiota, which alleviates the severity of the disease in patients suffering from IBD [27, 37]. The gut microbiome comprises of several trillion bacterial cells of varying genus' that line the walls of the human digestive tract. The microbiome plays several roles in maintaining human health including; digesting dietary molecules we are unable to digest normally into useable metabolites, and in preventing the colonization of the gut by potential pathogenic bacteria [38, 39]. In the context of IBD, imbalances in specific bacterial populations associated with the incidence of IBD. Patients presenting with IBD have frequently shown that increased populations of *Enterobacteriaceae* and decreases in *Clostridium*, *Firmicutes*, *Bifidobacterium*, and *Lactobacillus* populations [40]. Curcumin has been documented to promote the growth of *Lactobacillus* and *Bifidobacterium* [41]. In a mouse model of IBD, researchers demonstrated that nanoparticle delivery of curcumin increased butyrate concentrations in the gut, believed to be sourced from commensal *Clostridium* clusters, which in turn promoted the activity of gut T-regulatory lymphocytes and reductions in gut inflammatory markers [42].

The NFκB pathway has a critical role in cell survival and apoptosis, programmed cell death [28] and given the inhibitory effect of curcumin on this pathway, it has become a molecule of interest in the context of cancer treatment and prevention. Cancer in its simplest terms can be described as a division of abnormal cells associated with prolonged cell survival and an evasion of apoptosis, a natural process in an organism's growth and development [43]. Curcumin was shown to have a notable effect on a variety of cancer types [27]. It has been demonstrated *in vitro* that curcumin, in the presence of paclitaxel (a chemotherapy medication used to treat a number of types of cancer) has the ability to suppress antiapoptotic genes, proliferative genes and metastatic genes in a paclitaxel-resistant breast cell line via the NFκB signaling pathway. In the same study, using a human breast cancer xenograph model, researchers discovered that oral administration of curcumin decreased the occurrence of lung metastases in mice [44]. Further synergy with the chemotherapeutic docetaxel was also demonstrated *in vitro* using a metastatic prostate cancer cell line. Curcumin-docetaxel treatment increased the efficacy of docetaxel compared to curcumin and docetaxel treatment alone [45].

Curcumin can suppress angiogenesis, a key hallmark in cancer pathogenesis and therefore could preventing the formation of new blood vessels inhibiting the delivery of oxygen and nutrients to tumor tissue. VEGF-A is the principal mediator of angiogenesis and curcumin is thought to prevent angiogenesis through the modulation of the VGEF signaling pathway [30, 46]. In a model of murine Dalton's lymphoma, it was discovered that curcumin can promote the activity of the tumor suppressor gene p53, suggesting a role in cancer prevention [47]. A direct role in the killing of cancer cells has also been described *in vitro*, by causing cell cycle arrest, apoptosis, and inducing autophagy in pancreatic cancer cells [48]. This finding is significant given the poor prognosis associated with pancreatic cancer due to the lack of effective treatments [49].

Curcumin displays potential beneficial effects in protecting against the development of atherosclerotic plaques in human arteries. Atherosclerosis is a degenerative condition in which excess circulating lipids are deposited in the artery walls, causing the development of atherosclerotic plaques. These plaques cause narrowing of the artery, resulting in hypertension. Injury to the arterial wall and the presence of ectopic lipids in the blood leads to platelet aggregation and clotting, where the resulting clot blocks the flow of blood in the artery resulting in stroke or myocardial infarction [50]. Inflammatory M1 macrophages are the predominant immune cell during plaque formation. The chronic activation of M1 macrophages impairs the healing of the damaged cell wall, promoting the development of atherosclerotic lesions. The protective effects of curcumin in this context may be multifaceted. One mechanism by which curcumin may protect against atherosclerosis is by modulating the immune microenvironment, which can influence plaque formation. As previously mentioned, curcumin is believed to skew macrophage from an inflammatory M1 phenotype towards an anti-inflammatory M2 phenotype, which promotes wound healing [34, 51]. Furthermore, the serum lipid-lowering effects of curcumin have also been extensively documented. This effect may be two-fold as it may lead to a reduced presence of lipid activated M1 inflammatory macrophage in atherosclerotic lesions and the reduction of excess serum lipid concentrations reduces the risk of a major cardiac event [51, 52]. *In vivo* murine models of atherosclerosis appear to support these mechanisms, however further investigation in both human and animal models is required [53].

Curcumins therapeutic potential also extends as a possible intervention to improve the clinical symptoms of type 2 diabetes mellitus (T2DM). Inflammation is hypothesized to play an integral role in the development of T2DM, through the induction of adipose tissue and skeletal muscle insulin resistance, a symptom often seen in pre-diabetic individuals [54, 55]. By reducing chronic systemic inflammation, curcumin treatment may act as a strategy to prevent the onset of T2DM. This is exemplified in a human study, of pre-diabetic patients as supplementation with 250 mg/day of curcuminoids, prevented the onset of T2DM in the treatment group, whereas 16.4% of patients in the placebo group were diagnosed with T2DM. Those who received the Curcumin treatment exhibited improved sensitivity to insulin, improved pancreatic β-cell function, and improved inflammatory profiles [56]. In summary, the direct effects of curcumin in both *in vitro* and *in vivo* human and animal trials included, decreased levels of circulating LDL, and triglycerides, anti-hyperglycemic activity, improved fasting blood glucose concentrations, antioxidant potential, and decreased circulating inflammatory markers [57, 58].

To date, the recommended for daily intake for curcumin ranges from of 0 to 3 mg/kg of body weight, as approved by the European Food Safety Authority (EFSA) and the Joint FAO/WHO Expert Committee on Food Additives (JECFA) [59]. However, at high dose intakes can have several adverse effects of excessive intakes of the nutraceutical. For example, at daily doses ranging between 1000 mg and 12,000 mg can result in yellowing of the stool, diarrhea, and a rash [60].

6.2.2 OMEGA-3 POLYUNSATURATED FATTY ACIDS (PUFAS)

Perhaps one of the longest standing health boosting bioactives are the omega-3 PUFAs. Long believed to be of great benefit for cardiovascular health, omega-3 PUFAs are found predominantly in the flesh of oily fish such as salmon, trout, and mackerel. Fish oils are also abundant with high levels of omega-3 PUFAs, particularly cod liver oil, a popular over the counter supplement [61]. Docosahexaenoic acid (DHA) and eicosapentaenoic acid (EPA), are two of the main omega-3 PUFAs shown to have protective effects against a variety of chronic diseases, which are found in varying amounts depending on the source [62]. EPA and DHA are 20 and 22 carbon fatty acids, respectively, and are isolated primarily from marine sources, but can also be found in animal sources such as eggs [63]. However, small quantities of EPA and DHA can be synthesized endogenously through consumption of foods rich in another essential omega-3 PUFAs known as alpha-linoleic acid (ALA) [64], which in its own right has had noticeable benefits for several aspects of human health, most notably, brain health.

The most common omega-3 fatty acid is ALA, which can be sourced from vegetable oils, nuts (especially walnuts), flax seeds and flaxseed oil, leafy vegetables, and some animal fat, especially in grass-fed animals [65, 66]. Data suggests that ALA plasma status is inversely associated with the incidences of stroke in adult men [67]. Furthermore, animal models have demonstrated that ALA plays an active role in inducing recovery from stroke [68]. Interestingly, ALA supplementation has also been linked with potential applications for improving mental health [69]. In a mouse model of depression, diets high in ALA were capable of inducing anti-depressive behaviors. Phenotypic changes in the hippocampus and changes in gene expression in the brain were also observed, suggesting a potential therapeutic avenue for mental health disorders [70]. However, comprehensive human trials will need to be conducted to uncover the potential of ALA for neuroprotection and mental health applications.

Marine sourced omega-3 PUFAs have a long-documented history in the treatment of several risk factors linked to CVD. Documented modalities of omega-3 PUFAs include modulation of vascular endothelial cell function, anti-hypertensive activity, anti-hypercholesterolemia activity, reduction in platelet aggregation, and modulation of heart rate to reduce tachycardia [62]. A study conducted in an American cohort demonstrated an association between plasma concentrations of omega-3 PUFAs, including EPA and DHA, and a decreased likelihood of death as a result of a major cardiac event [71]. A meta-analysis study of randomized placebo-control clinical trials noted that considerable improvements in circulating triglyceride levels, improvements in blood pressure and heart rate were among the big indicators for improved cardiovascular health. A decrease in mortality as a result of major cardiac events was also observed following consumption of marine sourced omega-3 PUFAs [72]. Given the direct effects of CVD risk factors such as hypertension and hyperlipidemia, omega-3 PUFAs also display anti-inflammatory properties, which may indirectly improve cardiovascular health by modifying the release of inflammatory eicosanoids, which are lipid derived modulators that regulate several aspects of cellular signaling and immune function [62, 73].

Many studies have examined the effect of omega-3 PUFAs on immune function, and in summary, they conclude that that omega-3 PUFAs exert potent anti-inflammatory effects *in vitro* and *in vivo* [65, 74, 75]. The mechanisms by which omega-3 PUFAs influence inflammatory cell function has yet to be clarified, however it is believed to exert its effect through several mechanisms including; direct signaling through fatty acid receptors, incorporation into membrane phospholipids that consequently effects membrane fluidity and extracellular signaling, and influence hormonal pathways linked to inflammatory processes [74]. An *in vitro* study using THP-1 macrophage demonstrated that EPA exhibits potent anti-inflammatory properties, suppressing the expression of genes associated with inflammatory cytokines and chemokines, and genes involved in the NFκB signaling cascade. Furthermore, enhanced expression of mitochondrial tumor suppressor 1 (MTSG1), a candidate tumor suppressor protein, and an inverse suppression of NOS2, indicate that EPA is capable of relieving oxidative stress in inflammatory states [76]. A separate study, also using THP-1 macrophages showed that along with cytokine suppression, NF-κB p65 transcription factor was also suppressed in EPA and DHA treated cells [77]. DHA has also displayed similar activity in an *ex vivo* human study that compared macrophages from healthy control patients, and from patients suffering with small abdominal aortic aneurism. Isolated macrophage was treated with DHA and suppression of inflammatory signaling was observed with upregulation of free radical scavenging activity in both healthy and diseased states [78].

The *in vitro* trend of consistent anti-inflammatory activity by omega-3 PUFAs has been examined in numerous human studies [79, 80]. For example, while there may be several beneficial uses of omega-3 PUFAs, one that is being explored is its use in promoting healthy ageing. As we age, our bodies become increasingly inflammatory, with noted increases in circulating inflammatory biomarkers. This phenomenon known as "inflammaging," is a chronic, but low-grade upregulation of inflammation that can have a detrimental effect on the body leading to the onset of several chronic diseases [81, 82]. Inflammaging has been linked with the onset of a disease known as sarcopenia that causes a degenerative loss of muscle mass over time [83]. Sarcopenia is a deleterious condition that impacts DALYs [84–86]. It is believed that inflammation may directly impact the anabolic sensitivity of skeletal muscle, which is crucial for muscle health as it promotes anabolic pathways that allow muscle tissue to maintain its mass and function. Since, animal and human studies have observed a negative correlation between the anabolic capacity of muscle and the concentration of circulating inflammatory markers [84, 87], omega-3 PUFA supplementation is one of several strategies being considered to treat sarcopenia in older adults. Emerging data suggest that omega-3 PUFA ingestion is correlated with a decrease in circulating inflammatory markers in older adults [88, 89]. A direct mechanism for promoting the anabolic capacity of muscle by omega-3 PUFAs has also been described. In C2C12 myotubes, treatment with EPA and DHA was found to enhance phosphorylated activation of mTOR signaling [90]. mTOR is considered as the master regulator of protein synthesis in skeletal muscle [91]. Decreased activity of this pathway leads to reduced protein synthesis and subsequent reduced muscle mass and strength, which can significantly impact on an individual's quality of life [86]. *In vivo*, after eight weeks of consumption of an EPA and DHA rich supplement, researchers found that in older adults, mTOR signaling, the primary anabolic pathway in muscle, was upregulated suggesting that omega-3 PUFA ingestion could be an avenue to promote muscle protein synthesis in sarcopenic patients [92].

The anti-inflammatory nature of omega-3 PUFAs can be utilized to treat a variety of chronic immune disorders. IBD is an overarching term that describes several diseases, including CD and ulcerative colitis (UC), where chronic inflammation in the gut leads to adverse health conditions. Numerous studies have demonstrated that ingestion of omega-3 PUFAs reduce intestinal inflammation [93]. Furthermore, a study with a European cohort demonstrated an inverse relationship between the consumption of DHA and the occurrence of CD [94]. A case report concerning 22 females with IBD presented an imbalance of omega-6 PUFA and omega-3 PUFA with co-occurring vitamin D deficiency. Rebalancing the ratio of omega-6 PUFA and omega-3 PUFA with an EPA and DHA supplement with co-ingestion of vitamin D reduced disease severity and symptoms [95]. The imbalance of the Omega-6/3 PUFAs is related to several chronic disease states such as CVD, auto-immune conditions, and metabolic disorders [96]. Westernized diets are often deficient in omega-3 PUFAs, which are also associated with the increased prevalence of chronic diseases such as CVD and CID. Omega-3 PUFA supplements may be used to restore this balance to ameliorate and prevent the onset of chronic disease [97].

Omega-3 PUFA containing fish oil supplements are widely available in pharmacies and health food stores, commonly seen at 1000 mg. The WHO advises 250 mg/day of EPA and DHA for men and women, and it is recommended to increase intake to 300 mg/day for pregnant and lactating women [98]. Few adverse effects are associated with regular intake of marine-derived omega-3 PUFA. Although one study reported that over a four-week period, in patients diagnosed with CVD, intake of 1.7 g/day of an EPA and DHA containing capsule resulted in 12% of the participant groups suffering from adverse gastrointestinal (GI) symptoms such as abdominal pain, diarrhea, and GI bleeding [99]. Omega-3 PUFA supplements derived from fish oils are not suitable for certain groups such as vegans or some vegetarians. However, algae are potent sources of omega-3 PUFAs and could be a potential source for these cohorts [100, 101]. However, there appears to be a gap in the literature, comparing the efficacy of fish-derived omega-3 PUFAs and algae-derived omega-3 PUFAs, thus work is required in this regard.

6.2.3 CAROTENOIDS

Carotenoids are a class of pigmented phytochemical found almost exclusively in plants [102]. Carotenoids have a forty-carbon skeleton of isoprene units and maybe cyclized at one or both ends, with various hydrogenation levels or oxygen-containing functional groups, primarily in the transform occurring naturally [103]. Carotenoid-rich foods include carrots, tomatoes, spinach, and apricots [102]. The most commonly studied carotenoids are lycopene and β-carotene, and they have been shown to display beneficial properties inducing antioxidant and immunomodulatory properties.

6.2.3.1 LYCOPENE

Lycopene is an acylated carotenoid that is found in abundance in tomatoes and tomato-based processed foods [104]. As well as exhibiting cholesterol-lowering bioactivity, it displays potent antioxidant and anti-inflammatory properties and an ability to protect cells from oxidative damage [46]. Lycopene inhibited the inflammatory cytokines, IL-1β, IL-6, and TNFα, the activity of iNOS and disrupted NFκB signaling in LPS stimulated RAW264.7 macrophages [105, 106]. In RAW264.7 macrophage, treatment with lycopene and β-carotene suppressed reactive oxygen species (ROS) production *in vitro* [107]. Furthermore, a study in THP-1 macrophage demonstrated the negative regulation of oxidative pathways by lycopene as it reduced basal ROS production and ROS production in 7-ketocholesterol stimulated THP-1 macrophages [108]. The antioxidant and immunomodulatory properties of lycopene may protect against the development of cancer, CVD, and Met-S.

Lycopene has also been investigated as a nutraceutical intervention for the treatment of CVD. In rabbit models, lycopene supplementation has been found to decrease total circulating cholesterol levels, LDL levels, and triglyceride levels [109]. Furthermore, some research has associated lycopene intake with a reduction in the size of atherosclerotic lesions [110]. However, there are conflicting reports in the literature on the beneficial properties of lycopene, with one study associating increased lycopene intake with a decreased incidence of CVD [111] while conversely, a second study concerning 39,876 women over the age of 45, found no correlation between lycopene intake and CVD incidence. Researchers did note, however, that there was an inverse correlation between tomato product intake and CVD incidence, suggesting that a synergistic contribution with other phytochemicals found in tomato-based foodstuffs may be beneficial [112].

Lycopene is perhaps most recognized for its alleged role in the prevention of cancer with several studies linking lycopene intake to a decreased risk of prostate cancer [113]. In a randomized control trial, ingestion of a lycopene rich supplement twice daily was associated with a decreased incidence of prostate cancer in patients diagnosed with high-grade prostate intraepithelial neoplasia, a precursor condition to prostate cancer, compared to the placebo group [114]. Another study found that increased lycopene intake was associated with a reduced risk of severe prostate cancer development [115]. Despite its obvious bioactivity as a potent antioxidant and anti-inflammatory, there are questions as to whether lycopene alone, is capable of reducing the risk of cancer development. Studies investigating the anticancer potential of lycopene have either used purified lycopene or a tomato-based foodstuff as a vector to deliver the lycopene [116]. For example, in male rats, consumption of a tomato-based powder and not purified lycopene contributed to a reduced risk of death from prostate cancer [117]. There is no significant contribution of diets rich in lycopene in lowering the risk of ovarian cancer, whereas other carotenoids such as alpha-carotene, beta-carotene, lutein, and beta-cryptoxanthin were associated with a decreased risk [118]. It is possible that the effects of lycopene on cancer could be mechanistically specific to prostate cancer. However, there is some preliminary evidence that lycopene can inhibit the growth of endometrial cancer cell line [119] and therefore more longitudinal studies in the context of other cancer types is required to determine this.

6.2.3.2 Β-*CAROTENE*

Like its counterpart lycopene, β-carotene, historically was discussed extensively in the context of cancer prevention [120]. Recently, β-carotene containing creams have been documented to decrease the absorption and penetration of UV light into the epidermis of pig's ears, suggesting a potential protective effect against sunlight-induced melanomas [121]. Epidemiological studies that garnered support for β-carotene in cancer prevention were based primarily on retrospective and prospective questionnaires concerning dietary intake. However, a meta-analysis of randomized control studies has indicated that dietary intake of β-carotene had no effect on the incidence of pancreatic, colorectal, prostate, breast, melanoma, and non-melanoma cancers [122].

β-carotene is a pro-vitamin A molecule that possesses many of the immunoregulatory and antioxidant properties of vitamin A [123]. In RAW264.7 macrophage, treatment with β-carotene suppressed ROS production *in vitro* [107]. The antioxidant potential of β-carotene was demonstrated *in vivo* in humans suffering from chronic lead poisoning. Twelve weeks of supplementation with a β-carotene supplement called Beta Karoten® reduced oxidative profiles in a cohort of men frequently exposed to lead in their place of work [124].

However, there is evidence to suggest that direct β-carotene supplementation may be detrimental to human health under certain conditions. In the context of cardiovascular health, supplementation with β-carotene has no effect and even worsening effects on the likelihood of a major cardiac event in certain groups. In smokers, for example, β-carotene supplementation increased the likelihood of death or a cardiovascular event [125]. Some studies also suggest that β-carotene may increase the likelihood of lung cancer development in smokers [126]. However, the anticancer effects of β-carotene are well documented, believed to exert its effects through epigenetic modification [127, 128]. Thus, supplementation with β-carotene may not be suitable for specific cohorts, therefore ensuring that adequate dietary carotenoid levels are obtained from whole food sources such as tomatoes.

6.2.3.3 FLAVONOIDS

Flavonoids are a group of polyphenolic metabolites found primarily in edible fruits, vegetables, and plants [129]. Bioactive properties of flavonoids include antioxidant, anti-inflammatory, and anti-thrombotic effects [130]. Flavonoids can be divided into several groups such as flavanones, flavone, flavonols, isoflavones, and chalcones. These flavonoid subgroups differ based on their molecular structure, yet all exhibit similar bioactivities in the range of immunomodulatory, antioxidant, chemoprotective, and cardioprotective properties [131]. Dietary flavonoids come from a variety of plant-based sources, including red wine, tea, peppers, blueberries, and citrus fruits [132].

Several studies have identified the increased dietary intakes on flavonoids was correlated with decreased risk of mortality as a result of a cardiovascular event [133]. It has been suggested that the protective effects of flavonoids may be mediated through blocking the oxidation of LDL and improving vascular smooth muscle cell tolerance to oxidized-LDL [134]. Anti-hypertensive effects are also theorized to play a role in reducing CVD mortality [135, 136]. The ability of flavonoids to regulate oxidative metabolisms are also postulated to play an integral role in the prevention of cancer [137]. Furthermore, several flavonoids play a direct role in the destruction of cancer cells, through promoting apoptosis [138].

Flavonoids have a long-documented role in the attenuation of chronic inflammation, which enables flavonoids to be potentially employed to treat the symptoms of inflammatory disorders [139, 140]. Apigenin is a flavone derived molecule found in abundance in fruit skins [132]. Murine derived bone marrow dendritic cells activated with LPS, exhibited a reduction in inflammatory cytokines and cell surface expression of the key co-stimulatory molecules CD80, CD86, and CD40 when treated with apigenin. In the same study, researchers using the mouse model of collagen-induced arthritis demonstrated that apigenin supplementation improved the clinical symptoms of CIA employing a similar mechanism of action that targeted inflammatory cytokines and co-stimulatory marker expression [141]. There is a lack of clinical studies examining the beneficial effects of apigenin in human disease, therefore there is no evidence that these *in vitro* studies and *in vivo* findings translate to humans. However, there is promising data in a randomized control study examining the topical administration of apigenin-rich chamomile oil in osteoarthritic subjects for three weeks who displayed reduced pain levels as measured by a reduction in the need for pain-relieving medication [142].

The anti-inflammatory properties of flavonoids also mean that they may have an application in promoting healthy ageing, by modulating the phenomenon of "inflammaging" which has been previously discussed. Flavonoids are known to modulate the expression and activity of NFκB, COX-2 and iNOS, therefore reducing inflammatory activity [132, 143]. Retrospective studies have demonstrated that increased dietary intake of flavonoids is associated with an increased quality of life as we age [144]. This is also supported mechanistically, as diets high in flavonoids are associated with a decreased circulating levels of c-reactive proteins, IL-6, IL-8, IL-18, and TNF-R2 [145]. Through modulating inflammation, there is a decreased likelihood of developing chronic diseases such as cancers, CVD, and metabolic disease as we age.

There has also been much discussion around using flavonoids to treat the clinical symptoms of diabetes. Several flavonoids have been found to drive improvements in glycemic control, *in vitro* and *in vivo* through several mechanisms of action, such as; promoting GLUT4 activity, boosting insulin secretion, downregulating inflammation, and oxidation pathways, and modulating circulating lipid profiles [146]. In a diabetic rat model, researchers demonstrated that ingestion of an apigenin analog was able to enhance glucose uptake and reduce circulating glucose concentrations [147]. However, as outlined in Al-Ishaq et al. [146], there is a wealth of *in vivo* data concerning the effects of flavonoids supplementation in rodent models of diabetes and metabolic disease, however, there is a gap in the literature concerning the effects in humans. Comprehensive studies are required to determine the efficacy of flavonoid supplementation in the case of T2DM.

6.3 AMINO ACIDS

Amino acids are the fundamental building blocks of life, forming peptide bonds to create structural units that make up proteins. There are 20 amino acids that are incorporated into protein molecular structures during synthesis, despite many more been described, however only 9 amino acids are considered essential as they cannot be synthesized by the body [148]. Amino acids have an array of functions, both enabling cellular function and supporting cellular function through nutrition. With regards to important nutraceutical amino acids, studies have shown that L-glutamine and leucine amino acids have beneficial effects in diseases such as inflammatory bowel syndrome (IBS) and metabolic health.

6.3.1 L-GLUTAMINE

L-glutamine is one such example of a nutraceutical amino acid that plays a biological role beyond being a protein building block. Glutamine is the most abundant amino acid found in the human body. It is an L-α-amino acid, which is not one of the essential amino acids [149]. L-glutamine has many biological properties, which make it a potential health-boosting supplement. One such activity is the ability to modulate cellular and tissue integrity, which in the context of gut health has been widely explored. Tight junction proteins are essential in maintaining tissue integrity, and in the gut, sustained expression and activity of these proteins ensure that the gut lining remains selectively permeable, protecting the host from immunogenic microbiota and chronic bowel conditions [150]. In a study that encompassed several clinical symptoms of human IBS, researchers found that the expression of tight junction proteins, ZO-1, occludin, and claudin-1, in IBS presenting individuals were negatively regulated, increasing gut permeability [151]. IBS is very different to IBD in that both are chronic conditions that cause abdominal pains, cramping, and urgent bowel movements; however, IBS does not cause inflammation or destruction of the bowel wall, which leads to diseases such as CD and colitis [152]. Several *in vitro* studies using the colonic cell line CaCO-2, have remarked that L-glutamine supplementation improves tight junction protein expression and function [149]. Similar effects on claudin-1 expression were observed in colonic explants derived from patients suffering from diarrhea dominant IBS [153]. L-glutamine supplementation improved human gut permeability and reduced disease severity [154]. This information, combined with knowledge that L-glutamine supplementation was not associated with any side effects, suggests that L-glutamine supplementation may be a therapeutic strategy to improve gut layer integrity in disease states.

L-glutamine displays potent immunomodulatory potential. During an immune response, active immune cells require a high output energy source. Studies have shown that glutamine and glucose are required in equal measure to promote the appropriate activity of both innate and adaptive immune cell function [155]. Appropriate endogenous glutamine stores are therefore essential to promote effective immune function. In healthy individuals, glutamine synthase inhibition or poor plasma concentrations may result in impaired immune function and resolution of inflammation, which can have detrimental effects over time [156, 157]. Similar to other nutraceuticals, L-glutamine was shown to skew macrophage phenotypes; however, in this context, it switched cells from an M2 to Ml phenotype through the inhibition of glutamine synthase. In the context of this study, this inhibition was beneficial as the Ml macrophage induced T-cell activation and reduced cancer metastasis [158]. There is evidence to suggest that L-glutamine supplementation during chemotherapy treatment may reduce the severe side effects associated with chemotherapeutic infusions [159]. Studies have documented that oral supplementation of an L-glutamine rich source improves chemotherapy side effects such as poor gut function, mucositis, and weight loss [155].

The anti-inflammatory effect in the human gut was demonstrated *in vitro* using colonic explants. Researchers showed that pro-inflammatory cytokines were down regulated, following 2 weeks of supplementation [153]. Furthermore, dual supplementation with L-glutamine and L-alanine, reduced inflammation in rat skeletal muscle in response to injury, potentially protecting surrounding tissue from further damage [160]. Similar effects were also seen in an elderly cohort who undertook 30 days of glutamine supplementation combined with exercise. Compared to the control exercise group (no glutamine), those that exercised and ingested glutamine showed improved oxidative and inflammatory balance, suggesting an application to promote healthy aging [161].

6.3.2 LEUCINE

Leucine is a branched-chain amino acid and is one of the essential amino acids required by humans [148] that is of interest with regards to metabolic health in humans. Leucine is particularly important with regard to muscle metabolic, and subsequent locomotive functions. Leucine was identified to play an important role in modulating muscle protein synthesis in skeletal muscle. Leucine was identified to promote muscle protein synthesis through the activation and activity of mTOR and 4E-BP1 [162]. This was confirmed in a human exercise model, where subjects fed a leucine rich and carbohydrate mix supplement exhibited enhanced post-exercise activation of mTOR signaling [163]. In the context of human health, leucine supplementation was found to improve muscle strength and reduce baseline inflammation in patients diagnosed with cerebral palsy, improving quality of life [164]. In contrast, a study involving post-exercise supplementation with a leucine enriched whey protein source versus a standard whey protein source found that there was no difference in the ability of the enriched source to boost muscle protein post-exercise, despite an increase in plasma leucine concentrations. However, the leucine enriched whey protein source was associated with a more sustained activation of muscle protein synthesis pathways up to five hours post-exercise [165].

Leucine also plays a role in regulating glucose metabolism as it can modulate the activity of glucose transport proteins GLUT1 and GLUT4 that regulates insulin secretion and glucose intake pathways [166]. In a mouse model of obesity, ingestion of leucine improved the glycemic profile by improving insulin sensitivity and glucose tolerance [167]. The anabolic and glucoregulatory activity has made leucine a promising candidate for the treatment of Met-S such as sarcopenia. However, a long-term 3-month study conducted in older men, found that consumption of 7.5 g of leucine daily, had no significant impact on the regulation of muscle mass or glucoregulatory activity [168]. There is a lack of long-term studies concerning the effects of leucine on metabolic health and its suitability as a nutraceutical intervention, therefore more data and in-depth studies are required (Table 6.1) [169].

6.4 PROTEIN AND BIOACTIVE PEPTIDES

Proteins are three-dimensional macromolecular structures that are one of the most fundamental building blocks of life. They are multifunctional molecules, involved in every aspect of cellular function, including influencing gene expression, functioning as structural proteins, intercellular communication signals, and intracellular signal transducers. Proteins can be synthesized *de novo*, or essential amino acids are obtained from dietary sources. When consumed, whole-food proteins are digested, and peptide bonds hydrolyzed to form small peptide sequences or amino acids that are subsequently absorbed and used as the building blocks for endogenous protein synthesis. Deviation in protein structures because of protein gene mutations, errors in ribonucleic acid (RNA) translation, or protein misfolding in the endoplasmic reticulum can drastically affect how a protein functions, and is at the root of many acquired and inherited diseases.

Deficiency in dietary protein plays a considerable role in the development of human disease having major implications for human health. Consequently, the lack of protein and essential amino acid dietary intake can lead to a lowering of health quality including decreased strength, decreased muscle function, and immunodeficiency [177]. For example, in C57/Bl6 mice, a protein-deficient diet increased the susceptibility of infection by influenza, which could be the result of reduced immune function [178]. This observation was supported in a study in older women who consumed a protein-deficient diet exhibited decreased immune function compared to those who consumed a protein sufficient diet [179]. Conversely, excess intake of dietary protein can also have negative implications for human health as it is linked with the increased incidence of hepatic and renal disease [8].

TABLE 6.1 Summary of Key Beneficial Properties of Fatty Acid, Plant Derived Nutraceuticals and Amino Acids

Bioactive Food Molecule	Source	RDA (WHO)	Health Boosting Properties	*In Vitro*	*In Vivo*	Application to NCD Treatment
Curcumin	Turmeric	0–3 mg/kg of body weight	Anti-inflammatory Antioxidant Glucoregulatory Chemoprotective Adjunct chemotherapeutic	[31] [170] [171] [47] [48]	[36] [172] [56] [173] [45]	CID Caner T2DM CVD
Omega 3 PUFAs (EPA, DHA, ALA)	Salmon Trout Mackerel Fish oils	300 mg/day Combinations of EPA and DHA, of which should be predominately DHA	Cardioprotective Anti-inflammatory Antioxidant Pro-anabolic Anti-depressive behavior	n/a [77] [76] [90] n/a	[72] [89] [92] [70]	CID Sarcopenia CVD Cancer
Flavonoids	Edible fruits and veg	To be determined	Antioxidant Anti-inflammatory Cardioprotective Glycemic control	[134] [141] [134] [174]	[175] [145] [133] [147]	Cancer CID CVD T2DM
Carotenoids: Lycopene B-Carotene	Tomatoes Tomato-based products	To be determined	Antioxidant Anti-inflammatory Cardioprotective Chemopreventive	[108] [105] n/a n/a	[124] [95] [109] [113]	Cancer CID CVD
L-glutamine	Dietary Protein	To be determined	Immunomodulatory Tissue integrity	[151] [149]	[161] [154]	CID
Leucine	Dietary Protein	To be determined	Pro-anabolism Glucoregulatory	[91] [176]	[164] [167]	Sarcopenia T2DM

Balanced protein intake, based on physical activity is critical to human health in the long-term. Currently, the recommended daily intake of dietary protein for healthy adults is 0.8 g/kg of body weight [6]. This increases to 1.2 g/kg to 2.0 g/kg for those undertaking frequent intense aerobic and/or resistance exercise [180]. Bovine milk is one of the most protein dense nutrient sources available. There are two primary protein families in bovine milk, whey protein and casein protein, which account for 80% and 20% of the total milk protein content, respectively. When consumed, whey and casein proteins are hydrolyzed and broken down into peptides. It is believed that the subsequent release of these peptides, which are known to possess specific bioactivities, are the drivers of health boosting properties [181]. Although these bioactive peptide sequences can be released through the natural digestive process, there are questions as to whether they are sufficiently and frequently bioavailable to exert health boosting effects. Consequently, there is a worldwide push to mine protein-rich sources for bioactive health-boosting peptides using bioinformatic and laboratory-based approaches [182]. Mining protein-rich sources can produce two potential nutraceuticals; specific bioactive peptide sequences or hydrolysates, which are a mixture of several different bioactive peptides that work together to produce health benefits. Although bovine milk is highly protein-dense, only a small fraction of the milk protein is initially digestible. Whey protein is acid-soluble and is digested and absorbed rapidly by the body. Conversely, casein protein, which comprises a majority of the protein content of milk, is not acid-soluble and tend to coagulate in the stomach, reducing surface area digestibility. Due to this fact, it is believed that undigestible whey and casein components may harbor a wealth of bioactive peptide or hydrolysates that can beneficially modulate several aspects of human health [183, 184].

6.4.1 WHEY PROTEIN

Whey protein comprises 20% of the total protein content of milk and is a major by-product generated by the cheese making industry. Whey protein is comprised of a mixture of several proteins, peptides, and enzymes including; β-lactoglobulin, α-lactalbumin, bovine serum albumin, lactoferrin, immunoglobulins, and lysozyme with other growth factors such as TGFβ, IGF-I, and IGF-II [185–187].

A number of formulations exist on the international market that supply bioactive peptides derived from whey proteins that positively affect cardiovascular and metabolic health in humans [188]. For example, NOP-47™ is a bioactive peptide derived from whey protein that is produced by Glanbia Nutritionals. Two *in vivo* human studies, in both men and women concerning the use of NOP-47™, found that noticeable improvements in cardiovascular function in the groups that were fed NOP-47™ compared to the placebo groups [189, 190]. Whey protein hydrolysates also display angiotensin-converting enzyme (ACE) inhibitory properties, an important regulator of blood pressure. ACE converts Angiotensin I to Angiotensin II, which in turn exerts contractile effects on the vascular system, causing an increase in blood pressure which contributes to hypertension that subsequently has long-term implications on the cardiovascular system [191]. Several ACE inhibitory bioactive peptides have been isolated from the whey proteins α- and β-Lactoglobulin [192, 193]. BioZate® is a second whey protein-based product that in humans has shown to positively regulate blood pressure through ACE inhibition [188, 194].

Consumption of whole whey protein has been shown to increase the whole insulin sensitivity in obese, healthy, and insulin-resistant subjects [195]. Furthermore, whey protein hydrolysates have exhibited the capacity to drive the translocation of the primary insulin-stimulated glucose transporter GLUT-4 to a greater degree, in a rat exercise model [196]. It is understood that whey protein possesses the ability to suppress appetite, which, by default, may aid in weight loss due to reduced calorie intake, subsequently improving obesity, which is a significant risk factor for cardiovascular and metabolic health [195, 197].

6.4.2 CASEIN PROTEIN

Casein is composed of four protein subunits (αs1-, αs2-, β-, and κ-casein) constituting about 80% of the total protein content in bovine milk [181] and like whey, casein has been shown to display nutraceutical properties. Hydrolysates derived from these subunits have been extensively explored for their immunomodulatory and anti-inflammatory potential [181]. Several studies have cited the anti-proliferative effects of casein hydrolysates in lymphocytes [198]. A κ-casein fragment has also been shown to induce an M2-like phenotype in macrophages and suppress LPS induced cytokine signaling by abrogating the NFκB signaling pathway. Researchers also demonstrated that κ-casein treated macrophages and dendritic cells also displayed a reduced capacity to induce robust T lymphocyte responses [21].

Furthermore, unspecified casein hydrolysates of varying sizes, have been shown to suppress the expression of inflammatory cytokines on the colonic $CaCO_2$ cell line and *ex vivo* colonic explants [199, 200]. These findings suggest that an immunomodulatory casein hydrolysates or peptides could modulate the gut microenvironment through direct interaction with immune cells and the gut epithelial cells to suppress inflammation and restore gut homeostasis, potentially improving IBD/IBS symptoms [201].

Casein hydrolysates may also be beneficial in boosting skeletal muscle anabolic signaling via amplification of mTOR signaling. In a human exercise trial in trained cyclists, researchers showed that ingestion of a casein hydrolysate in combination with a carbohydrate source was able to significantly enhance the activity of an mTOR signal transducer, called 4E Binding protein 1 (4E-BP1), compared to the whole protein [202]. In an inactive state, 4E-BP1 is bound to the translation inhibitor eukaryotic translation initiation factor 4E (eIF4e), once phosphorylated 4E-BP1 and eIF4e dissociate and 4E-BP1 moves to initiate translation [203]. This upregulation of 4E-BP1 is therefore correlated with improved protein synthesis in skeletal muscle, helping to maintain and improve muscle mass and function. In the elderly, 4E-BP1 signaling is blunted and the anabolic capacity of muscle is down regulated, leading to syndromes such as sarcopenia [84]. As has been demonstrated with whey protein ingestion previously, combinatory interventions with resistance exercise and hydrolysate consumption may improve skeletal muscle and overall health in the elderly, drastically improving the quality of life [204].

Whole Casein protein, however, has not had any notable effects on glucoregulatory pathways; however, several studies concerning casein isolated peptides have demonstrated that they are capable of inducing insulin sensitivity. *In vitro*, a casein-derived macro peptide, was capable of inducing insulin sensitivity in the human hepatic cancer cell line, HepG2 via AMPK amplification [205]. Similarly, in an *in vivo* animal model, rats fed a high-fat diet exhibited superior glucose tolerance when administered a casein hydrolysate over the whole protein prior to exercise [206]. Given this preliminary evidence, further research into the glucoregulatory potential of casein hydrolysates is required.

6.4.3 PLANT-BASED PROTEIN

Despite animal products being the richest source of high-quality protein, plant-based protein is also a good source of quality protein. Although the pro-anabolic effects of plant protein are measurably less than animal protein, in terms of the whole food consumption, protein-rich plant foods, such a green leafy vegetable, often contain phytochemicals, and are rich in minerals and trace nutrients that have added health-boosting properties [207]. Nevertheless, several sources of plant-based proteins have exhibited promising health-boosting bioactivities. Vegetarians and vegans alike, despite social misconceptions, are capable of gathering enough dietary protein to sufficiently support their bodies. Legumes such as chickpeas, beans, lupins, and lentils are a rich source of high-quality protein [208]. Peas, soya beans, and lupins also have comparable levels of amino acids, such as leucine, lysine, and isoleucine, histidine, and phenylalanine, to animal-based sources, such as eggs [209]. Furthermore, lupin protein hydrolysates have been shown to be ACE inhibitory, suggesting a potential application to treat hypertension in humans [210].

In particular, pea is a commonly occurring food on dinner tables across the world that is a good source of plant protein. Similar to cow's milk proteins, after ingestion, the pea undergoes hydrolysis into amino acids and peptides, and this process is identical to all plant proteins. Peptides isolated from the yellow pea following enzymatic hydrolysis of whole protein yielded a mixture of peptides that we're able to suppress M1 macrophage function. These peptides suppressed IL-6 and TNFα and reduced iNOS activity *in vitro*. In the same study, female BALB/c mice were administered oral supplements of these peptides enhancing the phagocytic activity of gut peritoneal macrophage, whilst also increasing the number of IgA, IL-4, IL-10, and IFNγ producing immune cells in the gut. This study suggests that pea hydrolysates may also act as anti-inflammatory, antioxidant chemoprotective ingredients, and an immune-boosting ingredient to protect the host from gut pathogens [211].

As previously mentioned, the anabolic capacity of plant protein is less efficient than animal-based counterparts. However, pro-anabolic protein sources can be found in plant-based foods, and may offer an avenue to boost the anabolic potential of plant-based foods. In a recent study, involving young healthy female subjects, individuals were fed an isolated potato protein and its ability to induce muscle protein synthesis was examined. The sample group who consumed the protein isolate twice daily for two weeks exhibited elevated levels of muscle protein synthesis both at rest and after exercise [212]. These findings are very important, as it may offer a way to improve muscle health and function over time, even while resting.

Nuritas is an Irish-based company involved in mining different sources of food for functional bioactive peptides with health-boosting and pharmacological applications. Nuritas is unique as the company employs the use of a patented artificial intelligence (AI) model to identify peptides of interest. Touted as the world's first bioactive hydrolysates identified by AI, *PeptAIde* contains peptides isolated from brown rice. The effects of *PeptAIde* were examined in a kinetic study using healthy adults. This study demonstrated that *PeptAIde* was indeed immunomodulatory, having the ability to decrease inflammatory cytokine and chemokine secretion. Although, levels of circulating inflammatory markers returned to baseline after 24 hrs., it demonstrated that a single 20 g dose was able to suppress whole body inflammations [213]. These studies demonstrate that several NCDs can be prevented using an anti-inflammatory peptide and that these new technologies can fast track the identification of novel peptides (Table 6.2).

TABLE 6.2 Summary of Key Beneficial Properties of Proteins and Bioactive Peptides

Protein Source	Bioactive Content	Health Boosting Properties	Mechanisms	Application for NCD Treatment	Refs.
Bovine whey	Whey protein hydrolysate	Insulin sensitizing	Enhancing GLUT4 function	T2DM	[196]
	NOP-47 hydrolysate	Cardioprotective	Modulating vascular endothelial cell function	CVD	[189]
	BioZate hydrolysate	Anti-hypertensive	ACE inhibitory	CVD	[194]
Bovine casein	κ-casein Subunit	Immunomodulatory	Suppression of NFκB	CID	[21]
	Heterogenous hydrolysates	Anti-inflammatory	Suppression of inflammatory gene expression in colonic tissue	IBD	[199]
		Glucoregulatory	Improvements in whole-body glucose tolerance	T2DM Pre-diabetes	[206]
		Pro-anabolic	Enhanced mTOR signaling	Sarcopenia	[202]

TABLE 6.2 *(Continued)*

Protein Source	Bioactive Content	Health Boosting Properties	Mechanisms	Application for NCD Treatment	Refs.
Yellow pea	Hydrolyzed yellow pea protein	Immunomodulatory, antioxidant	Inhibition of M1 Macrophage function, improvements of pathogen/ monitoring immune function in the gut	CID Chemo-protective	[211]
Potato	Potato protein isolate	Pro-anabolic	Enhanced muscle protein synthesis post-exercise and at rest	Sarcopenia	[212]
Brown rice	*PeptAIde* (Nuritias)	Anti-inflammatory	Acute suppression of inflammatory cytokines and chemokines in healthy subjects	CID	[213]

6.5 COMPOUNDS CURRENTLY IN CLINICAL TRIALS

The bioactivity associated with the nutraceuticals discussed in this chapter, provides a wealth of information concerning the potential application of nutraceutical interventions to chronic human disease. However, as also mentioned, the efficacy of some of these nutraceuticals remains to be fully elucidated, and thus, their translation into a clinical setting cannot be fully determined until they have been examined robustly under clinical trial conditions. While there are many clinical trials currently underway to examine the benefits of nutraceuticals, this chapter will discuss a range of clinical trials to provide some insight into the activity that is ongoing in the clinical setting and the range of molecules that are currently being examined.

6.5.1 RESVERATROL

Plant-derived polyphenols and phytochemicals continue to be examined extensively for their clinical applications. There are several ongoing clinical trials investigating the potential application of resveratrol, a compound found in abundance in grapes. *In vitro* and *in vivo* studies have demonstrated the antioxidant, anti-inflammatory, and anti-tumorigenic properties of this compound [214]. The *in vitro* and *in vivo* anticancer properties of resveratrol are reviewed extensively in the following review chapter [215]. In brief, resveratrol is believed to be both a chemopreventative and a potential chemotherapeutic strategy in cancer as its potent antioxidant and free radical scavenging activity is believed to protect cells from oxidative damage and subsequent DNA damage that may result in carcinogenesis [216]. Furthermore, resveratrol was found to suppress the growth of neoblastoma xenographs in mice at serum concentrations of 2–10 μmol/L. Resveratrol treatment exhibited reduced tumor cell viability *in vitro,* [217] while in humans, resveratrol taken over a 10-year period reduced the likelihood of breast cancer development [218].

The antioxidant and anti-inflammatory properties of resveratrol are believed to exert protective effects in a variety of NCDs, including CVD, neurodegenerative disease, and metabolic disease [214, 219]. Several recruiting and ongoing clinical trials concerning resveratrol aim to uncover the potential application of the compound in chronic human disease states (Table 6.3). Resveratrol administration reduced disease severity in humans with mild to moderate Alzheimer's disease [220], while a clinical trial (NCT03762096), examined the benefits of resveratrol supplementation in individuals that suffer from diabetes that carry an increased risk of developing CVD. Individuals who received resveratrol supplementation of 2 g/day over six weeks, seemed to have improved cardiac function, cardiac metabolism, and immune function. A second study (NCT03525379) is examining whether a resveratrol supplement (500 mg per dose) taken twice daily for eight weeks has any improvement in blood flow, vascular function, and oxygen uptake in the skeletal muscle of patients with heart failure. The findings from this trial are yet to be published.

6.5.2 BIOACTIVE DIETARY POLYPHENOL PREPARATIONS (BDPP)

Bioactive dietary polyphenol preparations (BDPP) are also receiving much attention as another nutraceutical therapeutic. It consists of a combination of grape-derived bioactive polyphenolic compounds, one of which is resveratrol, touted to have an array of bioactivities consistent with other polyphenols. BDPP therapeutic strategy is to exploit this compound in order to treat metabolic and neurodegenerative disease. BDPP relieves pain in a rat model of intervertebral disc degeneration, suggesting a potential application as a strategy to treat chronic pain in humans [221]. Furthermore, it is believed that BDPP supplementation may delay the transition from mild cognitive impairment to fully active Alzheimer's disease [222, 223]. BDPP was found to improve brain synaptic brain function in mice, while also improving several aspects of metabolic disease, suggesting that Met-S may put individuals at risk of developing neurodegenerative disease [224].

Clinical translation to human disease models is lacking in the context of BDPP, however there are clinical trials currently recruiting to model the efficacy of BDPP in metabolic and neurodegenerative disease. In a phase 1 clinical trial (NCT02502253), researchers aim to answer several outstanding questions of BDPP supplementation in humans who display mild cognitive impairment. Over a four-month period, participants will receive low, medium, and high doses of BDPP. Side effects of BDPP ingestion at the varying doses will be monitored, which will be a critical point to determine BDPPs suitability for human consumption. Furthermore, due to the effect of BDPP in mouse models of Alzheimer's disease, several outputs of cognitive function will be monitored for any changes due to BDPP ingestion. Related to the first study, a second phase 1 study (NCT04421079) will examine the pharmacokinetics of BDPP ingestion. At low, medium, and high doses, researchers aim to assess the bioavailability of active BDPP metabolites over five weeks, specifically, dihydrocaffeic acid (DHCA). Researchers will subsequently assess whether there is any correlation between BDPP intake, DHCA bioavailability and blood serum concentrations of IL-6.

6.5.3 NUTRAFOL

Nutrafol is a commercially available nutraceutical that has been documented to improve hair growth in women [225]. One of the primary bioactive compounds in Nutrafol is a curcumin extract, of which bioactivity has been discussed in detail previously. The cosmetic benefits of hair growth extend beyond the surface and hair loss can often have negative psychological impacts in both men and women [226]. Clinical trials are currently ongoing with the Nutrafol product line and its application to promoting hair growth in menopausal and pre-menopausal women (NCT04048031). Participants in this study will receive either 4 x Nutrafol capsules daily or a placebo for a six-month period. During this time, researchers will monitor the effects of Nutrafol supplementation on hair growth and volume.

6.5.4 *KB220*

Addiction is a growing pandemic in the 21st century, and given the lack of effective treatments, any benefit that a nutraceutical could provide would be very novel. KB220 is a glutaminergic-dopaminergic compound that contains an array of amino acids, such as L-glutamine, L-phenylalanine, and tryptophan, in combination with several other compounds and trace minerals that act as dopamine and neurotransmitter precursors. It is believed that this nutraceutical cocktail can boost dopamine synthesis in the brain and balance the brain reward circuitry system, which is often imbalanced in cases of addiction [227, 228]. Potential applications of KB220 are believed to be helping to resolve alcohol, nicotine, and opioid addictions, of which widespread usages are linked with chronic diseases, such as CVD and cancer. Clinical trials are ongoing in an African American population with opioid addiction (NCT03861832) in the USA. Researchers hypothesize that there are genetic variants associated with low dopamine status, which causes imbalances within the brain reward circuitry system, resulting in increased incidences of addiction. This may be present in a higher frequency in African American population compared to European Americans. Researchers aim to both assess the genetic variations addressed above and whether administration with KB220 can improve the brain reward circuitry balance in opioid addicts to reduce dependencies and relapses in opioid use.

6.5.5 *OMEGA 3-PUFAS*

As discussed in detail previously, omega-3 PUFAs have a long history of health boosting properties in the context of several disease states, notably CVD. Despite the number of studies to date, many questions remain unanswered. One of which is whether or not genotype plays a role in how a person responds to omega-3 PUFAs. An active phase 1 clinical trial is attempting to answer whether or not the response of measurable risk factors for CVD and metabolic disease, such as blood pressure and glycemic status, are affected by the genetic variations in fatty acid sensor genes (NCT01343342). Over a six-week period, participants will ingest 1.9 g of EPA and 1.1 g of DHA in combination with 5 g of fish oils per day. Researchers will measure changes, if any, in blood pressure, serum lipid profiles and monitor changes in fatty acid sensor gene expression.

TABLE 6.3 Current Clinical Trials Examining the Benefits of Nutraceuticals

Compound	Sources	Bioactivity	Clinical Trial	Status	Clinical Trials Government ID
Resveratrol	Grape skin Seeds	Anticancer, Antioxidant, Neuroprotective Anti-inflammatory Cardioprotective	Short interval resveratrol trial in cardiovascular surgery	Recruiting, active	NCT03762096
			Evaluating the clinical efficacy of resveratrol improving metabolic and skeletal muscle function in patients with heart failure	Recruiting, active	NCT03525379
BDPP	Grapeseed	Anti-inflammatory Antioxidant, Neuroprotective	BDPP treatment for mild cognitive impairment (MCI) and prediabetes or Type 2 diabetes mellitus (T2DM)	Recruiting, active	NCT02502253
			Metabolism of bioactive dietary polyphenol preparation (BDPP)	Not yet recruiting, active	NCT04421079
Nutrafol	Phytochemical supplement (including curcumin)	Anti-inflammatory, antioxidant	Efficacy and safety of a nutraceutical supplement with standardized botanicals in peri-menopausal and menopausal women with thinning hair	Recruiting, active	NCT04048031
KB220	Glutaminergic-dopaminergic supplement	Neuroprotective Mood altering Anti-addiction	SMART brain health in African-Americans (SMART)	Recruiting, active	NCT03861832
Omega-3 fatty acids	EPA/DHA Cod liver oil	Cardioprotective	Genes, omega-3 fatty acids and CVD risk factors (FAS)	Not recruiting, active	NCT01343342

6.6 CONCLUSION AND FUTURE DIRECTIONS

This chapter highlighted the significant amount of data on commonly available nutraceuticals and the new generation of protein-derived nutraceuticals that are currently under development. It also provided evidence on the health protecting aspects of these nutraceuticals with regard to the prevention and treatment of NCDs that are prevalent in the 21st century [22]. The beneficial properties of these nutraceuticals include anti-inflammatory [31], anti-oxidant [124], pro-anabolic [164, 212], liporegulatory, [72] and glucoregulatory [171] that have the potential to promote metabolic [91, 147], cardiovascular [72], and tissue health [154] when consumed in the right amounts. Furthermore, nutraceuticals such as curcumin, have shown synergistic effects with established chemotherapeutic strategies to treat cancer [44, 45]. Many of the nutraceuticals are found in abundance in superfoods such as milk, fish oils, tomatoes, berries, and dark chocolate, which are promoted as health boosting foods [229]. Despite the strong evidence of their beneficial properties, several important questions remain unanswered, such as whether there is a need for regulation of nutraceuticals to the same extent as drug products and the role of these nutraceuticals in the management of NCDs.

One concern is that the consumption of nutraceuticals is not regulated in the same regard as pharmaceutical products, thus potentially anyone can freely consume these concentrated products without medical oversight. There is strong evidence that over consumption of vitamins leads to hypervitaminosis such as hypervitaminosis A that can lead to vitamin A toxicity, which caused symptoms such as changes to vision, bone pain, and skin changes. In extreme cases it can lead to liver damage and increased cranial pressure [230]. Although the adverse effects due to over consumption of nutraceuticals may not be as severe as drug therapy, they are still relevant, and the side effects observed can be considered mild to moderate. No clinical information is available that answers whether or not, supplementing a diet, with a given nutraceutical, has any long-term adverse effects if taken over a considerable length of time.

There are many over the counter food supplements that have health claims with no robust evidence to support these claims. However, the European Commission (EC) is working to address the issue of health claims associated with nutraceuticals. The EFSA is responsible for evaluating the scientific evidence supporting health claims ensuring that health claims provided are based on sound scientific evidence that can be easily understood by consumers. The suitability of certain nutraceuticals is also of concern, such as the increased risk of developing lung cancer with increased β-carotene consumption in smokers. Consequently, it may be necessary to regulate the sale and availability of concentrated nutraceuticals. There are no studies that have evaluated the prescribing of nutraceuticals by medical professionals to treat NCDs and if they are observing any beneficial effects. A recent clinical trial is currently recruiting volunteers that aims to address that question (NCT04161859). Researchers will aim to evaluate the clinical usage of several common nutraceutical compounds such as; Omega 3 PUFAs, resveratrol, curcumin, and alpha-linolenic acid, to name but a few. Specifically, researchers will be monitoring the clinical usage of these nutraceuticals with regard to the treatment of CVD.

The use of nutraceuticals in the context of disease management is a very important issue that needs to be addressed fully. This will require strong clinical evidence with medical oversight or clear communication and education from the manufacturer with regard to its use. In the same regard that athletes will periodize their training and nutrition over a calendar year based on competition schedule, a similar tactic could be employed with regard to prescribed usages of nutraceuticals. For example, rather than take Vitamin C supplements all year, periodizing consumption during periods of increased likelihood of contracting cold and influenza and decreasing supplemented consumption outside of this period. This is a strategy that several international athletes have employed when traveling abroad for competition concerning the use of probiotics, to reduce the likelihood of contracting GI infection abroad [231–233]. However, NCDs are very complex, and many can remain asymptomatic or exhibit vague symptoms until the disease enters the more chronic stages of disease. Thus, without constant health monitoring, this strategy or periodization may not be useful. Rather, ensuring that adequate dietary requirements and limits are adhered to through consumption of whole foods naturally containing these nutraceuticals may be a strategy for preventing the onset of chronic NCDs later in life.

Age is a major factor when it comes to the prevalence of NCD, and the use of nutraceuticals to prevent age-related diseases is the area where it can have the greatest impact. As we age, our metabolism slows, and our dietary calorie intake reduces [234], which makes it difficult to consume a diet that has optional nutrition. Supplementing food with nutraceuticals means that we can enhance the health benefits of food. As mentioned previously, "inflammaging" a phenomenon in which the basal immune state of an individual leans towards a more inflammatory state as the body ages [82]. This phenomenon over chronic low-grade inflammation appears to underly the development of several NCDs, including cancer, diabetes, and CVD [81, 235, 236]. The WHO estimates that by 2050, there will be over 2 billion people worldwide aged over 60, so it stands to reason that we may see an increase in NCD prevalence in the not-so-distant future. With this in mind, the beneficial effects of anti-inflammatory nutraceuticals are clear, in the potential to delay the onset and development of disease as we age. Future efforts at mining food sources for health-boosting bioactives should focus on the hunt for anti-inflammatory and antioxidant bioactives so we can support an ever-growing aging population.

KEYWORDS

- **cardiovascular vascular disease**
- **Crohn's disease**
- **metabolic syndromes**
- **nicotinic acid**
- **non-communicable diseases**
- **rheumatoid arthritis**
- **tumor necrosis factor**

REFERENCES

1. Witkamp, R. F., & Van, N. K., (2018). Let thy food be thy medicine when possible. *Eur. J. Pharmacol., 836*, 102–114. https://doi.org/10.1016/j.ejphar.2018.06.026.
2. Hunt, N., (2018). *Immunomodulatory Protein Hydrolysates for the Management of Intestinal Immune Disorders in Infants.* PhD thesis, Dublin City University.
3. Saunders, J., & Smith, T., (2010). Malnutrition: Causes and consequences. *Clin. Med. (Northfield. Il), 10*, 624–627. https://doi.org/10.7861/clinmedicine.10-6-624.
4. Chambial, S., Dwivedi, S., Shukla, K. K., John, P. J., & Sharma, P., (2013). Vitamin C in disease prevention and cure: An overview. *Indian J. Clin. Biochem., 28*, 314–328. https://doi.org/10.1007/s12291-013-0375-3.
5. Gani, L., & How, C., (2015). Vitamin D deficiency. *Singapore Med. J., 56*, 433–437. https://doi.org/10.11622/smedj.2015119.
6. Lonnie, M., Hooker, E., Brunstrom, J. M., Corfe, B. M., Green, M. A., Watson, A. W., Williams, E. A., et al., (2018). Protein for life: Review of optimal protein intake, sustainable dietary sources, and the effect on appetite in aging adults. *Nutrients, 10*, 1–18. https://doi.org/10.3390/nu10030360.

7. Carrera-Bastos, P., Fontes, O'Keefe, Lindeberg, & Cordain, (2011). The western diet and lifestyle and diseases of civilization. *Res. Reports Clin. Cardiol.,* 15. https://doi.org/10.2147/RRCC.S16919.
8. Martin, W. F., Armstrong, L. E., & Rodriguez, N. R., (2005). Dietary protein intake and renal function. *Nutr. Metab., 2,* 1–9. https://doi.org/10.1186/1743-7075-2-25.
9. Cali, A. M. G., & Caprio, S., (2008). Obesity in children and adolescents. *J. Clin. Endocrinol. Metab., 93,* s31–s36. https://doi.org/10.1210/jc.2008-1363.
10. Neovius, M., Janson, A., & Rossner, S., (2006). Prevalence of obesity in the United States. *Obes. Rev.,* 7, 1–3. https://doi.org/10.1111/j.1467-789x.2006.00190.x.
11. Ottova, V., Erhart, M., Rajmil, L., Dettenborn-Betz, L., & Ravens-Sieberer, U., (2012). Overweight and its impact on the health-related quality of life in children and adolescents: Results from the European KIDSCREEN survey. *Qual. Life Res., 21,* 59–69. https://doi.org/10.1007/s11136-011-9922-7.
12. Wabitsch, M., Moss, A., & Kromeyer-Hauschild, K., (2014). Unexpected plateauing of childhood obesity rates in developed countries. *BMC Med., 12,* 17. https://doi.org/10.1186/1741-7015-12-17.
13. Woods, C. B., Powell, C., Saunders, J. A., O'Brien, W., Murphy, M. H., Duff, C., Farmer, O., et al., (2018). *The Children's Sport Participation and Physical Activity Study 2018 (CSPPA 2018),* 1–108.
14. Pi-Sunyer, X., (2010). The medical risks of obesity. *PMC 121,* 21–33. https://doi.org/10.3810/pgm.2009.11.2074.
15. Metsälä, J., Lundqvist, A., Virta, L. J., Kaila, M., Gissler, M., Virtanen, S. M., & Nevalainen, J., (2018). The association between asthma and type 1 diabetes: A pediatric case-cohort study in Finland, years 1981–2009. *Int. J. Epidemiol., 47,* 409–416. https://doi.org/10.1093/ije/dyx245.
16. Miranda, J. J., Kinra, S., Casas, J. P., Davey, S. G., & Ebrahim, S., (2008). Non-communicable diseases in low- and middle-income countries: Context, determinants, and health policy. *Trop. Med. Int. Heal., 13,* 1225–1234. https://doi.org/10.1111/j.1365-3156.2008.02116.x.Non-communicable.
17. Miranda, J. J., Barrientos-Gutiérrez, T., Corvalan, C., Hyder, A. A., Lazo-Porras, M., Oni, T., & Wells, J. C. K., (2019). Understanding the rise of cardiometabolic diseases in low- and middle-income countries. *Nat. Med., 25,* 1667–1679. https://doi.org/10.1038/s41591-019-0644-7.
18. Strong, K., Mathers, C., Leeder, S., & Beaglehole, R., (2005). Preventing chronic diseases: How many lives can we save? *Lancet, 366,* 1578–1582. https://doi.org/10.1016/S0140-6736(05)67341-2.
19. Sidiropoulos, P. I., & Boumpas, D., (2006). Differential drug resistance to anti-tumor necrosis factor agents in rheumatoid arthritis. *Ann. Rheum. Dis., 65,* 701–703. https://doi.org/10.1136/ard.2005.049890.
20. Bodor, E. T., & Offermanns, S., (2009). Nicotinic acid: An old drug with a promising future. *Br. J. Pharmacol., 153,* S68–S75. https://doi.org/10.1038/sj.bjp.0707528.
21. Lalor, R., & O'Neill, S., (2019). Bovine κ-casein fragment induces hypo-responsive m2-like macrophage phenotype. *Nutrients, 11.* https://doi.org/10.3390/nu11071688.
22. El Sohaimy, S., (2012). Functional foods and nutraceuticals-modern approach to food science. *World Appl. Sci. J., 20,* 691–708. https://doi.org/10.5829/idosi.wasj.2012.20.05.66119.

23. Hartmann, R., & Meisel, H., (2007). *Food-Derived Peptides with Biological Activity: From Research to Food Applications* (pp. 163–169). https://doi.org/10.1016/j.copbio.2007.01.013.
24. Kiewiet, M., Faas, M., & De Vos, P., (2018). In: Kiewiet, M., Faas, M., & De Vos, P., (eds.), *University of Groningen Immunomodulatory Protein Hydrolysates and Their Application.* https://doi.org/10.3390/nu10070904.
25. Udenigwe, C. C., & Aluko, R. E., (2012). *Food Protein-Derived Bioactive Peptides: Production, Processing, and Potential Health Benefits.* https://doi.org/10.1111/j.1750-3841.2011.02455.x.
26. Souyoul, S. A., Saussy, K. P., & Lupo, M. P., (2018). Nutraceuticals: A review. *Dermatol. Ther. (Heidelb), 8*, 5–16. https://doi.org/10.1007/s13555-018-0221-x.
27. Kunnumakkara, A. B., Bordoloi, D., Padmavathi, G., Monisha, J., Roy, N. K., Prasad, S., & Aggarwal, B. B., (2017). Curcumin, the golden nutraceutical: Multitargeting for multiple chronic diseases. *Br. J. Pharmacol., 174*, 1325–1348. https://doi.org/10.1111/bph.13621.
28. Liu, T., Zhang, L., Joo, D., & Sun, S. C., (2017). NF-κB signaling in inflammation. *Signal Transduct. Target. Ther., 2.* https://doi.org/10.1038/sigtrans.2017.23.
29. Apostolakis, J., Ivantchenko, A. V., Ivanchenko, V. N., Kossov, M., Quesada, J. M., & Wright, D. H., (2014). NF-κB, an active player in human cancers. *Int. Top. Meet. Nucl. Res. Appl. Util. Accel., 2*, 823–830. https://doi.org/10.1158/2326-6066.CIR-14-0112. NF-.
30. Bose, S., Panda, A. K., Mukherjee, S., & Sa, G., (2015). Curcumin and tumor immune-editing: Resurrecting the immune system. *Cell Div., 10*, 6–8. https://doi.org/10.1186/s13008-015-0012-z.
31. Kim, G. Y., Kim, K. H., Lee, S. H., Yoon, M. S., Lee, H. J., Moon, D. O., Lee, C. M., et al., (2005). Curcumin inhibits immunostimulatory function of dendritic cells: MAPKs and translocation of NF-κB as potential targets. *J. Immunol., 174*, 8116–8124. https://doi.org/10.4049/jimmunol.174.12.8116.
32. Hofer, S., Rescigno, M., Granucci, F., Citterio, S., Francolini, M., & Ricciardi-Castagnoli, P., (2001). Differential activation of NF-κB subunits in dendritic cells in response to Gram-negative bacteria and to lipopolysaccharide. *Microbes Infect., 3*, 259–265. https://doi.org/10.1016/S1286-4579(01)01378-8.
33. McAleer, J. P., & Vella, A. T., (2008). Understanding how lipopolysaccharide impacts CD4 T-cell immunity. *Crit. Rev. Immunol., 28*, 281–299. https://doi.org/10.1615/critrevimmunol.v28.i4.20.
34. Mohammadi, A., Blesso, C. N., Barreto, G. E., Banach, M., Majeed, M., & Sahebkar, A., (2019). Macrophage plasticity, polarization, and function in response to curcumin, a diet-derived polyphenol, as an immunomodulatory agent. *J. Nutr. Biochem., 66*, 1–16. https://doi.org/10.1016/j.jnutbio.2018.12.005.
35. Dai, Q., Zhou, D., Xu, L., & Song, X., (2018). Curcumin alleviates rheumatoid arthritis-induced inflammation and synovial hyperplasia by targeting mTOR pathway in rats. *Drug Des. Devel. Ther., 12*, 4095–4105. https://doi.org/10.2147/DDDT.S175763.
36. Chandran, B., & Goel, A., (2012). A randomized, pilot study to assess the efficacy and safety of curcumin in patients with active rheumatoid arthritis. *Phyther. Res., 26*, 1719–1725. https://doi.org/10.1002/ptr.4639.

37. Burge, K., Gunasekaran, A., Eckert, J., & Chaaban, H., (2019). Curcumin and intestinal inflammatory diseases: Molecular mechanisms of protection. *Int. J. Mol. Sci., 20*. https://doi.org/10.3390/ijms20081912.
38. Harris, V. C., Haak, B. W., Boele, V. H. M., & Wiersinga, W. J., (2017). The intestinal microbiome in infectious diseases: The clinical relevance of a rapidly emerging field. *Open Forum Infect. Dis., 4,* 1–8. https://doi.org/10.1093/ofid/ofx144.
39. Malmuthuge, N., & Guan, L. L., (2016). Gut microbiome and omics: A new definition to ruminant production and health. *Anim. Front., 6*, 8–12. https://doi.org/10.2527/af.2016-0017.
40. Khan, I., Ullah, N., Zha, L., Bai, Y., Khan, A., Zhao, T., Che, T., & Zhang, C., (2019). Alteration of gut microbiota in inflammatory bowel disease (IBD): Cause or consequence? IBD treatment targeting the gut microbiome. *Pathogens, 8*, 126. https://doi.org/10.3390/pathogens8030126.
41. Di Meo, Margarucci, Galderisi, Crispi, & Peluso, (2019). Curcumin, gut microbiota, and neuroprotection. *Nutrients, 11*, 2426. https://doi.org/10.3390/nu11102426.
42. Ohno, M., Nishida, A., Sugitani, Y., Nishino, K., Inatomi, O., Sugimoto, M., Kawahara, M., & Andoh, A., (2017). Nanoparticle curcumin ameliorates experimental colitis via modulation of gut microbiota and induction of regulatory T-cells. *PLoS One 12*, 1–16. https://doi.org/10.1371/journal.pone.0185999.
43. Yousef, A. F. C. A., (2017). Revisiting the hallmarks of cancer. *Am. J. Cancer Res., 26*, 62.
44. Aggarwal, B. B., Shishodia, S., Takada, Y., Banerjee, S., Newman, R. A., Bueso-Ramos, C. E., & Price, J. E., (2005). Curcumin suppresses the paclitaxel-induced nuclear factor-κB pathway in breast cancer cells and inhibits lung metastasis of human breast cancer in nude mice. *Clin. Cancer Res., 11*, 7490–7498. https://doi.org/10.1158/1078-0432.CCR-05-1192.
45. Banerjee, S., Singh, S. K., Chowdhury, I., Lillard, J. W., & Singh, R., (2017). Combinatorial effect of curcumin with docetaxel modulates apoptotic and cell survival molecules in prostate cancer. *Front. Biosci.-Elit., 9*, 235–245. https://doi.org/10.2741/e798.
46. Calvani, M., Pasha, A., & Favre, C., (2020). Nutraceutical boom in cancer: Inside the labyrinth of reactive oxygen species. *Int. J. Mol. Sci., 21*. https://doi.org/10.3390/ijms21061936.
47. Das, L., & Vinayak, M., (2015). Long term effect of curcumin in restoration of tumor suppressor p53 and phase-II antioxidant enzymes via activation of Nrf2 signaling and modulation of inflammation in prevention of cancer. *PLoS One, 10*, 1–22. https://doi.org/10.1371/journal.pone.0124000.
48. Zhu, Y., & Bu, S., (2017). Curcumin induces autophagy, apoptosis, and cell cycle arrest in human pancreatic cancer cells. *Evidence-Based Complement. Altern. Med., 2017*, 1–13. https://doi.org/10.1155/2017/5787218.
49. Garcea, G., Neal, C. P., Pattenden, C. J., Steward, W. P., & Berry, D. P., (2005). Molecular prognostic markers in pancreatic cancer: A systematic review. *Eur. J. Cancer, 41*, 2213–2236. https://doi.org/10.1016/j.ejca.2005.04.044.
50. Lusis, A. J., (2000). Atherosclerosis. *Nature 407*, 233–241. https://doi.org/10.1038/35025203.Atherosclerosis.
51. Momtazi-Borojeni, A. A., Abdollahi, E., Nikfar, B., Chaichian, S., & Ekhlasi-Hundrieser, M., (2019). Curcumin as a potential modulator of M1 and M2 macrophages:

New insights in atherosclerosis therapy. *Heart Fail. Rev., 24*, 399–409. https://doi.org/10.1007/s10741-018-09764-z.
52. Wongcharoen, W., & Phrommintikul, A., (2009). The protective role of curcumin in cardiovascular diseases. *Int. J. Cardiol., 133*, 145–151. https://doi.org/10.1016/j.ijcard.2009.01.073.
53. Lin, K., Chen, H., Chen, X., Qian, J., Huang, S., & Huang, W., (2020). Efficacy of curcumin on aortic atherosclerosis: A systematic review and meta-analysis in mouse studies and insights into possible mechanisms. *Oxid. Med. Cell. Longev. 2020*. https://doi.org/10.1155/2020/1520747.
54. Shoelson, S. E., Lee, J., & Goldfine, A. B., (2006). Inflammation and insulin resistance. *J. Clin. Invest., 116*, 1793–1801. https://doi.org/10.1172/JCI29069.and.
55. Wei, Y., Chen, K., Whaley-Connell, A. T., Stump, C. S., Ibdah, J. A., & Sowers, J. R., (2008). Skeletal muscle insulin resistance: Role of inflammatory cytokines and reactive oxygen species. *Am. J. Physiol. Regul. Integr. Comp. Physiol., 294*, R673–680. https://doi.org/10.1152/ajpregu.00561.2007.
56. Chuengsamarn, S., Rattanamongkolgul, S., Luechapudiporn, R., Phisalaphong, C., & Jirawatnotai, S., (2012). Curcumin extract for prevention of type 2 diabetes. *Diabetes Care, 35*, 2121–2127. https://doi.org/10.2337/dc12-0116.
57. Azhdari, M., Karandish, M., & Mansoori, A., (2019). Metabolic benefits of curcumin supplementation in patients with metabolic syndrome: A systematic review and meta-analysis of randomized controlled trials. *Phyther. Res., 33*, 1289–1301. https://doi.org/10.1002/ptr.6323.
58. Pivari, F., Mingione, A., Brasacchio, C., & Soldati, L., (2019). Curcumin and type 2 diabetes mellitus: prevention and treatment. *Nutrients, 11*, 204–208. https://doi.org/10.1177/146642405007000307.
59. Hewlings, S., & Kalman, D., (2017). Curcumin: A review of its' effects on human health. *Foods, 6*, 92. https://doi.org/10.3390/foods6100092.
60. Lao, C. D., Ruffin, M. T., Normolle, D., Heath, D. D., Murray, S. I., Bailey, J. M., Boggs, M. E., et al., (2006). Dose escalation of a curcuminoid formulation. *BMC Complement Altern. Med., 6*, 10. https://doi.org/10.1186/1472-6882-6-10.
61. Calder, P. C., (2017a). New evidence that omega-3 fatty acids have a role in primary prevention of coronary heart disease. *J. Public Heal. Emerg., 1*, 1155–66. https://doi.org/10.21037/jphe.2017.03.03.
62. Innes, J. K., & Calder, P. C., (2020). Marine omega-3 (N-3) fatty acids for cardiovascular health: An update for 2020. *Int. J. Mol. Sci., 21*, 1–21. https://doi.org/10.3390/ijms21041362.
63. Khan, S. A., Khan, A., Khan, S. A., Beg, M. A., Ali, A., & Damanhouri, G., (2015). Comparative study of fatty-acid composition of table eggs from the Jeddah food market and effect of value addition in omega-3 bio-fortified eggs. *Saudi J. Biol. Sci., 24*, 929–935. https://doi.org/10.1016/j.sjbs.2015.11.001.
64. Mohebi-Nejad, A., & Bikdeli, B., (2014). Omega-3 supplements and cardiovascular diseases. *Tanaffos, 13*, 6–14.
65. Calder, P. C., (2017b). Omega-3 fatty acids and inflammatory processes: From molecules to man. *Biochem. Soc. Trans., 45*, 1105–1115. https://doi.org/10.1042/BST20160474.
66. Lemahieu, C., Bruneel, C., Ryckebosch, E., Muylaert, K., Buyse, J., & Foubert, I., (2015). Impact of different omega-3 polyunsaturated fatty acid (n-3 PUFA) sources (flaxseed, Isochrysis galbana, fish oil and DHA Gold) on n-3 LC-PUFA enrichment

(efficiency) in the egg yolk. *J. Funct. Foods, 19*, 821–827. https://doi.org/10.1016/j.jff.2015.04.021.
67. Blondeau, N., Lipsky, R. H., Bourourou, M., Duncan, M. W., Gorelick, P. B., & Marini, A. M., (2015). Alpha-linolenic acid: An omega-3 fatty acid with neuroprotective properties—ready for use in the stroke clinic? *Biomed Res. Int., 2015*, 1–8. https://doi.org/10.1155/2015/519830.
68. Blondeau, N., Nguemeni, C., Debruyne, D. N., Piens, M., Wu, X., Pan, H., Hu, X. Z., et al., (2009). Subchronic alpha-linolenic acid treatment enhances brain plasticity and exerts an antidepressant effect: A versatile potential therapy for stroke. *Neuropsychopharmacology, 34*, 2548–2559. https://doi.org/10.1038/npp.2009.84.
69. Lucas, M., Mirzaei, F., O'Reilly, E. J., Pan, A., Willett, W. C., Kawachi, I., Koenen, K., & Ascherio, A., (2011). Dietary intake of n-3 and n-6 fatty acids and the risk of clinical depression in women: A 10-y prospective follow-up study. *Am. J. Clin. Nutr., 93*, 1337–1343. https://doi.org/10.3945/ajcn.111.011817.
70. Venna, V. R., Deplanque, D., Allet, C., Belarbi, K., Hamdane, M., & Bordet, R., (2009). PUFA induce antidepressant-like effects in parallel to structural and molecular changes in the hippocampus. *Psychoneuroendocrinology, 34*, 199–211. https://doi.org/10.1016/j.psyneuen.2008.08.025.
71. Mozaffarian, D., Lemaitre, R. N., King, I. B., Song, X., Huang, H., Sacks, F. M., Rimm, E. B., et al., (2013). Plasma phospholipid long-chain omega-3 fatty acids and total and cause-specific mortality in older adults: The cardiovascular health study. *Ann. Intern. Med., 158*, 515–525. https://doi.org/10.1145/1080754.1080765.
72. AbuMweis, S., Jew, S., Tayyem, R., & Agraib, L., (2018). Eicosapentaenoic acid and docosahexaenoic acid containing supplements modulate risk factors for cardiovascular disease: A meta-analysis of randomized placebo-control human clinical trials. *J. Hum. Nutr. Diet, 31*, 67–84. https://doi.org/10.1111/jhn.12493.
73. Dennis, E. A., & Norris, P. C., (2015). Eicosanoid storm in infection and inflammation. *Nat. Rev. Immunol., 15*, 511–523. https://doi.org/10.1038/nri3859.
74. Calder, P. C., (2010). Omega-3 fatty acids and inflammatory processes. *Nutrients, 2*, 355–374. https://doi.org/10.3390/nu2030355.
75. Layé, S., Nadjar, A., Joffre, C., & Bazinet, R. P., (2018). Anti-inflammatory effects of omega-3 fatty acids in the brain: Physiological mechanisms and relevance to pharmacology. *Pharmacol. Rev., 70*, 12–38. https://doi.org/10.1124/pr.117.014092.
76. Allam-Ndoul, B., Guénard, F., Barbier, O., & Vohl, M. C., (2016). Effect of n-3 fatty acids on the expression of inflammatory genes in THP-1 macrophages. *Lipids Health Dis., 15*, 1–7. https://doi.org/10.1186/s12944-016-0241-4.
77. Mullen, A., Loscher, C. E., & Roche, H. M., (2010). Anti-inflammatory effects of EPA and DHA are dependent upon time and dose-response elements associated with LPS stimulation in THP-1-derived macrophages. *J. Nutr. Biochem., 21*, 444–450. https://doi.org/10.1016/j.jnutbio.2009.02.008.
78. Meital, L. T., Windsor, M. T., Perissiou, M., Schulze, K., Magee, R., Kuballa, A., Golledge, J., et al., (2019). Omega-3 fatty acids decrease oxidative stress and inflammation in macrophages from patients with small abdominal aortic aneurysm. *Sci. Rep., 9*, 1–11. https://doi.org/10.1038/s41598-019-49362-z.
79. Spencer, M., Finlin, B. S., Unal, R., Zhu, B., Morris, A. J., Shipp, L. R., Lee, J., et al., (2013). Omega-3 fatty acids reduce adipose tissue macrophages in human subjects with insulin resistance. *Diabetes, 62*, 1709–1717. https://doi.org/10.2337/db12-1042.

80. Tortosa-Caparrós, E., Navas-Carrillo, D., Marín, F., & Orenes-Piñero, E., (2017). Anti-inflammatory effects of omega 3 and omega 6 polyunsaturated fatty acids in cardiovascular disease and metabolic syndrome. *Crit. Rev. Food Sci. Nutr., 57*, 3421–3429. https://doi.org/10.1080/10408398.2015.1126549.
81. Ferrucci, L., & Fabbri, E., (2018). Inflammageing: Chronic inflammation in ageing, cardiovascular disease, and frailty. *Physiol. Behav., 176*, 139–148. https://doi.org/10.1016/j.physbeh.2017.03.040.
82. Franceschi, C., Garagnani, P., Parini, P., Giuliani, C., & Santoro, A., (2018). Inflammaging: A new immune-metabolic viewpoint for age-related diseases. *Nat. Rev. Endocrinol., 14*, 576–590. https://doi.org/10.1038/s41574-018-0059-4.
83. Kalyani, R. R., Corriere, M., & Ferrucci, L., (2014). Age-related and disease-related muscle loss: The effect of diabetes, obesity, and other diseases. *Lancet Diabetes Endocrinol., 2*, 819–829. https://doi.org/10.1016/S2213-8587(14)70034-8.
84. Breen, L., & Phillips, S. M., (2011). Skeletal muscle protein metabolism in the elderly: Interventions to counteract the 'anabolic resistance' of ageing. *Nutr. Metab., 6*, 1–11.
85. Dardevet, D., Rémond, D., Peyron, M. A., Papet, I., Savary-Auzeloux, I., & Mosoni, L., (2012). Muscle wasting and resistance of muscle anabolism: The "anabolic threshold concept" for adapted nutritional strategies during sarcopenia. *Sci. World J., 2012*. https://doi.org/10.1100/2012/269531.
86. Haran, P. H., Rivas, D. A., & Fielding, R. A., (2012). Role and potential mechanisms of anabolic resistance in sarcopenia. *J. Cachexia. Sarcopenia Muscle, 3*, 157–162. https://doi.org/10.1007/s13539-012-0068-4.
87. Toth, M. J., (2004). Age-related differences in skeletal muscle protein synthesis: Relation to markers of immune activation. *AJP Endocrinol. Metab., 288*, E883–E891. https://doi.org/10.1152/ajpendo.00353.2004.
88. Dupont, J., Dedeyne, L., Dalle, S., Koppo, K., & Gielen, E., (2019). The role of omega-3 in the prevention and treatment of sarcopenia. *Aging Clin. Exp. Res., 31*, 825–836. https://doi.org/10.1007/s40520-019-01146-1.
89. Tan, A., Sullenbarger, B., Prakash, R., & McDaniel, J. C., (2018). Supplementation with eicosapentaenoic acid and docosahexaenoic acid reduces high levels of circulating proinflammatory cytokines in aging adults: A randomized, controlled study. *Prostaglandins, Leukot. Essent. Fat. Acids, 132*, 23–29. https://doi.org/10.1016/j.plefa.2018.03.010.
90. Kamolrat, T., & Gray, S. R., (2013). The effect of eicosapentaenoic and docosahexaenoic acid on protein synthesis and breakdown in murine C2C12 myotubes. *Biochem. Biophys. Res. Commun., 432*, 593–598. https://doi.org/10.1016/j.bbrc.2013.02.041.
91. Son, S. M., Park, S. J., Lee, H., Siddiqi, F., Lee, J. E., Menzies, F. M., & Rubinsztein, D. C., (2019). Leucine signals to mTORC1 via its metabolite acetyl-coenzyme A. *Cell Metab., 29*, 192–201.e7. https://doi.org/10.1016/j.cmet.2018.08.013.
92. Smith, G. I., Atherton, P., Reeds, D. N., Mohammed, B. S., Rankin, D., Rennie, M. J., & Mittendorfer, B., (2011). Dietary omega-3 fatty acid supplementation increases the rate of muscle protein synthesis in older adults: A randomized controlled trial. *Am. J. Clin. Nutr., 93*, 402–412. https://doi.org/10.3945/ajcn.110.005611.
93. Marton, L. T., Goulart, R. D. A., Carvalho, A. C. A. D., & Barbalho, S. M., (2019). Omega fatty acids and inflammatory bowel diseases: An overview. *Int. J. Mol. Sci., 20*. https://doi.org/10.3390/ijms20194851.

94. Chan, S. S. M., Luben, R., Olsen, A., Tjonneland, A., Kaaks, R., Lindgren, S., Grip, O., et al., (2014). Association between high dietary intake of the n-3 polyunsaturated fatty acid docosahexaenoic acid and reduced risk of Crohn's disease. *Aliment. Pharmacol. Ther., 39*, 834–842. https://doi.org/10.1111/apt.12670.
95. Hozawa, A., Jacobs, D. R., Steffes, M. W., Gross, M. D., Steffen, L. M., & Lee, D. H., (2007). Relationships of circulating carotenoid concentrations with several markers of inflammation, oxidative stress, and endothelial dysfunction: The coronary artery risk development in young adults (CARDIA)/young adult longitudinal trends in antioxidants (YALT). *Clin. Chem., 53*, 447–455. https://doi.org/10.1373/clinchem.2006.074930.
96. Simopoulos, A. P., (2004). Omega-6/omega-3 essential fatty acid ratio and chronic diseases. *Food Rev. Int., 20*, 77–90. https://doi.org/10.1081/FRI-120028831.
97. Simopoulos, A., (2002). The importance of the ratio of omega-6/omega-3 essential fatty acids. *Biomed. Pharmacother., 56*, 365–379. https://doi.org/10.1016/S0753-3322(02)00253-6.
98. WHO, (2008). *Interim Summary of Conclusions and Dietary Recommendations on Total Fat & Fatty Acids*. WHO Expert Consultation.
99. Svensson, M., Schmidt, E. B., Jørgensen, K. A., & Christensen, J. H., (2006). N-3 fatty acids as secondary prevention against cardiovascular events in patients who undergo chronic hemodialysis: A randomized, placebo-controlled intervention trial. *Clin. J. Am. Soc. Nephrol., 1*, 780–786. https://doi.org/10.2215/CJN.00630206.
100. Lenihan-Geels, G., Bishop, K., & Ferguson, L., (2013). Alternative sources of omega-3 fats: Can we find a sustainable substitute for fish? *Nutrients, 5*, 1301–1315. https://doi.org/10.3390/nu5041301.
101. Peltomaa, E., Johnson, M. D., & Taipale, S. J., (2018). Marine cryptophytes are great sources of EPA and DHA. *Mar. Drugs, 16*, 1–11. https://doi.org/10.3390/md16010003.
102. Rao, A. V., & Rao, L. G., (2007). Carotenoids and human health. *Pharmacol. Res., 55*, 207–216. https://doi.org/10.1016/j.phrs.2007.01.012.
103. Liu, R. H., (2013). Dietary bioactive compounds and their health implications. *J. Food Sci., 78*, A18–A25. https://doi.org/10.1111/1750-3841.12101.
104. Agarwal, S., & Rao, A. V., (2000). Tomato lycopene and its role in human health and chronic diseases. *CMAJ, 163*, 739–744. https://doi.org/10.1016/j.postharvbio.2004.05.023.
105. Feng, D., Ling, W. H., & Duan, R. D., (2010). Lycopene suppresses LPS-induced NO and IL-6 production by inhibiting the activation of ERK, p38MAPK, and NF-κB in macrophages. *Inflamm. Res., 59*, 115–121. https://doi.org/10.1007/s00011-009-0077-8.
106. Marcotorchino, J., Romier, B., Gouranton, E., Riollet, C., Gleize, B., Malezet-Desmoulins, C., & Landrier, J. F., (2012). Lycopene attenuates LPS-induced TNF-α secretion in macrophages and inflammatory markers in adipocytes exposed to macrophage-conditioned media. *Mol. Nutr. Food Res., 56*, 725–732. https://doi.org/10.1002/mnfr.201100623.
107. Kawata, A., Murakami, Y., Suzuki, S., & Fujisawa, S., (2018). Anti-inflammatory activity of β-carotene, lycopene and tri-n-butyl borane, a scavenger of reactive oxygen species. *In Vivo (Brooklyn), 32*, 255–264. https://doi.org/10.21873/invivo.11232.
108. Palozza, P., Simone, R., Catalano, A., Boninsegna, A., Böhm, V., Fröhlich, K., Mele, M. C., et al., (2010). Lycopene prevents 7-ketocholesterol-induced oxidative stress, cell cycle arrest and apoptosis in human macrophages. *J. Nutr. Biochem., 21*, 34–46. https://doi.org/10.1016/j.jnutbio.2008.10.002.

109. Kulczyński, B., Gramza-Michałowska, A., Kobus-Cisowska, J., & Kmiecik, D., (2017). The role of carotenoids in the prevention and treatment of cardiovascular disease: Current state of knowledge. *J. Funct. Foods, 38*, 45–65. https://doi.org/10.1016/j.jff.2017.09.001.
110. Hu, M. Y., Li, Y. L., Jiang, C. H., Liu, Z. Q., Qu, S. L., & Huang, Y. M., (2008). Comparison of lycopene and Fluvastatin effects on atherosclerosis induced by a high-fat diet in rabbits. *Nutrition, 24*, 1030–1038. https://doi.org/10.1016/j.nut.2008.05.006.
111. Jacques, P. F., Lyass, A., Massaro, J. M., Vasan, R. S., & D'Agostino, S. R. B., (2013). Relationship of lycopene intake and consumption of tomato products to incident CVD. *Br. J. Nutr., 110*, 545–551. https://doi.org/10.1017/S0007114512005417.
112. Sesso, H. D., Liu, S., Gaziano, J. M., & Buring, J. E., (2003). Dietary lycopene, tomato-based food products and cardiovascular disease in women. *J. Nutr., 133*, 2336–2341. https://doi.org/10.1093/jn/133.7.2336.
113. Holzapfel, N., Holzapfel, B., Champ, S., Feldthusen, J., Clements, J., & Hutmacher, D., (2013). The potential role of lycopene for the prevention and therapy of prostate cancer: From molecular mechanisms to clinical evidence. *Int. J. Mol. Sci., 14*, 14620–14646. https://doi.org/10.3390/ijms140714620.
114. Mohanty, N. K., Saxena, S., Singh, U. P., Goyal, N. K., & Arora, R. P., (2005). Lycopene as a chemo preventive agent in the treatment of high-grade prostate intraepithelial neoplasia. *Urol. Oncol. Semin. Orig. Investig., 23*, 383–385. https://doi.org/10.1016/j.urolonc.2005.05.012.
115. Zu, K., Mucci, L., Rosner, B. A., Clinton, S. K., Loda, M., Stampfer, M. J., & Giovannucci, E., (2014). Dietary lycopene, angiogenesis, and prostate cancer: A prospective study in the prostate-specific antigen era. *J. Natl. Cancer Inst.*, 106. https://doi.org/10.1093/jnci/djt430.
116. Hwang, E. S., & Bowen, P. E., (2003). Can the consumption of tomatoes or lycopene reduce cancer risk? *Integr. Cancer Ther., 1*, 121–132. https://doi.org/10.1177/1534735402001002003.
117. Boileau, T. W. M., Liao, Z., Kim, S., Lemeshow, S., Erdman, J. W., & Clinton, S. K., (2003). Prostate carcinogenesis in N-methyl-N-nitrosourea (NMU)-testosterone-treated rats fed tomato powder, lycopene, or energy-restricted diets. *J. Natl. Cancer Inst., 95*, 1578–1586. https://doi.org/10.1093/jnci/djg081.
118. Zhang, M., Holman, C. D. A. J., & Binns, C. W., (2007). Intake of specific carotenoids and the risk of epithelial ovarian cancer. *Br. J. Nutr., 98*, 187–193. https://doi.org/10.1017/S0007114507690011.
119. Levy, J., Bosin, E., Feldman, B., Giat, Y., Miinster, A., Danilenko, M., & Sharoni, Y., (1995). Lycopene is a more potent inhibitor of human cancer cell proliferation than either A-carotene or β-carotene. *Nutr. Cancer, 24*, 257–266. https://doi.org/10.1080/01635589509514415.
120. Hercberg, S., (2005). The history of β-carotene and cancers: From observational to intervention studies. What lessons can be drawn for future research on polyphenols? *Am. J. Clin. Nutr., 81*, 218S–222S. https://doi.org/10.1093/ajcn/81.1.218S.
121. Freitas, J. V., Praça, F. S. G., Bentley, M. V. L. B., & Gaspar, L. R., (2015). Trans-resveratrol and beta-carotene from sunscreens penetrate viable skin layers and reduce cutaneous penetration of UV-filters. *Int. J. Pharm., 484*, 131–137. https://doi.org/10.1016/j.ijpharm.2015.02.062.

122. Druesne-Pecollo, N., Latino-Martel, P., Norat, T., Barrandon, E., Bertrais, S., Galan, P., & Hercberg, S., (2010). Beta-carotene supplementation and cancer risk: A systematic review and meta-analysis of randomized controlled trials. *Int. J. Cancer, 127*, 172–184. https://doi.org/10.1002/ijc.25008.
123. Kaulmann, A., & Bohn, T., (2014). Carotenoids, inflammation, and oxidative stress—implications of cellular signaling pathways and relation to chronic disease prevention. *Nutr. Res., 34,* 907–929. https://doi.org/10.1016/j.nutres.2014.07.010.
124. Kasperczyk, S., Dobrakowski, M., Kasperczyk, J., Ostałowska, A., Zalejska-Fiolka, J., & Birkner, E., (2014). Beta-carotene reduces oxidative stress, improves glutathione metabolism, and modifies antioxidant defense systems in lead-exposed workers. *Toxicol. Appl. Pharmacol., 280*, 36–41. https://doi.org/10.1016/j.taap.2014.07.006.
125. Voutilainen, S., Nurmi, T., Mursu, J., & Rissanen, T. H., (2006). Carotenoids and cardiovascular health. *Am. J. Clin. Nutr., 83*, 1265–1271. https://doi.org/10.1093/ajcn/83.6.1265.
126. Middha, P., Weinstein, S. J., Männistö, S., Albanes, D., & Mondul, A. M., (2018). β-carotene supplementation and lung cancer incidence in the alpha-tocopherol, beta-carotene cancer prevention study: The role of tar and nicotine. *Nicotine Tob. Res., 21*, 1045–1050. https://doi.org/10.1093/ntr/nty115.
127. Kim, D., Kim, Y., & Kim, Y., (2019). Effects of β-carotene on expression of selected microRNAs, histone acetylation, and DNA methylation in colon cancer stem cells. *J. Cancer Prev., 24*, 224–232. https://doi.org/10.15430/jcp.2019.24.4.224.
128. Mezzomo, N., & Ferreira, S. R. S., (2016). Carotenoid's functionality, sources, and processing by supercritical technology: A review. *J. Chem., 2016.* https://doi.org/10.1155/2016/3164312.
129. Panche, A. N., Diwan, A. D., & Chandra, S. R., (2016). Flavonoids: An overview. *J. Nutr. Sci., 5*, e47. https://doi.org/10.1017/jns.2016.41.
130. Ponzo, V., Goitre, I., Fadda, M., Gambino, R., De Francesco, A., Soldati, L., Gentile, L., et al., (2015). Dietary flavonoid intake and cardiovascular risk: A population-based cohort study. *J. Transl. Med., 13,* 1–13. https://doi.org/10.1186/s12967-015-0573-2.
131. Lim, H., Heo, M. Y., & Kim, H. P., (2019). Flavonoids: Broad spectrum agents on chronic inflammation. *Biomol. Ther., 27*, 241–253. https://doi.org/10.4062/biomolther.2019.034.
132. Ginwala, R., Bhavsar, R., Chigbu, D. G. I., Jain, P., & Khan, Z. K., (2019). Potential role of flavonoids in treating chronic inflammatory diseases with a special focus on the anti-inflammatory activity of apigenin. *Antioxidants, 8*, 1–28. https://doi.org/10.3390/antiox8020035.
133. Peterson, J. J., Dwyer, J. T., Jacques, P. F., & McCullough, M. L., (2012). Do flavonoids reduce cardiovascular disease incidence or mortality in US and European populations? *Nutr. Rev., 70*, 491–508. https://doi.org/10.1111/j.1753-4887.2012.00508.x.
134. Vaya, J., Mahmood, S., Goldblum, A., Aviram, M., Volkova, N., Shaalan, A., Musa, R., & Tamir, S., (2003). Inhibition of LDL oxidation by flavonoids in relation to their structure and calculated enthalpy. *Phytochemistry, 62*, 89–99. https://doi.org/10.1016/S0031-9422(02)00445-4.
135. Clark, J. L., Zahradka, P., & Taylor, C. G., (2015). Efficacy of flavonoids in the management of high blood pressure. *Nutr. Rev., 73*, 799–822. https://doi.org/10.1093/nutrit/nuv048.

136. Juan, D., PÉrez-VizcaÍno, F., JimÉnez, J., Tamargo, J., & Zarzuelo, A., (2001). Flavonoids and cardiovascular diseases. In: *Studies in Natural Products Chemistry* (pp. 565–605). https://doi.org/10.1016/S1572-5995(01)80018-1.
137. Rodríguez-García, C., Sánchez-Quesada, C., Gaforio, J. J., & Gaforio, J. J., (2019). Dietary flavonoids as cancer chemopreventive agents: An updated review of human studies. *Antioxidants, 8*, 1–23. https://doi.org/10.3390/antiox8050137.
138. Abotaleb, M., Samuel, S. M., Varghese, E., Varghese, S., Kubatka, P., Liskova, A., & Büsselberg, D., (2019). Flavonoids in cancer and apoptosis. *Cancers (Basel), 11*. https://doi.org/10.3390/cancers11010028.
139. Choy, K. W., Murugan, D., Leong, X. F., Abas, R., Alias, A., & Mustafa, M. R., (2019). Flavonoids as natural anti-inflammatory agents targeting nuclear factor-kappa B (NFκB) signaling in cardiovascular diseases: A mini review. *Front. Pharmacol., 10*, 1–8. https://doi.org/10.3389/fphar.2019.01295.
140. Serafini, M., Peluso, I., & Raguzzini, A., (2010). Flavonoids as anti-inflammatory agents. *Proc. Nutr. Soc., 69*, 273–278. https://doi.org/10.1017/S002966511000162X.
141. Li, X., Han, Y., Zhou, Q., Jie, H., He, Y., Han, J., He, J., Jiang, Y., & Sun, E., (2016). Apigenin, a potent suppressor of dendritic cell maturation and migration, protects against collagen-induced arthritis. *J. Cell. Mol. Med., 20*, 170–180. https://doi.org/10.1111/jcmm.12717.
142. Shoara, R., Hashempur, M. H., Ashraf, A., Salehi, A., Dehshahri, S., & Habibagahi, Z., (2015). Efficacy and safety of topical *Matricaria chamomilla* L. (chamomile) oil for knee osteoarthritis: A randomized controlled clinical trial. *Complement. Ther. Clin. Pract., 21*, 181–187. https://doi.org/10.1016/j.ctcp.2015.06.003.
143. Raso, G. M., Meli, R., Di Carlo, G., Pacilio, M., & Di Carlo, R., (2001). Inhibition of inducible nitric oxide synthase and cyclooxygenase-2 expression by flavonoids in macrophage J774A.1. *Life Sci., 68*, 921–931. https://doi.org/10.1016/S0024-3205(00)00999-1.
144. Samieri, C., Sun, Q., Townsend, M. K., Rimm, E. B., & Grodstein, F., (2014). Dietary flavonoid intake at midlife and healthy aging in women. *Am. J. Clin. Nutr., 100*, 1489–1497. https://doi.org/10.3945/ajcn.114.085605.
145. Landberg, R., Sun, Q., Rimm, E. B., Cassidy, A., Scalbert, A., Mantzoros, C. S., Hu, F. B., & Van, D. R. M., (2011). Selected dietary flavonoids are associated with markers of inflammation and endothelial dysfunction in U.S. *Women. J. Nutr., 141*, 618–625. https://doi.org/10.3945/jn.110.133843.
146. AL-Ishaq, Abotaleb, Kubatka, Kajo, & Büsselberg, (2019). Flavonoids and their anti-diabetic effects: Cellular mechanisms and effects to improve blood sugar levels. *Biomolecules, 9*, 430. https://doi.org/10.3390/biom9090430.
147. Cazarolli, L. H., Folador, P., Moresco, H. H., Brighente, I. M. C., Pizzolatti, M. G., & Silva, F. R. M. B., (2009). Mechanism of action of the stimulatory effect of apigenin-6-C-(2″-O-α-l-rhamnopyranosyl)-β-l-fucopyranoside on 14C-glucose uptake. *Chem. Biol. Interact., 179*, 407–412. https://doi.org/10.1016/j.cbi.2008.11.012.
148. Akram, M., Asif, H. M., Uzair, M., Akhtar, N., Madni, A., Ali, S. S. M., Hasan, Z. U., & Ullah, A., (2011). Amino acids: A review article. *J. Med. Plants Res., 5*, 3997–4000.
149. Kim, M. H., & Kim, H., (2017). The roles of glutamine in the intestine and its implication in intestinal diseases. *Int. J. Mol. Sci., 18*. https://doi.org/10.3390/ijms18051051.

150. Van, D. H. R. R. W. J., Von, M. M. F., Deutz, N. E. P., Soeters, P. B., Brummer, R. J. M., Von, K. B. K., & Arends, J. W., (1993). Glutamine and the preservation of gut integrity. *Lancet, 341*, 1363–1365. https://doi.org/10.1016/0140-6736(93)90939-E.
151. Bertiaux-Vandaële, N., Youmba, S. B., Belmonte, L., Lecleire, S., Antonietti, M., Gourcerol, G., Leroi, A. M., et al., (2011). The expression and the cellular distribution of the tight junction proteins are altered in irritable bowel syndrome patients with differences according to the disease subtype. *Am. J. Gastroenterol., 106*, 2165–2173. https://doi.org/10.1038/ajg.2011.257.
152. Abdul, R. R., Raja, A. R. A., & Lee, Y. Y., (2016). Irritable bowel syndrome and inflammatory bowel disease overlap syndrome: Pieces of the puzzle are falling into place. *Intest. Res., 14*, 297. https://doi.org/10.5217/ir.2016.14.4.297.
153. Bertrand, J., Ghouzali, I., Guérin, C., Bôle-Feysot, C., Gouteux, M., Déchelotte, P., Ducrotté, P., & Coëffier, M., (2016). Glutamine restores tight junction protein claudin-1 expression in colonic mucosa of patients with diarrhea-predominant irritable bowel syndrome. *J. Parenter. Enter. Nutr., 40*, 1170–1176. https://doi.org/10.1177/0148607115587330.
154. Zhou, Q. Q., Verne, M. L., Fields, J. Z., Lefante, J. J., Basra, S., Salameh, H., & Verne, G. N., (2018). Randomized placebo-controlled trial of dietary glutamine supplements for postinfectious irritable bowel syndrome. *Gut., 68*, 996–1002. https://doi.org/10.1136/gutjnl-2017-315136.
155. Cruzat, V., Rogero, M. M., Keane, K. N., Curi, R., & Newsholme, P., (2018). Glutamine: Metabolism and immune function, supplementation, and clinical translation. *Nutrients, 10*, 1–31. https://doi.org/10.3390/nu10111564.
156. Fillmann, H., Kretzmann, N. A., San-Miguel, B., Llesuy, S., Marroni, N., González-Gallego, J., & Tuñón, M. J., (2007). Glutamine inhibits over-expression of pro-inflammatory genes and down-regulates the nuclear factor kappa-B pathway in an experimental model of colitis in the rat. *Toxicology, 236*, 217–226. https://doi.org/10.1016/j.tox.2007.04.012.
157. Palmieri, E. M., Menga, A., Lebrun, A., Hooper, D. C., Butterfield, D. A., Mazzone, M., & Castegna, A., (2017a). Blockade of glutamine synthetase enhances inflammatory response in microglial cells. *Antioxid. Redox Signal, 26*, 351–363. https://doi.org/10.1089/ars.2016.6715.
158. Palmieri, E. M., Menga, A., Martín-Pérez, R., Quinto, A., Riera-Domingo, C., De Tullio, G., Hooper, D. C., et al., (2017b). Pharmacologic or genetic targeting of glutamine synthetase skews macrophages toward an M1-like phenotype and inhibits tumor metastasis. *Cell Rep., 20*, 1654–1666. https://doi.org/10.1016/j.celrep.2017.07.054.
159. Ziegler, T. R., (2001). Glutamine supplementation in cancer patients receiving bone marrow transplantation and high dose chemotherapy. *J. Nutr., 131*, 2578S–2584S. https://doi.org/10.1093/jn/131.9.2578s.
160. Raizel, R., Leite, J. S. M., Hypólito, T. M., Coqueiro, A. Y., Newsholme, P., Cruzat, V. F., & Tirapegui, J., (2016). Determination of the anti-inflammatory and cytoprotective effects of l-glutamine and l-alanine, or dipeptide, supplementation in rats submitted to resistance exercise. *Br. J. Nutr., 116*, 470–479. https://doi.org/10.1017/S0007114516001999.
161. Almeida, E. B., Santos, J. M. B., Paixaõ, V., Amaral, J. B., Foster, R., Sperandio, A., Roseira, T., et al., (2020). L-glutamine supplementation improves the benefits of

combined-exercise training on oral redox balance and inflammatory status in elderly individuals. *Oxid. Med. Cell. Longev., 2020*. https://doi.org/10.1155/2020/2852181.

162. Layman, D. K., & Walker, D. A., (2006). Potential importance of leucine in treatment of obesity and the metabolic syndrome. *J. Nutr., 136*, 319S–323S. https://doi.org/10.1093/jn/136.1.319S.
163. Dreyer, H. C., Drummond, M. J., Pennings, B., Fujita, S., Glynn, E. L., Chinkes, D. L., Dhanani, S., et al., (2008). Leucine-enriched essential amino acid and carbohydrate ingestion following resistance exercise enhances mTOR signaling and protein synthesis in human muscle. *Am. J. Physiol. Endocrinol. Metab., 294*, 392–400. https://doi.org/10.1152/ajpendo.00582.2007.
164. Theis, N., Brown, M. A., Wood, P., & Waldron, M., (2020). Leucine supplementation increases muscle strength and volume, reduces inflammation, and affects wellbeing in adults and adolescents with cerebral palsy. *J. Nutr.,* 6–11. https://doi.org/10.1093/jn/nxaa006.
165. Hamarsland, H., Nordengen, A. L., Nyvik, A. S., Holte, K., Garthe, I., Paulsen, G., Cotter, M., et al., (2017). Native whey protein with high levels of leucine results in similar post-exercise muscular anabolic responses as regular whey protein: A randomized controlled trial. *J. Int. Soc. Sports Nutr., 14*, 43. https://doi.org/10.1186/s12970-017-0202-y.
166. Zhang, S., Zeng, X., Ren, M., Mao, X., & Qiao, S., (2017). Novel metabolic and physiological functions of branched chain amino acids: A review. *J. Anim. Sci. Biotechnol., 8*, 4–15. https://doi.org/10.1186/s40104-016-0139-z.
167. Guo, K., Yu, Y. H., Hou, J., & Zhang, Y., (2010). Chronic leucine supplementation improves glycemic control in etiologically distinct mouse models of obesity and diabetes mellitus. *Nutr. Metab. (Lond), 7*, 57. https://doi.org/10.1186/1743-7075-7-57.
168. Verhoeven, S., Vanschoonbeek, K., Verdijk, L. B., Koopman, R., Wodzig, W. K. W. H., Dendale, P., & Van, L. L. J., (2009). Long-term leucine supplementation does not increase muscle mass or strength in healthy elderly men. *Am. J. Clin. Nutr., 89*, 1468–1475. https://doi.org/10.3945/ajcn.2008.26668.
169. Leenders, M., & Van, L. L. J., (2011). Leucine as a pharmaconutrient to prevent and treat sarcopenia and type 2 diabetes. *Nutr. Rev., 69*, 675–689. https://doi.org/10.1111/j.1753-4887.2011.00443.x.
170. Swatson, W. S., Katoh-Kurasawa, M., Shaulsky, G., & Alexander, S., (2017). Curcumin affects gene expression and reactive oxygen species via a PKA dependent mechanism in dictyostelium discoideum. *PLoS One, 12*, e0187562. https://doi.org/10.1371/journal.pone.0187562.
171. Mohiti-Ardekani, J., Asadi, S., Ardakani, A. M., Rahimifard, M., Baeeri, M., & Momtaz, S., (2019). Curcumin increases insulin sensitivity in C2C12 muscle cells via AKT and AMPK signaling pathways. *Cogent Food Agric., 5*. https://doi.org/10.1080/23311932.2019.1577532.
172. Dall'Acqua, S., Stocchero, M., Boschiero, I., Schiavon, M., Golob, S., Uddin, J., Voinovich, D., et al., (2016). New findings on the *in vivo* antioxidant activity of Curcuma longa extract by an integrated 1H NMR and HPLC–MS metabolomic approach. *Fitoterapia, 109*, 125–131. https://doi.org/10.1016/j.fitote.2015.12.013.
173. Phillips, J. M., Clark, C., Herman-Ferdinandez, L., Moore-Medlin, T., Rong, X., Gill, J. R., Clifford, J. L., et al., (2011). Curcumin inhibits skin squamous cell carcinoma tumor growth *in vivo*. *Otolaryngol. Neck Surg., 145*, 58–63. https://doi.org/10.1177/0194599811400711.

174. Huang, Y., Hao, J., Tian, D., Wen, Y., Zhao, P., Chen, H., Lv, Y., & Yang, X., (2018). Antidiabetic activity of a flavonoid-rich extract from *Sophora davidii* (Franch.) skeels in KK-Ay mice via activation of AMP-activated protein kinase. *Front. Pharmacol., 9,* 1–15. https://doi.org/10.3389/fphar.2018.00760.
175. Wu, P., Ma, G., Li, N., Deng, Q., Yin, Y., & Huang, R., (2015). Investigation of *in vitro* and *in vivo* antioxidant activities of flavonoids rich extract from the berries of *Rhodomyrtus tomentosa* (Ait.) Hassk. *Food Chem., 173,* 194–202. https://doi.org/10.1016/j.foodchem.2014.10.023.
176. Yagasaki, K., (2014). Anti-diabetic phytochemicals that promote GLUT4 translocation via AMPK signaling in muscle cells. *Nutr. Aging, 2,* 35–44. https://doi.org/10.3233/NUA-130032.
177. Wu, G., (2016). Dietary protein intake and human health. *Food Funct., 7,* 1251–1265. https://doi.org/10.1039/c5fo01530h.
178. Taylor, A. K., Cao, W., Vora, K. P., De La Cruz, J., Shieh, W. J., Zaki, S. R., Katz, J. M., et al., (2013). Protein energy malnutrition decreases immunity and increases susceptibility to influenza infection in mice. *J. Infect. Dis., 207,* 501–510. https://doi.org/10.1093/infdis/jis527.
179. Castaneda, C., Charnley, J. M., Evans, W. J., & Crim, M. C., (1995). Elderly women accommodate to a low-protein diet with losses of body cell mass, muscle function, and immune response. *Am. J. Clin. Nutr., 62,* 30–39. https://doi.org/10.1093/ajcn/62.1.30.
180. Egan, B., (2016). Protein intake for athletes and active adults: Current concepts and controversies. *Nutr. Bull., 41,* 202–213. https://doi.org/10.1111/nbu.12215.
181. Marcone, S., Belton, O., & Fitzgerald, D. J., (2017). Milk-derived bioactive peptides and their health promoting effects: A potential role in atherosclerosis. *Br. J. Clin. Pharmacol., 83,* 152–162. https://doi.org/10.1111/bcp.13002.
182. Li-chan, E. C. Y., (2015). Bioactive peptides and protein hydrolysates: Research trends and challenges for application as nutraceuticals and functional food ingredients. *Curr. Opin. Food Sci., 1,* 28–37. https://doi.org/10.1016/j.cofs.2014.09.005.
183. Dalziel, J., Young, W., McKenzie, C., Haggarty, N., & Roy, N., (2017). Gastric emptying and gastrointestinal transit compared among native and hydrolyzed whey and casein milk proteins in an aged rat model. *Nutrients, 9,* 1351. https://doi.org/10.3390/nu9121351.
184. Wilborn, C. D., Taylor, L. W., Outlaw, J., Williams, L., Campbell, B., Foster, C. A., Smith-Ryan, A., et al., (2013). The effects of pre- and post-exercise whey vs. Casein protein consumption on body composition and performance measures in collegiate female athletes. *J. Sport. Sci. Med., 12,* 74–79.
185. De Wit, J. N., (1998). Nutritional and functional characteristics of whey proteins in food products. *J. Dairy Sci., 81,* 597–608. https://doi.org/10.3168/jds.S0022-0302(98)75613-9.
186. Edwards, P. J., & Jameson, G. B., (2014). Structure and stability of whey proteins. In: *Milk Proteins from Expression to Food* (pp. 201–242). Elsevier Inc. https://doi.org/10.1177/0738894281006001O4.
187. Qi, P. X., & Onwulata, C. I., (2011). Physical properties, molecular structures, and protein quality of texturized whey protein isolate: Effect of extrusion moisture content. *J. Dairy Sci., 94,* 2231–2244. https://doi.org/10.3168/jds.2010-3942.

188. Dullius, A., Inês, M., Fernanda, C., & Souza, V. D., (2018). Whey protein hydrolysates as a source of bioactive peptides for functional foods-biotechnological facilitation of industrial scale-up. *J. Funct. Foods, 42,* 58–74. https://doi.org/10.1016/j.jff.2017.12.063.
189. Ballard, K. D., Bruno, R. S., Seip, R. L., Quann, E. E., Volk, B. M., Freidenreich, D. J., Kawiecki, D. M., et al., (2009). Acute ingestion of a novel whey-derived peptide improves vascular endothelial responses in healthy individuals: A randomized, placebo-controlled trial. *Nutr. J., 8,* 1–11. https://doi.org/10.1186/1475-2891-8-34.
190. Ballard, K. D., Kupchak, B. R., Volk, B. M., Mah, E., Shkreta, A., Liptak, C., Ptolemy, A. S., et al., (2013). Acute effects of ingestion of a novel whey-derived extract on vascular endothelial function in overweight, middle-aged men and women. *Br. J. Nutr., 109,* 882–893. https://doi.org/10.1017/S0007114512002061.
191. Silva, M. R., Silvestre, M. P. C., Silva, V. D. M., Souza, M. W. S., Lopes, J. C. O., Afonso, W. O., Lana, F. C., & Rodrigues, D. F., (2014). Production of ace-inhibitory whey protein concentrate hydrolysates: Use of pancreatin and papain. *Int. J. Food Prop., 17,* 1002–1012. https://doi.org/10.1080/10942912.2012.685821.
192. Mullally, M. M., Meisel, H., & Fitzgerald, R. J., (1997). Identification of a novel angiotensin-I-converting enzyme inhibitory peptides corresponding to a tryptic fragment of bovine β-lactoglobulin. *FEBS Lett., 402,* 99–101. https://doi.org/10.1016/S0014-5793(96)01503-7.
193. Pihlanto-Leppälä, A., Koskinen, P., Phlola, K., Tupasela, T., & Korhonen, H., (2000). Angiotensin I-converting enzyme inhibitory properties of whey protein digests: Concentration and characterization of active peptides. *J. Dairy Res., 67,* 53–64. https://doi.org/10.1017/S0022029999003982.
194. Iwaniak, A., Minkiewicz, P., & Darewicz, M., (2014). Food-originating ACE inhibitors, including antihypertensive peptides, as preventive food components in blood pressure reduction. *Compr. Rev. Food Sci. Food Saf., 13,* 114–134. https://doi.org/10.1111/1541-4337.12051.
195. Sousa, G. T. D., Lira, F. S., Rosa, J. C., Oliveira, E. P. D., Oyama, L. M., Santos, R. V., & Pimentel, G. D., (2012). *Dietary Whey Protein Lessens Several Risk Factors for Metabolic Diseases: A Review.?, 11,* 1. https://doi.org/10.1186/1476-511X-11-67.
196. Morato, P. N., Lollo, P. C. B., Moura, C. S., Batista, T. M., Camargo, R. L., Carneiro, E. M., & Amaya-Farfan, J., (2013). Whey protein hydrolysate increases translocation of GLUT-4 to the plasma membrane independent of insulin in Wistar rats. *PLoS One, 8.* https://doi.org/10.1371/journal.pone.0071134.
197. Mollahosseini, M., Shab-Bidar, S., Rahimi, M. H., & Djafarian, K., (2017). Effect of whey protein supplementation on long- and short-term appetite: A meta-analysis of randomized controlled trials. *Clin. Nutr. ESPEN, 20,* 34–40. https://doi.org/10.1016/j.clnesp.2017.04.002.
198. Reyes-Diaz, A., F Gonzalez-Cordova, A., Hernandez-Mendoza, A., Reyes-Diaz, R., & Vallejo-Cordoba, B., (2018). Immunomodulation by hydrolysates and peptides derived from milk proteins. *Int. J. Dairy Technol., 71,* 1–9. https://doi.org/10.1111/1471-0307.12421.
199. Mukhopadhya, A., Noronha, N., Bahar, B., Ryan, M. T., Murray, B. A., Kelly, P. M., Loughlin, I. B. O., et al., (2014). Anti-inflammatory effects of a casein hydrolysate and its peptide-enriched fractions on TNF-α-challenged Caco-2 cells and LPS-challenged porcine colonic explants. *Food Sci. Nutr., 2,* 712–723. https://doi.org/10.1002/fsn3.153.

200. Mukhopadhya, A., Noronha, N., Bahar, B., Ryan, M. T., Murray, B. A., Kelly, P. M., O'Loughlin, I. B., et al., (2015). The anti-inflammatory potential of a moderately hydrolyzed casein and its 5 kDa fraction in *in vitro* and ex vivo models of the gastrointestinal tract. *Food Funct., 6*, 612–621. https://doi.org/10.1039/c4fo00689e.
201. Martínez-Augustin, O., Rivero-Gutiérrez, B., Mascaraque, C., & Sánchez, D. M. F., (2014). Food derived bioactive peptides and intestinal barrier function. *Int. J. Mol. Sci., 15*, 22857–22873. https://doi.org/10.3390/ijms151222857.
202. Cogan, K. E., Evans, M., Iuliano, E., Melvin, A., Susta, D., Neff, K., De Vito, G., & Egan, B., (2018). Co-ingestion of protein or a protein hydrolysate with carbohydrate enhances anabolic signaling, but not glycogen resynthesis, following recovery from prolonged aerobic exercise in trained cyclists. *Eur. J. Appl. Physiol., 118*, 349–359. https://doi.org/10.1007/s00421-017-3775-x.
203. Musa, J., Orth, M. F., Dallmayer, M., Baldauf, M., Pardo, C., Rotblat, B., Kirchner, T., et al., (2016). Eukaryotic initiation factor 4E-binding protein 1 (4E-BP1): A master regulator of mRNA translation involved in tumorigenesis. *Oncogene 35*, 4675–4688. https://doi.org/10.1038/onc.2015.515.
204. Hulmi, J. J., Tannerstedt, J., Selänne, H., Kainulainen, H., Kovanen, V., & Mero, A. A., (2009). Resistance exercise with whey protein ingestion affects mTOR signaling pathway and myostatin in men. *J. Appl. Physiol., 106*, 1720–1729. https://doi.org/10.1152/japplphysiol.00087.2009.
205. Song, J. J., Wang, Q., Du, M., Li, T. G., Chen, B., & Mao, X. Y., (2017). Casein glycomacropeptide-derived peptide IPPKKNQDKTE ameliorates high glucose-induced insulin resistance in HepG2 cells via activation of AMPK signaling. *Mol. Nutr. Food Res., 61*, 1–12. https://doi.org/10.1002/mnfr.201600301.
206. Matsunaga, Y., Tamura, Y., Sakata, Y., Nonaka, Y., Saito, N., Nakamura, H., Shimizu, T., et al., (2017). Comparison between pre-exercise casein peptide and intact casein supplementation on glucose tolerance in mice fed a high-fat diet. *Appl. Physiol. Nutr. Metab., 43*, 355–362. https://doi.org/10.1139/apnm-2017-0485.
207. Berrazaga, I., Micard, V., Gueugneau, M., & Walrand, S., (2019). The role of the anabolic properties of plant-versus animal-based protein sources in supporting muscle mass maintenance: A critical review. *Nutrients, 11*. https://doi.org/10.3390/nu11081825.
208. Bouchenak, M., & Lamri-Senhadji, M., (2013). Nutritional quality of legumes, and their role in cardiometabolic risk prevention: A review. *J. Med. Food, 16*, 185–198. https://doi.org/10.1089/jmf.2011.0238.
209. Gorissen, S. H. M., Crombag, J. J. R., Senden, J. M. G., Waterval, W. A. H., Bierau, J., Verdijk, L. B., & Van, L. L. J. C., (2018). Protein content and amino acid composition of commercially available plant-based protein isolates. *Amino Acids, 50*, 1685–1695. https://doi.org/10.1007/s00726-018-2640-5.
210. Boschin, G., Scigliuolo, G. M., Resta, D., & Arnoldi, A., (2014). ACE-inhibitory activity of enzymatic protein hydrolysates from lupin and other legumes. *Food Chem., 145*, 34–40. https://doi.org/10.1016/j.foodchem.2013.07.076.
211. Ndiaye, F., Vuong, T., Duarte, J., Aluko, R. E., & Matar, C., (2012). Anti-oxidant, anti-inflammatory and immunomodulating properties of an enzymatic protein hydrolysate from yellow field pea seeds. *Eur. J. Nutr., 51*, 29–37. https://doi.org/10.1007/s00394-011-0186-3.
212. Oikawa, S. Y., Bahniwal, R., Holloway, T. M., Lim, C., Mcleod, J. C., Mcglory, C., Baker, S. K., & Phillips, S. M., (2020). *Potato Protein Isolate Stimulates Muscle*

Protein Synthesis at Rest and with Resistance Exercise in Young Women, 1. https://doi.org/10.3390/nu12051235.
213. Rein, D., Ternes, P., Demin, R., Gierke, J., Helgason, T., & Schön, C., (2019). Artificial intelligence identified peptides modulate inflammation in healthy adults. *Food Funct., 10*, 6030–6041. https://doi.org/10.1039/c9fo01398a.
214. Salehi, B., Mishra, A., Nigam, M., Sener, B., Kilic, M., Sharifi-Rad, M., Fokou, P., et al., (2018). Resveratrol: A double-edged sword in health benefits. *Biomedicines, 6*, 91. https://doi.org/10.3390/biomedicines6030091.
215. Sinha, D., Sarkar, N., Biswas, J., & Bishayee, A., (2016). Resveratrol for breast cancer prevention and therapy: Preclinical evidence and molecular mechanisms. *Semin. Cancer Biol., 40, 41*, 209–232. https://doi.org/10.1016/j.semcancer.2015.11.001.
216. Pezzuto, J. M., (2008). Resveratrol as an inhibitor of carcinogenesis. *Pharm. Biol., 46*, 443–573. https://doi.org/10.1080/13880200802116610.
217. Van, G. P. R., Sareen, D., Subramanian, L., Walker, Q., Darjatmoko, S. R., Lindstrom, M. J., Kulkarni, A., et al., (2007). Resveratrol inhibits tumor growth of human neuroblastoma and mediates apoptosis by directly targeting mitochondria. *Clin. Cancer Res., 13*, 5162–5169. https://doi.org/10.1158/1078-0432.CCR-07-0347.
218. Levi, F., Pasche, C., Lucchini, F., Ghidoni, R., Ferraroni, M., & La Vecchia, C., (2005). Resveratrol and breast cancer risk. *Eur. J. Cancer Prev., 14*, 139–142. https://doi.org/10.1097/00008469-200504000-00009.
219. Pasinetti, G. M., Wang, J., Ho, L., Zhao, W., & Dubner, L., (2015). Roles of resveratrol and other grape-derived polyphenols in Alzheimer's disease prevention and treatment. *Biochim. Biophys. Acta-Mol. Basis Dis., 1852*, 1202–1208. https://doi.org/10.1016/j.bbadis.2014.10.006.
220. Zhu, C. W., Grossman, H., Neugroschl, J., Parker, S., Burden, A., Luo, X., & Sano, M., (2018). A randomized, double-blind, placebo-controlled trial of resveratrol with glucose and malate (RGM) to slow the progression of Alzheimer's disease: A pilot study. Alzheimer's dement. *Transl. Res. Clin. Interv., 4*, 609–616. https://doi.org/10.1016/j.trci.2018.09.009.
221. Lai, A., Ho, L., Evashwick-Rogler, T. W., Watanabe, H., Salandra, J., Winkelstein, B. A., Laudier, D., et al., (2019). Erratum: Dietary polyphenols as a safe and novel intervention for modulating pain associated with intervertebral disc degeneration in an *in-vivo* rat model (*Plos One, 14*, 10 (E0223435)). doi: 10.1371/journal.pone.0223435). *PLoS One 14*, 1–24. https://doi.org/10.1371/journal.pone.0225674.
222. Ibrahim, F. G., & Zaki, R. M., (2019). Possible neuromodulating role of different grape (*Vitis vinifera* L.) derived polyphenols against Alzheimer's dementia: Treatment and mechanisms. *Bull. Natl. Res. Cent., 43*, 108. https://doi.org/10.1186/s42269-019-0149-z.
223. Wang, J., Bi, W., Cheng, A., Freire, D., Vempati, P., Zhao, W., Gong, B., Janle, E. M., et al., (2014). Targeting multiple pathogenic mechanisms with polyphenols for the treatment of Alzheimer's disease-experimental approach and therapeutic implications. *Front. Aging Neurosci., 6*, 1–10. https://doi.org/10.3389/fnagi.2014.00042.
224. Wang, J., Tang, C., Ferruzzi, M. G., Gong, B., Song, B. J., Janle, E. M., Chen, T. Y., et al., (2013). Role of standardized grape polyphenol preparation as a novel treatment to improve synaptic plasticity through attenuation of features of metabolic syndrome in a mouse model. *Mol. Nutr. Food Res., 57*, 2091–2102. https://doi.org/10.1002/mnfr.201300230.

225. Ablon, G., & Kogan, S., (2018). A six-month, randomized, double-blind, placebo-controlled study evaluating the safety and efficacy of a nutraceutical supplement for promoting hair growth in women with self-perceived thinning hair. *J. Drugs Dermatology, 17*, 558–565.
226. Hunt, N., & McHale, S., (2007). The psychological impact of alopecia. *Psychologist, 20*, 362–364.
227. Blum, K., Febo, M., & Badgaiyan, R. D., (2016). Fifty years in the development of a glutaminergic-dopaminergic optimization complex (KB220) to balance brain reward circuitry in reward deficiency syndrome: A pictorial. *Austin Addict. Sci., 1*, 1–34.
228. Blum, K., Febo, M., Fried, L., Baron, D., Braverman, E. R., Dushaj, K., Li, M., et al., (2017). Pro-dopamine regulator-(KB220) to balance brain reward circuitry in reward deficiency syndrome (RDS). *J. Reward Defic. Syndr. Addict. Sci., 03*. https://doi.org/10.17756/jrdsas.2017-034.
229. Proestos, C., (2018). Superfoods: Recent data on their role in the prevention of diseases. *Curr. Res. Nutr. Food Sci. J., 6*, 576–593. https://doi.org/10.12944/CRNFSJ.6.3.02.
230. Marcinowska-Suchowierska, E., Kupisz-Urbańska, M., Łukaszkiewicz, J., Płudowski, P., & Jones, G., (2018). Vitamin D toxicity-a clinical perspective. *Front. Endocrinol. (Lausanne). 9*, 1–7. https://doi.org/10.3389/fendo.2018.00550.
231. McFarland, L. V., & Goh, S., (2019). Are probiotics and prebiotics effective in the prevention of travelers' diarrhea: A systematic review and meta-analysis. *Travel Med. Infect. Dis., 27*, 11–19. https://doi.org/10.1016/j.tmaid.2018.09.007.
232. O' Donovan, C. M., Connor, B., Madigan, S. M., Cotter, P. D., & O' Sullivan, O., (2020). Instances of altered gut microbiomes among Irish cricketers over periods of travel in the lead up to the 2016 world cup: A sequencing analysis. *Travel Med. Infect. Dis., 35*, 101553. https://doi.org/10.1016/j.tmaid.2020.101553.
233. Pyne, D. B., West, N. P., Cox, A. J., & Cripps, A. W., (2015). Probiotics supplementation for athletes-Clinical and physiological effects. *Eur. J. Sport Sci., 15*, 63–72. https://doi.org/10.1080/17461391.2014.971879.
234. Amarya, S., Singh, K., & Sabharwal, M., (2015). Changes during aging and their association with malnutrition. *J. Clin. Gerontol. Geriatr., 6*, 78–84. https://doi.org/10.1016/j.jcgg.2015.05.003.
235. Kalinkovich, A., & Livshits, G., (2017). Sarcopenic obesity or obese sarcopenia: A cross talk between age-associated adipose tissue and skeletal muscle inflammation as a main mechanism of the pathogenesis. *Ageing Res. Rev., 35*, 200–221. https://doi.org/10.1016/j.arr.2016.09.008.
236. Noy, R., & Pollard, J. W., (2015). Tumor-associated macrophages: From mechanisms to therapy. *Immunity, 41*, 49–61. https://www.ncbi.nlm.nih.gov/pmc/articles/PMC4137410/ (accessed 10 August 2021).
237. Lalor, R., (2019). *Immunomodulatory Properties of Bovine Caseins on Innate Immune Cells*. Dublin City University.

CHAPTER 7

Bioactive Proteins and Peptides as Functional Foods

DEEPA THOMAS[1] and M. S. LATHA[2,3]

[1]*Research and Post Graduate Department of Chemistry, Bishop Moore College, Mavelikara, Alappuzha, Kerala, India*

[2]*Department of Chemistry, Sree Narayana College, Chathannur, Kollam, Kerala, India*

[3]*Department of Chemistry, Sree Narayana College, Kollam, Kerala, India*

ABSTRACT

The quality of food plays an important role in the prevention of disease. Functional food describes "food that can provide health benefits beyond basic nutrition." Bioactive proteins and peptides from an important fortifying ingredient for functional food. Apart from providing nutritional benefits, they resist invasion of disease, inhibit pathophysiological pathways, and suppress pathogenic molecular activity. They also exhibit antioxidant, antihypertensive, antimicrobial, antimutagenic, anti-inflammatory, immune, and cytomodulatory and mineral-binding property. Dairy products, eggs, fish, wheat, maize, soy, rice, and mushrooms are the major sources of these functional ingredients. Fortification with bioactive proteins and peptides enhances the desirable physiological and immunological effects of the food system. Development of fully integrated bioprocesses for the large-scale manufacture and refinement of these significant biomolecules would accelerate their introduction to the main consumer markets. Therefore, effective measures are required to implement economically viable methods for the production of these functional foods on an industrial scale.

7.1 INTRODUCTION

Functional food is a concept defined as "foods that may provide health benefits beyond basic nutrition." A functional food may be derived from nature, a food from which an element has been eliminated or added, or a food in which the formulation of one or more constituents has been altered, or a food wherein the bioavailability of one or more elements or some combination thereof has been altered. 'Functional food science' helps to foster the development of functional food. The essential strategies for the production of functional foods are fortification, enrichment, modification, and enhancement through new feed composition, unique growing conditions, or genetic manipulation [1, 2]. Bioactive proteins and peptides form an essential ingredient of functional foodstuffs. They not only provide nutritional benefits but also help to resist the development of disease, inhibit pathophysiological pathways, or suppress pathogenic molecular activity. They exhibit specific bioactivities such as improving nutrient intake, growth enhancement, enzyme inhibition, protection against pathogenic agents and immune system modulation. They are capable of exhibiting local effects in the gastrointestinal (GI) system or having systemic effects following intestinal absorption and circulation. They display numerous functions, including antioxidant, antihypertensive, antimicrobial, antimutagenic, antioxidative, anti-inflammatory, immune, and cytomodulatory and mineral binding activity. They are able to produce beneficial effects on major body systems such as digestive, immune, cardiovascular, nervous, and endocrine systems. Biologically active peptides may be generated from precursor proteins in the following ways: (a) digestive enzymatic hydrolysis, (b) fermentation with proteolytic starter cultures, (c) proteolysis by microorganism or plant-derived enzymes. An arrangement of the above-mentioned methods has proven successful in many studies in generating short functional peptides [3].

7.2 INGREDIENTS IN FUNCTIONAL FOOD

There are many methods for producing functional foods, such as food processing modification, genetic engineering, etc., which enables the food industry to produce new products with added market value. Probiotics, prebiotics, biogenic, and nutrients are the most important components which can be added to food.

7.2.1 PROBIOTICS

They are living microorganisms such as lactobacilli and bifidobacteria that have several immune-enhancing effects on host health and these microbial supplements assistant naturally with the intestinal mucosa, build up the intestinal microbial balance. Probiotic bacteria have been gradually introduced into a variety of items, including milk powders, yogurts, cheeses, ice cream, frozen dairy desserts, fermented vegetables, and meats because of their perceived health benefits. The survival and multiplication of probiotic microorganisms in the host significantly influence their advantages. In addition to the immunological benefits and the prevention, defense, and elimination of pathogenic bacteria, probiotic bacteria are also associated with cancer treatment and are found to be beneficial in patients with elevated cholesterol rates [4].

7.2.2 PREBIOTICS

They are nondigestible food elements, such as dietary fibers and oligosaccharides that are capable of stimulating the growth of beneficial intestinal bacteria in the colon, by offering growth enhancers and nutrients to probiotic bacteria. They are short-chain carbohydrates that may be fermented in the broad bowel and promote the production of significantly valuable probiotics [5].

7.2.3 NUTRIENTS

They include fatty acids, minerals, and vitamins that are specific and targeted action. A lot of these nutrients are used in foodstuffs such as cereals, beverages, dairy products, bakery items, etc., due to their significant role in preventing disease and wellness promotion [5].

7.2.4 BIOGENICS

These are biologically active molecules includes peptides, proteins, enzymes, carotenoids, phenolic acids, and flavonoids that benefit the host through direct immunostimulation, suppression of mutagenesis, peroxidation, tumorigenesis, or intestinal putrefaction. Such functional components operate biologically within the GI tract and are capable of modifying the gut microbiota, influencing endotoxin translocation and eventual immune activation, and promoting host nutrition [6, 7]. They are also related to the neutralization of reactive species that target cell molecules and in the prevention of several oxidative and nitrosative stress-related diseases such as cardiovascular diseases (CVD), cancer, hypertension, atherosclerosis, diabetes mellitus, and neurological disorders. To improve functional activities, they are included in various foods such as pasta, fish products, ice cream, yogurt, and cheese. Among biogenic, proteins, and peptides are important class of ingredients for functional foods. They activate physiological intrinsic behaviors which make them useful as therapeutic agents. Such bio functionalities can be exploited in therapeutic products and for immune-nutrition as well. They are obtainable from the most varied natural sources.

7.3 BIOLOGICALLY ACTIVE PEPTIDES AND PROTEINS: FUNCTIONS

Bioactive proteins and peptides have the potential to arrest certain diseases. It is found that certain proteins and peptides extracted from foods such as milk, egg, meat, pulses, algae, and fungi help to delay the development of disease and inhibit the process of pathophysiology. They possess antidisease characteristics such as: antihypertensive activity; antimicrobial activity; anticancerous and antitumorigenic activity; antiobesity activity; anti-inflammatory activity; mineral binding activity; immune- and cytomodulatory activity; and antioxidant activity which are depicted in Figure 7.1.

1. **Antihypertensive Activity:** Hypertension is a medical disorder that refers to chronic high blood pressure in the arteries and initiates substantial risk aspects for heart disease and stroke. Angiotensin I-enzyme (ACE), a dipeptidyl carboxypeptidase plays a pivotal role in the control of blood pressure. The switching of angiotensin-I, a decapeptide to angiotensin-II, an octapeptide is catalyzed by ACE. Inactivating ACE is known to be the first step of therapy to treat hypertension. Therefore, ACE-inhibitory components are benefited to reduce the blood pressure in hypertensive patients [7, 8]. In addition to promote improvements in the general lifestyle, attempts have been engendered to develop functional foods that possess elements that help to reduce blood pressure and maybe a supplement or alternative to the prescribed hypertension treatment. Bioactive proteins and peptides with ACE inhibitory and antihypertensive activity have been the subject of special attention. They are isolated from plants, animals, aquatic, or microbial sources [9].

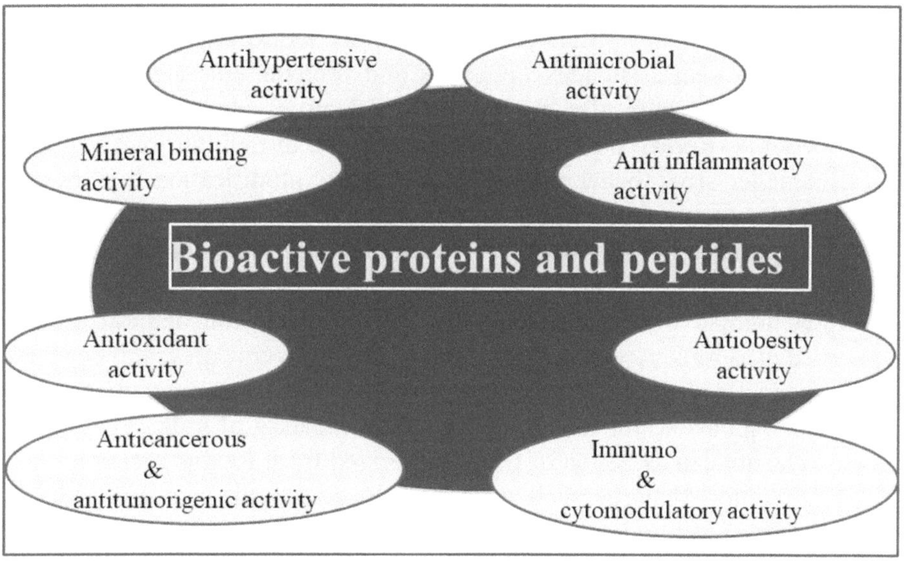

FIGURE 7.1 Functions of bioactive proteins and peptides.

2. **Antimicrobial Activity:** Antimicrobial peptides and proteins are an integral part of living organisms in providing natural protection against foreign pathogenic substances. The great advantage of antimicrobial peptides produced from food proteins is that they are derived from harmless substances, so their safety can be expected for use in medicines and the food industry. They show a wide range of activity against a vast array of microorganisms, including Gram-positive and Gram-negative bacteria, yeast, protozoa, and fungi. They protect the GI tract from invasive viruses and bacteria. They function meanderingly by promoting the development of advantageous microorganisms in the gut or explicitly by performing an antimicrobial operation or neutralizing the attachment or penetration pathways of pathogenic substances. The antimicrobial peptides work by disintegrating the microorganism's cell membrane. Many antimicrobial peptides are cationic and have an alpha-helical structure. The cationic properties of peptides allow attachment with the anionic cell membrane, which is in the lipid-rich nature and the initiation of cell membrane lysis through three possible mechanisms includes the formation of toroidal pores, barrel stave formation and carpet formation. Numerous species of antimicrobial proteins and peptides have been reported from insects to plants and animals [10].

3. **Anticancerous and Antitumorigenic Activity:** Cancer is a chronic disease which appears to be one of the world's leading causes of human death. Bioactive peptides and proteins extracted from food have been clinically proven to be effective substitutes for cancer control. Research results showed that some of their attributes such as smaller sizes, better synthesis, specificity, modification facility, and better penetration into cell membranes make them suitable candidates for cancer management. Pharmaceutical research and development related to peptide anticancer is likely to attract considerable attention. A number of anticancer protein and peptides from natural sources such as milk, rice, corn, mushrooms, soybean, chickpea, and egg have been reported [11, 12]. The phosphate and calcium content of casein is responsible for the anticarcinogenic capacity of milk [11, 13–15]. The anticancer action of peptides derived from food is rooted in their structural features such as composition, length, sequence, overall charge, and hydrophobicity of amino acids. The positive charge and hydrophobicity are responsible for the amphipathic attachment of target cells to the membrane. Phosphatidylserine is possessed by the outer layer of cancer cells. It is a phospholipid possesses a negative charge and is responsible for electrostatic attraction between cancer cells and peptides. Improving the hydrophobicity of peptides has been demonstrated to be good in making them extra stable in the serum and boosting their anticancer efficacy. Hydrophobic amino acids such as proline, glycine, alanine, leucine, and one or more residues of arginine, serine, lysine, glutamic acid, threonine, and tyrosine are the prevailing amino acids of food protein-derived anticancer peptides [13, 16, 17].
4. **Antiobesity Activity:** Research has shown that eating foods with functional benefits as part of a balanced diet on a daily basis can help to minimize the risk or control a variety of health problems. Over the years, it has been observed that diet and physical activity remain the best and most efficient options for maintain body weight in overweight and obese individuals. Dietary maintenance can be accomplished by recognizing bioactive functional food ingredients that could be useful in modulating molecular pathways and functions of genes and proteins along with calorie restriction and exercise [18]. Moderate protein consumption displays a key role in reducing and maintaining the body weight. Studies have shown that functional peptides derived from food proteins play a significant role in the loss of body weight and regulation of lipid metabolism.

5. **Anti-Inflammatory Activity:** The therapeutic uses of natural compounds and their derivatives are becoming increasingly important as healthier alternatives to the currently available anti-inflammatory drugs. Because of their food sources and the perceived lack of severe side effects, bioactive peptides and proteins can potentially provide a safer alternative to conventional pharmaceuticals for inhibition and treatment of inflammation. Different studies have revealed that inflammatory markers include IL -1, IL -6, IL -8, TNF -α and CRP and various transcription elements such as STAT and NF -π B are critical components regulating inflammatory diseases. Studies show that bioactive peptides derived from food proteins show significant anti-inflammatory activity by suppressing or decreasing the expression of the inflammatory biomarkers and/or by reducing the function of those transcription factors [19].
6. **Mineral Binding Activity:** Some minerals, like Zn, Fe, and Ca are vital to life, and their deficiencies may result in a variety of health problems. Since bioactive peptides have mineral binding potential, they may be used as a functional ingredient to improve mineral availability and open up new food supplementation opportunities. Metals exist in a soluble form during complexation between metals and peptides that is readily accessible to the organism. The mineral-binding activity is correlated with the peptide molecular weight and amino acids in the peptide sequence. The peptide-metal bond is formed by the interaction between the electron donator group of the ligand surface (peptide) and electron receptor (metallic ion) with one or more available coordination sites. The complex making specificity is regulated by the spatial arrangement of the ligands and hence the sequence of peptide. The interactions can be altered by varying the amino acid residues sequences. Peptide surface charge also plays a crucial role for determining the metal complex stability [20].
7. **Immuno- and Cytomodulatory Activity:** The immune system performs an essential role in protecting the body from invading pathogens [21–24]. Immunomodulator is any material that can control or adjust the functions of the immune response or of the immune system. There is various recombinant, artificial, and natural preparations as immunomodulators. Dietary strategies involving the intake of essential nutrients and promising functional foods are an efficient and successful technique for the modulation of the immune

system. They provide a more convenient and cost-effective source of specific antibodies. Studies show that both immunosuppression and immunostimulation can indeed be essential in the prevention and control of various pathological states of the organism and that the biopeptides can act against inflammation and autoimmune diseases, prevent transplant refusal, and improve overall health. Cytomodulatory peptides obstruct the growth of cancer cells or promote the development of neonatal intestinal cells and immune cells.

8. **Antioxidant Activity:** Antioxidants can prevent or impede oxidation by either inhibiting or inactivating the production of reactive oxygen species (ROS) in the metabolism. Antioxidants are therefore of great significance in the human diet because they can enable the body to minimize oxidative damage. Bioactive proteins and peptides from food sources have emerged as a new source of natural antioxidants. They act as dietary antioxidant supplements and as food preservatives. Their antioxidant activity may be attributed to the radical scavenging and chelation properties with metal ions and inhibition of lipid peroxidation. It has also been proposed that the peptide structure and its sequence of amino acids can affect its antioxidant properties. Amino acids with aromatic residues strengthen the radical-scavenging abilities of peptide. An increase in peptide hydrophobicity is also believed to increase their lipid solubility and thus improve their antioxidant activity [25, 26].

7.4 SOURCES OF BIOLOGICALLY ACTIVE PEPTIDES AND PROTEINS

Bovine milk and dairy products are considered as the major source of food-derived bioactive proteins and peptides. However, bioactive peptides and proteins are also derived from other sources, including animals and plants. Bioactive proteins are specifically detected or observed in bovine milk, meat, eggs, and diverse fish species such as tuna, herring, sardine, and salmon and also in wheat, maize, rice, mushrooms, soy, pumpkin, and sorghum. Thus, it is known that peptides and proteins extracted from these sources exhibit an overwhelming capacity that can be used in feed and therapeutic sectors. The coming section discusses the major sources of bioactive proteins and peptides and is shown in Figure 7.2.

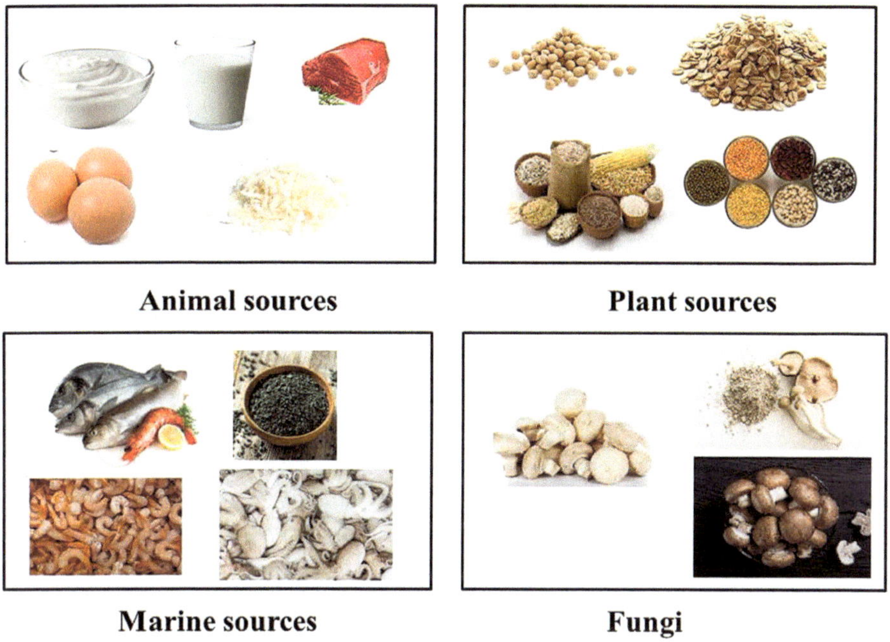

FIGURE 7.2 Sources of bioactive proteins and peptides.

7.4.1 ANIMAL PRODUCTS

Milk and dairy products are the main source of health protecting bioactive proteins and peptides. Mammalian milk contains over 60 unique enzymes including digestive enzymes and antioxidant and antimicrobial enzymes which are essential in terms of milk stability and mammalian defense against pathogenic agents. Fermented dairy products play a functional role either through a direct probiotic effect (action of microorganisms) or through an indirect biogenic effect (action of microbial metabolites formed during the fermentation process). The health-promoting mechanisms of probiotic action are largely focused on the beneficial impact of cytokines and antimicrobial peptides on the immune response due to activation of natural immunity. Whey proteins have anticarcinogenic, immunostimulatory, health-promoting, and antimicrobial functions, which can limit fat accumulation and consequently increase insulin sensitivity [28]. Processes such as fermentation or cheese maturation induce the release of bioactive peptides during the manufacture of milk products. As a result, fermented dairy products have been contained in a wide range of these bioactive peptides. Milk-derived bioactive peptides have shown *in vivo* and *in vitro* health-promoting activities. The milk proteins have gained growing attention as ingredients of health-promoting functional foods which aimed at diet-related chronic diseases such as cardiovascular disease, type two diabetes, and obesity,

Milk protein is associated with reducing the risk of hypertension. Multiple ACE-inhibitory peptides have been found in fermented milk, cheese, and yogurt. Casein, the main milk protein, may generate multiple ACE inhibitory peptides when hydrolyzed by trypsin in the intestinal tract. This has been a topic of increasing commercial interest with greater awareness and scientific credibility [30]. The presence of the VPP and IPP tripeptides in the milk that was fermented with *L. helveticus* and Saccharomyces cerevisiae has been identified since 1995. Numerous animal studies have shown that the use of fermented milk containing Val-Pro-Pro (VPP) and Ile-Pro-Pro (IPP) results in a decrease of blood pressure. These studies are the basement for the production of hypotensive milk-drink products such as Ameal S™ (Calpis Company, Japan) and Evolus® (Valio, Finland) [31, 32].

Milk and dairy products are the main source of antimicrobial activated bioactive proteins and peptides. Fresh milk comprises a unique combination of antimicrobial activity. Many researchers have been paying much attention to antimicrobial peptides released from milk which are known to be non-toxic to mammalian cells as they are extracted from a benign origin. Fermented dairy products play a functional role either actively via contact with ingested microbes or passively via the action of microbial metabolites such as proteins, oligosaccharides, peptides, organic acids, and vitamins formed during the fermentation step. Thus, milk-derived antimicrobial peptides have been considered to possess an overwhelming capacity for use in medical industries and feed.

It is worth noting that proteins and peptides extracted from milk have immunoregulatory characteristics. In addition to its function as a growth factor and its antimicrobial activity, lactoferrin is found to exhibit various immunomodulatory effects. Lactoferrin is a very active protein obtained from milk to hinder microbial growth, and it has been suggested that this function is because of its potential to bind iron and get rid of microorganisms. Bovine lactoferrin (bLF) works well against viral infections. Clinical and animal trials have shown the therapeutic effects of these bLF-containing products. bLF's antimicrobial and antioxidant properties support its use as a preservative in foods and add in commercial foods include yogurt, milk-based beverages, nutritional supplements, skim milk, and pet foods. In addition, it is used in oral care products and cosmetics. The beneficial effect of pet food combined with LF on dermatitis has also been seen in dogs and cats. It is often used as a spray on the surface of raw beef carcasses to minimize microbial contamination and as a component of edible coatings.

K-casein has also been shown to be a class of milk proteins and a significant source to the use of antibacterial peptides in food preservation and health care. This peptide is shown to be inhibitory against bacterium and yeast and is therefore suggested for use in infant nutrition to activate new-born host defense mechanisms. Kappacin (genetic variant of K-casein) shows antimicrobial activity and is used for oral therapy as a pharmaceutical supplement. It has also been commercially available for use for dental care and is a suitable and safe food-grade bio preservative with high potential for use in the food industry.

Whey protein-derived peptides have demonstrated capability to bind calcium, iron, and zinc. Caseinophosphopeptides (CCPs) are bioactive peptides derived from tryptic casein digestion that can bind and solubilize metals such as Ca, Fe, Mg, Zn, Ba, Ni, Co, Cr, and Se. CCPs are known to be additives in functional foods and medicinal formulations and are used in confectionery products. CPPs benefit in reducing anemia, high blood pressure, and osteoporosis [27–29]. Iron peptide complexes are seen as an alternative to reducing iron fortification problems and are considered as an alternative for iron supplements [30, 31].

Milk-derived proteins and peptides with substantial nutritional and therapeutic benefits received growing interest as potential constituents of health-promoting functional foods aimed at diet-related chronic diseases such as obesity [39, 65]. The anti-inflammatory effects of the milk-derived proteins and peptides have been demonstrated on the basis of many *in vitro* and *in vivo* studies [40]. Bioactive peptides derived from milk are considered prominent candidates for numerous health-promoting functional foods aimed at the health of the heart, bone, and digestive system, as well as enhancing immune defense, mood, and stress control. The sour milk products Calpis^{TM} and Evolus^{R}, which contain antihypertensive tripeptides are available in the market and have been clinically proven to reduce blood pressure in human studies.

For centuries, the egg has been known as a high-value source of food for humans, because it is a rich and healthy source of essential amino acids, proteins, minerals, and vitamins. In addition, egg has also found significant applications as additives in functional food preparations as well as in cosmeceutical and pharmaceutical products, due to their gel-forming, emulsifying, and bioactive properties. A number of these properties are associated with the respective protein and peptide components found therein. Additionally, eggs have properties that promote health; others are preventive in nature, and others have therapeutic potential. There is constant research into the potential use of particular egg yolk-derived antibodies in healthcare products and clinical medicine, and formulations are being tested against dental caries or gastritis and rotavirus infections [32, 33].

The plentiful essential protein lysozyme in hen's white egg exerts antibacterial action that is used in semi-hard and hard cheeses to prevent the late blowing defect of Clostridia. It is also used in frozen foods to inhibit the growth of pathogens and in oral health care products to protect against periodontal bacteria and to avoid oral mucosal infections. Eggs are a cheap and low-calorie source of high-quality proteins and other beneficial nutrients. Egg is also abundant in ovalbumin and phosvitin that have multifunctional properties. The phosphoprotein, phosvitin has antimicrobial activity. Ovalbumin and ovotransferrin, derived from egg white, are important sources of ACE inhibitory peptides [45]. The ovalbumin-derived peptide displays antihypertensive activity *in vivo*. From multiple *in vitro* and *in vivo* studies, it is obvious that proteins and peptides extracted from eggs are a healthier option for preserving muscle mass and weight loss and can be considered as a natural nutraceutical [34, 35]. Egg yolk also contains bioactive peptides with anti-inflammatory activities [48–50].

Meat is a highly protein-rich food and contains amino acids, minerals, and vitamins. Some bioactive peptides have also been found to be generated during the processing of meat, such as fermentation and hydrolysis, so the production of these compounds and eventual enhancement in meat products may be advantageous to human health. Meat and its derivatives can also be considered functional foods in so far as they contain a range of bioactive proteins and peptides that are known to work. The meat industry must pursue various possibilities beyond traditional shows, including manipulating the formulation of raw and processed products by attempting to change fatty acid profiles or adding antioxidants to them [36, 37].

7.4.2 PLANT SOURCES

Plants are also potential sources of bioactive proteins and peptides that are produced primarily from peas, wheat, rice, soybeans, pumpkins, oats, hemp seeds, canola, and flaxseed and have functional properties. Such functional bioactive proteins and peptides have wide applications in human nutrition includes components in energy drinks, weight management, and sports nutrition products; nutritious sources for elderly people and immune-compromised patients. Protein hydrolysates from agricultural crops like rapeseed, sunflower, soy, barley, and wheat have been explored for their antioxidant property [52, 55–58].

Soybean is a potential source of bioactive proteins and peptides among plant sources. Soy proteins have a high nutritional value, excellent functional features, and relatively inexpensive. In addition to being an excellent source of dietary protein, they do have antihypertensive, anticholesterolemic, antioxidant, antiobesity, and anticancer activity. The precursor of most peptides is glycine and β-conglycinine, the essential soy proteins. Lunasin, a bioactive peptide derived from soy protein shows anticancer, antioxidant, immunomodulatory, anti-inflammatory, and cholesterol-reducing activities. It is used as a dietary supplement in capsules or powder and as an ingredient of soy drinks. Within soy protein, all of the essential amino acids found in animal protein are present. The *in vivo* studies in rats demonstrate the ACE-inhibitory and blood pressure lowering capacity of soy protein-derived biopeptides [38, 39]. Studies of dietary activity in animals and humans indicate that proteins soybean play a vital role in loss and maintenance of body weight [60–66, 70]. Soybean is also a source of immunomodulatory peptides [68, 69].

Rice, wheat, maize, and millets are recognized worldwide as essential functional foods and provide good health benefit and health-promoting impact. Cereals are considered as a major source of ACE inhibitors. Bioactive peptides extracted from rice may be produced by enzymatic hydrolysis from bran and endosperm has functional properties. These can serve as direct scavengers of various free radicals and have beneficial effects, including antihypertensive, immunomodulatory, and anti-inflammatory activities. Functional protein supplements derived from Thai rice have become increasingly common among people who are health-conscious, athletes, and elderly. Brown rice protein fractionation hydrolysate has the efficacy to function as a versatile food component in nutraceutical foods and beverage products that can offer good taste and health benefits [40–43]. Rice dreg hydrolysate inhibitory peptides have shown important antihypertensive action. *In vivo* studies revealed that kurosu, a product from unpolished rice shows antihypertensive activity [8].

Studies show that the defatted wheat germ is an important source of protein that can be processed into value-added products like protein hydrolysates or bioactive peptides using suitable processing techniques. In a study, Zohreh et al. demonstrated the antioxidant, ACE-inhibitory, and antitumor activities of bioactive peptides derived from wheat germ protein hydrolysates [25]. In another study, Cian et al. reported the antioxidant and ACE inhibitory activity of wheat gluten hydrolysate peptides [44]. Durum wheat bran protein concentrate contains albumin and globulin proteins enriched in essential amino acids have good functional properties and are recommended in cereal-based foods such as pasta and bread as a fortification ingredient [45]. Bioactive peptides extracted from the rice display anti-inflammatory activities [42, 46].

Corn peptides actively prevent the generation of free radicals and are used as nutritional regulators and stabilizers in drinks, dairy products, and grains. It also acts as antioxidants for functional and medicinal nutritious foods. Corn peptides minimize subcutaneous fat and ingestion of corn peptides ensures weight loss. The experimental results showed that the food enriched with corn peptides generated the highest amount of heat, suggesting that corn peptides have high effect than other proteins on promoting the energy metabolism. Corn peptides are suitable diet for obese people and as a nutritional supplement for weight loss treatment [47].

Nutritional pulses are an important source of protein and have higher lysine, arginine, glutamic, and aspartic acid levels compared with cereals. In addition to their proven nutritional benefits, recent pulse intake has had preventive or therapeutic effects on chronic health conditions, such as CVD, diabetes, and cancer. The use of pulses is also associated with therapeutic or protective effects on health conditions such as overweight and obesity. The potential of pulse seed hydrolysates and BPs for cancer, inflammation, hypertension, cardiovascular disease, and high cholesterol is identified in various *in vitro* studies. Purified BPs can be used in functional foods as health-enhancing ingredients. Pulses flours are already used for enhancing the functionality and nutritional consistency of baked products and snacks. Pulse-based hydrolysates and bioactive peptides are suggested as suitable sources for the production of new protein-derived products [46, 48, 49].

Oat is a multifunctional crop considered superior to many other unfortified cereals in nutritional terms. Oats are widely used as whole grains, containing essential nutrients such as proteins, vitamins, unsaturated fatty acids, and minerals. Oat protein is high in quality and low in cost. Protein content in oats (11–15% of the grain) can be divided into four fractions. Water-soluble albumins (1–12% of the total protein), salt water soluble globulins (80% of total protein), alcohol soluble prolamins or avenins (10 to 15% of total proteins) and acid or base soluble glutelins (5 to 66% of total proteins). Owing to the higher lysine content, which is the key limiting amino acid in cereals, protein contained in oat is considered to be nutritionally superior to that of wheat. With the growing demand for gluten-free foods, oat is seen as a good option for diversifying the diet of patients with celiac disease. Multiple experimental and clinical trials have shown that oat-based products intake can reduce serum cholesterol levels, decrease glucose absorption, and lower plasma insulin response. Such health benefits of oat have drawn the wide interest of scientists and the public. Proper use of oat in food applications can help to prevent cardiovascular disease, obesity, diabetes, and many other diet-related diseases. Oat-based porridge, oat flour, oat bread, biscuits, and cookies, flakes, and infant foods are receiving a lot of interest due to their high nutritional value. Oat proteins have been used in food products such as heat-resistant chocolates, due to their emulsifying and viscous properties. Studies have reported using the oat bran as a fat replacement in meatballs. These oat-bran meatballs display high sensory acceptability. Several oats-based probiotic beverages such as Proviva, Yosa, Adavena M40 and Biovessina are launched in the market based on increased awareness of high nutritional value oats and increasing demand for healthy foods [50–54].

7.4.3 MARINE SOURCES

In recent times, the usage of aquatic food has increased globally, owing to a deeper insight of their health benefits and the good outlook of seafood among consumers. Marine organisms have evolved specific properties and bioactive compounds in contrast to terrestrial sources, owing to the large variety of their living environments. They are rich in beneficial nutrients. Several bioactive proteins and peptides are developed from marine resources, namely fish, oysters, algae, squids, salmon, sea urchins, shrimps, snow crabs, and seahorses. The extraction of functional food ingredients, value-added nutraceuticals, and natural health products from marine sources has been well recognized in conjunction with health promotion, mitigation of infection risk, and cost savings in health care.

Fish protein hydrolysates (FPHs) have become notable over the years as the major source of protein hydrolysates and bioactive peptides. Large variety food formulations may use the various properties of FPH such as good water-holding capability, solubility, emulsion capacity, foaming potential, heat tolerance ability, and gelling potential. They can be used in a wide range of products as stabilizing and emulsifying additives and can help to shape and stabilize foam-based products, including mayonnaise, salad dressings, sausages, beverages, creams, etc. FPHs are also said to have efficacy for pharmaceutical applications. Studies have shown that peptides derived from FPHs display antioxidants, anti-proliferation, antihypertension, anti-inflammatory, and antidiabetic efficacy. Fish is also a great source of anticancer peptides [86, 89]. Furthermore, the separation of effective anticancer components from fish tissue has made the argument for considering fish by-products as sources of chemo-preventive and anticancer components. The nutritional benefit of FPH makes it suitable food that can promote the growth and survival of aquatic life. Due to their efficacy in the prevention and treatment of hypertension, fish derived bioactive peptides have potential as nutraceuticals and pharmaceuticals and are commercially available as a dietary supplement under the brand names of Vasotensin^{R}, PeptACER, and Levenorm^{TM}. Fish proteins are also displayed anti-inflammatory effects *in vitro* as well as in animal studies [91]. In addition, the food items fortified with Omega-3 oil offer a way of achieving the desired biochemical effects of these nutrients without consumption of nutritional supplements, medications, or a significant shift in dietary habits [55–57].

Seaweeds with antibacterial, antiviral, and antifungal properties are known for their abundance in polysaccharides, minerals, and other vitamins, proteins, lipids, and polyphenols. This gives great potential to marine algae as a substitute in functional food. Multiple marine organisms generate biologically active proteins with antimicrobial, anticancer, anticoagulant, hypocholesterolemic, and immunostimulatory activities. Bioactive peptides derived from spirulina, the cyanobacterium (blue-green algae), shows antitumor activity. Most species of seaweed possess all the essential amino acids and are a significant source of acidic amino acids such as glutamic acid and aspartic acid. Numerous studies demonstrated the *in vivo* therapeutical potentials of red, green, and brown seaweeds. Bioactive lectins, carbohydrate-binding proteins of non-immune origin with antibacterial, anti-inflammatory, and anticancer activities are found in macroalgal species. Red seaweed-derived biliproteins are used as fluorescent markers. Phycobiliproteins displayed antioxidant properties that are helpful for the prevention and treatment of neurological disorders, tumors, and stomach ulcers. As examined by *in vitro* and *in vivo* assays, the sulfated heteropolysaccharide compounds, fucoidan found in seaweeds, are able to inhibit the growth of different cell lines. Food products supplemented with seaweeds and extracts of seaweed showed beneficial effects on the numerous lifestyle diseases such as obesity, diabetes, and hypertension [23, 94, 99, 100]. Studies on marine organisms such as seaweed and algae revealed their anti-inflammatory effect. Spirulina is a good source of phycobiliprotein that is examined for its anti-inflammatory properties.

The *in vitro* and *in vivo* studies performed in acetates Chinensis, a marine shrimp, usually used as a flavoring agent in shrimp sauce demonstrated antihypertensive activity. Many marine species, including mollusks and crustaceans, display calcium-binding, antimicrobial, and appetite suppressing activities, thereby encouraging human wellbeing, and avoiding chronic illness. Shrimp-derived peptides have a major effect on cholecystokinin, a hormone that controls appetite and gastric emptying. Specific foods containing these peptides have the potential to regulate the disorders associated with appetite. For their antioxidant and radical scavenging properties, seaweeds such as alginate, carrageenan, and agar are widely used in food. They have also been introduced to many products, including salad dressings, drinks, and baked goods to boost their protein content, or sold as protein supplements. Given their many possible health benefits, food products, supplements or natural health products that contain marine bioactive are expected to dominate a huge market. Collactive™, a marine source of collagen and elastin, may be used as an anti-wrinkle ingredient, and Nutripeptin™, another marine bioactive compound, has been found to be effective in enhancing satiety and weight loss response as examples of commercially available marine-food items. Foods containing FPHs/peptides are believed to be appropriate for consumption by people with mild hypertension. Examples of two such products include Lapis Support™ (beverage) and Valtyron® (additive for soft drinks, jelly, powdered soup, dietary supplements, etc.), [58–62].

7.4.4 FUNGI

Because of their special properties and nutrient content, fungi have already known applications in the medical and food industry. Mushrooms are a distinct category of edible macrofungi capable of providing good taste and nutritional value with high protein content and low fat. They are known as an alternate source of protein of good quality and are capable of providing the maximum protein amount. Moreover, it contains biologically active compounds having antifungal, anti-inflammatory, antitumor, antiviral, antibacterial, hepatoprotective, antidiabetic, hypolipidemic, antithrombotic, antihypertensive, immunomodulatory, and hypocholesterolemic properties. The protein in the mushrooms comprises the nine essential amino acids that humans require and is particularly rich in lysine and leucine that are deficient in most staple cereal food. In addition, several mushroom proteins demonstrate significant pH and thermal stability. 'Mushroom nutraceuticals' are the conventional preparations used from older days in the form of extracts, health tonics, fermented drinks, and soups. Due to their high nutritional value, edible items can be fortified with mushrooms, and such food serves as a reservoir of nutrients for undernourished populations. Canned mushrooms are commercialized and used to make soup and pizza. Powdered mushroom has been added to food items such as noodles, pasta, rice porridge, and bakery items. Studies reveal that the high protein content of mushroom help to build a better gluten network and provide better elasticity in bakery products noodles and pasta. The addition of mushrooms also enhances its antioxidant content. Thus, mushroom fortification leads to improve nutritional values, physical properties, and food quality [62–69].

7.5 CONCLUSION

Food-derived proteins and peptides have gained significant attention as chronic disease prevention agents because of their exceptional multifunctional properties related to the maintenance of general health. Such proteins and peptides are a profoundly fascinating commodity in health-promoting foods for future use as active ingredients. They possess multifunctional activities such as antioxidant, antihypertensive, antimicrobial, antimutagenic, antioxidative, anti-inflammatory, immune, and cytomodulatory and mineral binding activity. Depending on their activities, they may be sold as nutraceutical products or functional ingredients. Fortification with bioactive proteins and peptides leads to beneficial effects in food systems in terms of health implications and functionality. The creation of fully integrated bioprocesses which can be transferred to large-scale operations for the production and purification of these essential biomolecules would help to accelerate their categorization in the major consumer markets. In recent years, the market for functional ingredients and foods has grown astonishingly because of the awareness of consumers and the interest in promoting healthy diets and their lifestyles. It is possible to successfully integrate natural ingredients, such as bioactive proteins and peptides, into foods, creating new natural product categories and new business opportunities. In addition, steps need to be taken to incorporate effective and economically viable development methods on an industrial scale.

KEYWORDS

- **angiotensin I-enzyme**
- **caseinophosphopeptides**
- **fish protein hydrolysates**
- **functional foods**
- **Ile-Pro-Pro**
- **peptides**
- **proteins**
- **Pro-Val-Pro**

REFERENCES

1. Roberfroid, M. B., (2002). Global view on functional foods: European perspectives. *British Journal of Nutrition, 88*(S2), S133–S138. https://doi.org/10.1079/bjn2002677.
2. Siró, I., Kápolna, E., Kápolna, B., & Lugasi, A., (2008). Functional food. Product development, marketing, and consumer acceptance: A review. *Appetite, 51*, 456–467. https://doi.org/10.1016/j.appet.2008.05.060.
3. Korhonen, H., (2009). Milk-derived bioactive peptides: From science to applications. *Journal of Functional Foods, 1*(2), 177–187. https://doi.org/10.1016/j.jff.2009.01.007.
4. Govender, M., Choonara, Y. E., Kumar, P., Du Toit, L. C., Van, V. S., & Pillay, V., (2014). A review of the advancements in probiotic delivery: Conventional vs. Non-conventional formulations for intestinal flora supplementation. *AAPS PharmSciTech, 15*, 29–43. https://doi.org/10.1208/s12249-013-0027-1.
5. López-Varela, S., González-Gross, M., & Marcos, A., (2002). Functional foods and the immune system: A review. *European Journal of Clinical Nutrition, 56*(3), S29–S33. https://doi.org/10.1038/sj.ejcn.1601481.
6. Aslam, H., Green, J., Jacka, F. N., Collier, F., Berk, M., Pasco, J., & Dawson, S. L., (2018). Fermented foods, the gut and mental health: A mechanistic overview with implications for depression and anxiety. *Nutritional Neuroscience,* 1–13. https://doi.org/10.1080/1028415x.2018.1544332.
7. Yamamoto, N., Ejiri, M., & Mizuno, S., (2005). Biogenic peptides and their potential use. *Current Pharmaceutical Design, 9*(16), 1345–1355. https://doi.org/10.2174/1381612033454801.

8. Correia-da-Silva, Sousa, Pinto, and Kijjoa., (2017). Phenolic Compounds as Nutraceuticals or Functional Food Ingredients. *Current Pharmaceutical Design,23(19), 2787-2806.* https://doi.org/10.2174/1381612822666161227153906.
9. Huang, W. Y., Davidge, S. T., & Wu, J., (2013). Bioactive natural constituents from food sources-potential use in hypertension prevention and treatment. *Critical Reviews in Food Science and Nutrition, 53*, 615–630. https://doi.org/10.1080/10408398.2010.550071.
10. Walther, B., & Sieber, R., (2011). Bioactive proteins and peptides in foods. *International Journal for Vitamin and Nutrition Research, 81*(2, 3), 181–192. https://doi.org/10.1024/03009831/a000054.
11. Pellegrini, A., (2005). Antimicrobial peptides from food proteins. *Current Pharmaceutical Design, 9*(16), 1225–1238. https://doi.org/10.2174/1381612033454865.
12. Ayyash, M., Al-Nuaimi, A. K., Al-Mahadin, S., & Liu, S. Q., (2018). *In vitro* investigation of anticancer and ACE-inhibiting activity, α-amylase and α-glucosidase inhibition, and antioxidant activity of camel milk fermented with camel milk probiotic: A comparative study with fermented bovine milk. *Food Chemistry, 239*, 588–597. https://doi.org/10.1016/j.foodchem.2017.06.149.
13. Xu, X., Yan, H., Chen, J., & Zhang, X., (2011). Bioactive proteins from mushrooms. *Biotechnology Advances, 29*(6), 667–674. https://doi.org/10.1016/j.biotechadv.2011.05.003.
14. Chalamaiah, M., Yu, W., & Wu, J., (2018). Immunomodulatory and anticancer protein hydrolysates (peptides) from food proteins: A review. *Food Chemistry, 245*, 205–222. https://doi.org/10.1016/j.foodchem.2017.10.087.
15. Hernández-Ledesma, B., & Hsieh, C. C., (2017). Chemopreventive role of food-derived proteins and peptides: A review. *Critical Reviews in Food Science and Nutrition, 57*(11), 2358– 2376. https://doi.org/10.1080/10408398.2015.1057632.
16. Reynolds, E. C., (1998). Anticariogenic complexes of amorphous calcium phosphate stabilized by casein phosphopeptides: A review. *Special Care in Dentistry, 18*(1), 8–16. https://doi.org/10.1111/j.1754-4505.1998.tb01353.x.
17. Karami, Z., & Akbari-Adergani, B., (2019). Bioactive food derived peptides: A review on correlation between structure of bioactive peptides and their functional properties. *Journal of Food Science and Technology, 56*, 535–547. https://doi.org/10.1007/s13197-018-3549-4.
18. Sharma, P., Kaur, H., Kehinde, B. A., Chhikara, N., Sharma, D., & Panghal, A., (2020). Foodderived anticancer peptides: A review. *International Journal of Peptide Research and Therapeutics,* 1–16. https://doi.org/10.1007/s10989-020-10063-1.
19. Baboota, R. K., Bishnoi, M., Ambalam, P., Kondepudi, K. K., Sarma, S. M., Boparai, R. K., & Podili, K., (2013). Functional food ingredients for the management of obesity and associated co-morbidities: A review. *Journal of Functional Foods, 5*, 997–1012. https://doi.org/10.1016/j.jff.2013.04.014.
20. Majumder, K., Mine, Y., & Wu, J., (2016). The potential of food protein-derived antiinflammatory peptides against various chronic inflammatory diseases. *Journal of the Science of Food and Agriculture, 96*, 2303–2311. https://doi.org/10.1002/jsfa.7600.
21. Caetano-Silva, M. E., Netto, F. M., Bertoldo-Pacheco, M. T., Alegría, A., & Cilla, A., (2020). Peptide-metal complexes: Obtention and role in increasing bioavailability and decreasing the pro-oxidant effect of minerals. *Critical Reviews in Food Science and Nutrition*. https://doi.org/10.1080/10408398.2020.1761770.

22. Cai, J., Li, X., Du, H., Jiang, C., Xu, S., & Cao, Y., (2020). Immunomodulatory significance of natural peptides in mammalians: Promising agents for medical application. *Immunobiology*, 151936. https://doi.org/10.1016/j.imbio.2020.151936.
23. Chalamaiah, M., Hemalatha, R., Jyothirmayi, T., Diwan, P. V., Uday, K. P., Nimgulkar, C., & Dinesh, K. B., (2014). Immunomodulatory effects of protein hydrolysates from rohu (*Labeo rohita*) egg (roe) in BALB/c mice. *Food Research International, 62*, 1054–1061. https://doi.org/10.1016/j.foodres.2014.05.050.
24. Möller, N. P., Scholz-Ahrens, K. E., Roos, N., & Schrezenmeir, J., (2008). Bioactive peptides and proteins from foods: Indication for health effects. *European Journal of Nutrition, 47*(4), 171–182. https://doi.org/10.1007/s00394-008-0710-2.
25. Segerstrom, S. C., & Miller, G. E., (2004). Psychological stress and the human immune system: A meta-analytic study of 30 years of inquiry. *Psychological Bulletin, 130*(4), 601–630. https://doi.org/10.1037/0033-2909.130.4.601.
26. Karami, Z., Peighambardoust, S. H., Hesari, J., Akbari-Adergani, B., & Andreu, D., (2019). Antioxidant, anticancer and ACE-inhibitory activities of bioactive peptides from wheat germ protein hydrolysates. *Food Bioscience, 32*, 100450. https://doi.org/10.1016/j.fbio.2019.100450.
27. Giromini, Cheli, Rebucci, and Baldi, (2019). Dairy proteins a,nd bioactive peptides: Modeling digestion and the intestinal barrier. *Journal of dairy science, 102(2):*929-942. https://doi.org/ 10.3168/jds.2018-15163.
28. Benkerroum, (2010). Antimicrobial peptides generated from milk proteins: a survey and prospects for application in the food industry. A review. *International Journal of Dairy Technology, 63(3):*320-338. https://doi.org/10.1111/j.1471-0307.2010.00584.x
29. Zou, T. B., He, T. P., Li, H. B., Tang, H. W., & Xia, E. Q., (2016). The structure-activity relationship of the antioxidant peptides from natural proteins. *Molecules, 21*, 72. https://doi.org/10.3390/molecules21010072.
30. Beltrán-Barrientos, L. M., Hernández-Mendoza, A., Torres-Llanez, M. J., González-Córdova, A. F., Vallejo-Córdoba., B. (2016). Fermented milk as antihypertensive functional food. *Journal of Dairy Technology, 99(6),* 4099–4110. https://doi.org/10.3168/jds.2015-10054.
31. Korhonen, H., & Pihlanto, A. (2006). Bioactive peptides: production and functionality. *International Dairy Journal, 16(9):*945-960. https://doi.org/10.1016/j.idairyj.2005.10.012
32. Ohsawa, I., Nishimaki, K., Murakami, Y., Suzuki, Y., Ishikawa, M., & Ohta, S..(2008). Age-dependent neurodegeneration accompanying memory loss in transgenic mice defective in mitochondrial aldehyde dehydrogenase 2 activity.*The Journal of Neuroscience.28(24):*6239-49. https://doi.org/10.1523/JNEUROSCI.4956-07.2008
33. Meisel, H., & Fitzjerald, J., (2005). Biofunctional peptides from milk proteins: Mineral binding and cytomodulatory effects. *Current Pharmaceutical Design, 9*(16), 1289–1295. https://doi.org/10.2174/1381612033454847.
34. Lee, J. S., Yu, X., Wagoner, J. A. J., & Murphy, W. L., (2017). Mineral binding peptides with enhanced binding stability in serum. *Biomaterials Science, 5*(4), 663–668. https://doi.org/10.1039/c6bm00928j.
35. Sultan, S., Huma, N., Butt, M. S., Aleem, M., & Abbas, M., (2018). Therapeutic potential of dairy bioactive peptides: A contemporary perspective. *Critical Reviews in Food Science and Nutrition, 58*(1), 105–115. https://doi.org/10.1080/10408398.2015.1136590.
36. Caetano-Silva, M. E., Bertoldo-Pacheco, M. T., Paes-Leme, A. F., & Netto, F. M., (2015). Ironbinding peptides from whey protein hydrolysates: Evaluation, isolation,

and sequencing by LC-MS/MS. *Food Research International, 71*, 132–139. https://doi.org/10.1016/j.foodres.2015.01.008.
37. Caetano-Silva, M. E., Cilla, A., Bertoldo-Pacheco, M. T., Netto, F. M., & Alegría, A., (2018). Evaluation of *in vitro* iron bioavailability in free form and as whey peptide-iron complexes. *Journal of Food Composition and Analysis, 68*, 95–100. https://doi.org/10.1016/j.jfca.2017.03.010.
38. Mudgil, Kamal, Yuen, and Maqsood, (2018). Characterization and identification of novel antidiabetic and anti-obesity peptides from camel milk protein hydrolysates. *Food Chemistry, 259*:46-54. https://doi.org/10.1016/j.foodchem.2018.03.082.
39. Zimecki and Kruzel,(2007). Milk-derived proteins and peptides of potential therapeutic and nutritive value, *Journal of Experimental Therapeutics and Oncology 6(2),* 89–106. https://pubmed.ncbi.nlm.nih.gov/17407968.
40. Aguilar-Toalá JE, Santiago-López L, Peres CM, Peres C, Garcia HS, Vallejo-Cordoba B, González-Córdova AF, Hernández-Mendoza A (2017). Assessment of multifunctional activity of bioactive peptides derived from fermented milk by specific Lactobacillus plantarum strains. *Journal of Dairy Science. 100(1),* 65–75. doi: 10.3168/jds.2016-11846.
41. Campbell, L., Raikos, V., & Euston, S. R., (2003). Modification of functional properties of eggwhite proteins. *Nahrung/Food, 47*(6), 369–376. https://doi.org/10.1002/food.200390084.
42. Miranda, J., Anton, X., Redondo-Valbuena, C., Roca-Saavedra, P., Rodriguez, J., Lamas, A., & Cepeda, A., (2015). Egg and egg-derived foods: Effects on human health and use as functional foods. *Nutrients, 7*(1), 706–729. https://doi.org/10.3390/nu7010706.
43. Martinez, C. S., Alterman, C. D. C., Vera, G., Márquez, A., Uranga, J. A., Peçanha, F. M., & Wiggers, G. A., (2019). Egg white hydrolysate as a functional food ingredient to prevent cognitive dysfunction in rats following long-term exposure to aluminum. *Scientific Reports, 9*(1), 1–13. https://doi.org/10.1038/s41598-018-38226-7.
44. Zhang, B., Wang, H., Wang, Y., Yu, Y., Liu, J., Liu, B., & Zhang, T., (2019). Identification of antioxidant peptides derived from egg-white protein and its protective effects on H_2O_2-induced cell damage. *International Journal of Food Science & Technology, 54*(6), 2219–2227. https://doi.org/10.1111/IJFS.14133.
45. Hartmann and Meisel, (2007). Food-derived peptides with biological activity: from research to food applications.*Current Opinion in Biotechnology, 18(2),* 163–169. https://doi.org 10.1016/j.copbio.2007.01.013.
46. Khan, M. I., Arshad, M. S., Anjum, F. M., Sameen, A., Aneeq-ur-Rehman, & Gill, W. T., (2011). Meat as a functional food with special reference to probiotic sausages. *Food Research International, 44*, 3125–3133. https://doi.org/10.1016/j.foodres.2011.07.033.
47. Olmedilla-Alonso, B., Jiménez-Colmenero, F., & Sánchez-Muniz, F. J., (2013). Development and assessment of healthy properties of meat and meat products designed as functional foods. *Meat Science, 95*(4), 919–930. https://doi.org/10.1016/j.meatsci.2013.03.030.
48. Bitencourt, R. M., Pamplona, F. A., Takahashi, R. N. (2008). Facilitation of contextual fear memory extinction and anti-anxiogenic effects of AM404 and cannabidiol in conditioned rats. *European Neuropsychopharmacology. 18(12):*849-59. doi: 10.1016/j.euroneuro.2008.07.001.
49. Khan, M. J., Drochner, W., Steingass, H., Islam, K. M. S., (2008). Nutritive evaluation of some tree leaves from Bangladesh for feeding ruminant animals. *Indian Journal of Animal. Science., 78 (11),* 1273–1277. https://www.feedipedia.org/node/21450.

50. Vo, Ryu, & Kim, (2013). Purification of novel anti-inflammatory peptides from enzymatic hydrolysate of the edible microalgal Spirulina maxima. *Journal of Functional Foods 5(3),* 1336–1346. https://doi.org/10.1016/j.jff.2013.05.001.
51. Singh, B. P., Vij, S., & Hati, S., (2014). Functional significance of bioactive peptides derived from soybean. *Peptides, 54,* 171–179. https://doi.org/10.1016/j.peptides.2014.01.022.
52. Gibbs, B. F., Zougman, A., Masse, R. and Mulligan, C.(2004), Production and characterization of bioactive peptides from soy hydrolysate and soy-fermented food, *Food research international, 37 (2),* 123-131. . https://doi.org/ 10.1016/j.foodres.2003.09.010
53. Taniguchi, M., Aida, R., Saito, K., Kikura, T., Ochiai, A., Saitoh, E., & Tanaka, T., (2020). Identification and characterization of multifunctional cationic peptides from enzymatic hydrolysates of soybean proteins. *Journal of Bioscience and Bioengineering, 129*(1), 59–66. https://doi.org/10.1016/j.jbiosc.2019.06.016.
54. Dei Piu', L., Tassoni, A., Serrazanetti, D. I., Ferri, M., Babini, E., Tagliazucchi, D., & Gianotti, A., (2014). Exploitation of starch industry liquid by-product to produce bioactive peptides from rice hydrolyzed proteins. *Food Chemistry, 155,* 199–206. https://doi.org/10.1016/j.foodchem.2014.01.055.
55. Ma, Xiong, Zhai, Zhu, & Dziubla,(2010). Fractionation and evaluation of radical-scavenging peptides from in vitro digests of buckwheat protein. *Food Chemistry.118 (3)*:582-588. https://doi.org/10.1016/j.foodchem.2009.05.024.
56. Girón-Calle, J., Vioque, J., Pedroche, J., Alaiz, M., Yust, M. M., Megías, C., & Millán, F. (2008). Chickpea protein hydrolysate as a substitute for serum in cell culture. *Cytotechnology, 57*(3), 263–272. https://doi.org/10.1007/s10616-008-9170-z
57. Suetsuna, K., Ukeda, H., & Ochi, H. (2000). Isolation and characterization of free radical scavenging activities peptides derived from casein. *The Journal of nutritional biochemistry, 11*(3), 128–131. https://doi.org/10.1016/s0955-2863(99)00083-2.
58. Vioque, J., Sánchez-Vioque, R., Clemente, A. et al. (2000). Partially hydrolyzed rapeseed protein isolates with improved functional properties. *J Amer Oil Chem Soc 77,* 447–450 https://doi.org/10.1007/s11746-000-0072-y.
59. Liu, Y. Q., Strappe, P., Shang, W. T., & Zhou, Z. K., (2019). Functional peptides derived from rice bran proteins. *Critical Reviews in Food Science and Nutrition, 59,* 349–356. https://doi.org/10.1080/10408398.2017.1374923.
60. Adeneye, A. A., Adeyemi, O. O., Agbaje, E. O., & Banjo, A. A. (2010). Evaluation of the toxicity and reversibility profile of the aqueous seed extract of Hunteria umbellata (K. Schum.) Hallier f. in rodents. African journal of traditional, complementary, and alternative medicines: *AJTCAM, 7*(4), 350–369. https://doi.org/10.4314/ajtcam.v7i4.56704.
61. Aoyama, T., Fukui, K., Takamatsu, K., Hashimoto, Y., & Yamamoto, T. (2000). Soy protein isolate and its hydrolysate reduce body fat of dietary obese rats and genetically obese mice (yellow KK). *Nutrition (Burbank, Los Angeles County, Calif.), 16*(5), 349–354. https://doi.org/10.1016/s0899-9007(00)00230-6.
62. Bhandari, R., Xiao, J., & Shankar, A. (2013). Urinary bisphenol A and obesity in U.S. children. *American Journal of Epidemiology, 177*(11), 1263–1270. https://doi.org/10.1093/aje/kws391.
63. Cani, P. D., Neyrinck, A. M., Maton, N., & Delzenne, N. M. (2005). Oligofructose promotes satiety in rats fed a high-fat diet: involvement of glucagon-like Peptide-1. *Obesity Research, 13*(6), 1000–1007. https://doi.org/10.1038/oby.2005.117.

64. Ko, J.-Y.; Kang, N.; Lee, J.-H.; Kim, J.-S.; Kim, W.-S.; Park, S.-J.; Kim, Y.-T.; Jeon, Y.-J (2016). Angiotensin I-converting enzyme inhibitory peptides from an enzymatic hydrolysate of flounder fish (*Paralichthys olivaceus*) muscle as a potent anti-hypertensive agent. *Process Biochem., 51*, 535–541 https://doi.org/10.1016/j.procbio.2016.01.009.
65. Mudgil, P., Kamal, H., Yuen, G. C., & Maqsood, S. (2018). Characterization and identification of novel antidiabetic and anti-obesity peptides from camel milk protein hydrolysates. *Food chemistry, 259*, 46–54. https://doi.org/10.1016/j.foodchem.2018.03.082.
66. Velasquez, M. T., & Bhathena, S. J. (2007). Role of dietary soy protein in obesity. *International Journal of Medical Sciences, 4*(2), 72–82. https://doi.org/10.7150/ijms.4.72.
67. Zhu, Z., Jiang, W., & Thompson, H. J. (2012). Edible dry bean consumption (Phaseolus vulgaris L.) modulates cardiovascular risk factors and diet-induced obesity in rats and mice. *The British Journal of Nutrition, 108* Suppl 1, S66–S73. https://doi.org/10.1017/S0007114512000839.
68. Ashaolu, T. J. (2020). Immune boosting functional foods and their mechanisms: A critical evaluation of probiotics and prebiotics. *Biomedicine & Pharmacotherapy, 130,* 110625. https://doi.org/10.1016/j.biopha.2020.110625.
69. Gibbs, B. F., Zougman, A., Masse, R., & Mulligan, C. (2004). Production and characterization of bioactive peptides from soy hydrolysate and soy-fermented food. *Food Research International, 37(2)*, 123–131. https://doi.org/110.1016/j.foodres.2003.09.010.
70. Zhao, B., Cui, Y., Fan, X., Qi, P., Liu, C., Zhou, X., & Zhang, X. (2019). Anti-obesity effects of Spirulina platensis protein hydrolysate by modulating brain-liver axis in high-fat diet fed mice. *PloS one, 14*(6), e0218543. https://doi.org/10.1371/journal.pone.0218543.
71. Selamassakul, O., Laohakunjit, N., Kerdchoechuen, O., Yang, L., & Maier, C. S., (2020). Bioactive peptides from brown rice protein hydrolyzed by bromelain: Relationship between bio-functional activities and flavor characteristics. *Journal of Food Science, 85*(3), 707–717. https://doi.org/10.1111/1750-3841.15052.
72. Taniguchi, M., & Ochiai, A., (2017). Characterization and production of multifunctional cationic peptides derived from rice proteins. *Bioscience, Biotechnology and Biochemistry, 81*, 634–650. https://doi.org/10.1080/09168451.2016.1277944.
73. Cian, R. E., Vioque, J., & Drago, S. R., (2015). Structure-mechanism relationship of antioxidant and ACE I inhibitory peptides from wheat gluten hydrolysate fractionated by pH. *Food Research International, 69*, 216–223. https://doi.org/10.1016/j.foodres.2014.12.036.
74. Alzuwaid, N. T., Sissons, M., Laddomada, B., & Fellows, C. M., (2020). Nutritional and functional properties of durum wheat bran protein concentrate. *Cereal Chemistry, 97*(2), 304– 315. https://doi.org/10.1002/cche.10246.
75. López-Barrios, L., Gutiérrez-Uribe, J. A., & Serna-Saldívar, S. O., (2014). Bioactive peptides and hydrolysates from pulses and their potential use as functional ingredients. *Journal of Food Science, 79*(3). https://doi.org/10.1111/1750-3841.12365.
76. Li, G., Liu, W., Wang, Y., Jia, F., Wang, Y., Ma, Y., & Lu, J., (2019). Functions and applications of bioactive peptides from corn gluten meal. In: *Advances in Food and Nutrition Research* (Vol. 87, pp. 1–41). https://doi.org/10.1016/bs.afnr.2018.07.001.
77. Bessada, S. M. F., Barreira, J. C. M., & Oliveira, M. B. P. P., (2019). Pulses and food security: Dietary protein, digestibility, bioactive and functional properties. *Trends in Food Science and Technology, 93*, 53–68. https://doi.org/10.1016/j.tifs.2019.08.022.

78. Clark, S., & Duncan, A. M., (2017). The role of pulses in satiety, food intake and body weight management. *Journal of Functional Foods, 38*, 612–623. https://doi.org/10.1016/j.jff.2017.03.044.
79. Angelov, A., Yaneva-Marinova, T., & Gotcheva, V., (2018). Oats as a matrix of choice for developing fermented functional beverages. *Journal of Food Science and Technology, 55*, 2351–2360. https://doi.org/10.1007/s13197-018-3186-y.
80. Bekers, M., Marauska, M., Laukevics, J., Grube, M., Vigants, A., Karklina, D., & Viesturs, U., (2001). Oats and fat-free milk based functional food product. *Food Biotechnology, 15*(1), 1–12. https://doi.org/10.1081/FBT-100103890.
81. Rasane, P., Jha, A., Sabikhi, L., Kumar, A., & Unnikrishnan, V. S., (2013). Nutritional advantages of oats and opportunities for its processing as value-added foods: A review. *Journal of Food Science and Technology, 52*, 662–675. https://doi.org/10.1007/s13197-013-1072-1.
82. Sang, S., & Chu, Y. F., (2017). Whole grain oats, more than just a fiber: Role of unique phytochemicals. *Molecular Nutrition and Food Research, 61*, 1600715. https://doi.org/10.1002/mnfr.201600715.
83. Zheng, Z., Li, J., & Liu, Y., (2020). Effects of partial hydrolysis on the structural, functional, and antioxidant properties of oat protein isolate. *Food & Function, 11*(4), 3144–3155. https://doi.org/10.1039/c9fo01783f.
84. Calanche, J., Beltrán, H., Marquina, P., Roncalés, P., & Beltrán, J. A., (2019). Eating fish in another way: Development of functional pasta with added concentrates of farmed sea bass (*Dicentrarchus labrax*). *Cereal Chemistry, 96*(5), 856–865. https://doi.org/10.1002/cche.10186.
85. Tahergorabi, R., Beamer, S. K., Matak, K. E., & Jaczynski, J., (2012). Functional food products made from fish protein isolate recovered with isoelectric solubilization/precipitation. *LWT Food Science and Technology, 48*(1), 89–95. https://doi.org/10.1016/j.lwt.2012.02.018.
86. Correia-da-Silva, M., Sousa, E., Pinto, M., & Kijjoa, A. (2017). Anticancer and cancer preventive compounds from edible marine organisms. *Seminars in Cancer Biology, 46*, 55–64. https://doi.org/10.1016/j.semcancer.2017.03.011'
87. Zamora-Sillero, J., Gharsallaoui, A., & Prentice, C., (2018). Peptides from fish by-product protein hydrolysates and its functional properties: An overview. *Marine Biotechnology, 20*, 118–130. https://doi.org/10.1007/s10126-018-9799-3.
88. Ganesan, A. R., Mohanram, M. S. G., Balasubramanian, B., Kim, I. H., Seedevi, P., Mohan, K., & Ignacimuthu, S., (2020). Marine invertebrates' proteins: A recent update on functional property. *Journal of King Saud University-Science, 32*, 1496–1502. https://doi.org/10.1016/j.jksus.2019.12.003.
89. Palaniappan Seedevi, Meivelu Moovendhan, Shanmugam Vairamani, & Annaian Shanmugam (2017). Evaluation of antioxidant activities and chemical analysis of sulfated chitosan from Sepia prashadi, *International Journal of Biological Macromolecules, 99*, 519–529, https://doi.org/10.1016/j.ijbiomac.2017.03.012.
90. Halim, N. R. A., Yusof, H. M. & Sarbon, N. M. (2016) Functional and bioactive properties of fish protein hydolysates and peptides: A comprehensive review, *Trends in Food Science & Technology, 51*, 24–33, https://doi.org/10.1016/j.tifs.2016.02.007.
91. Harnedy, P. A., & FitzGerald, R. J., (2012). Bioactive peptides from marine processing waste and shellfish: A review. *Journal of Functional Foods, 4*, 6–24. https://doi.org/10.1016/j.jff.2011.09.001.

92. Mohamed, S., Hashim, S. N., & Rahman, H. A., (2012). Seaweeds: A sustainable functional food for complementary and alternative therapy. *Trends in Food Science and Technology, 23*, 83–96. https://doi.org/10.1016/j.tifs.2011.09.001.
93. Holdt, S.L., & Kraan, S. (2011). Bioactive compounds in seaweed: functional food applications and legislation. *J Appl Phycol 23*, 543–597. https://doi.org/10.1007/s10811-010-9632-5
94. Shahidi, F., & Ambigaipalan, P., (2015). Novel functional food ingredients from marine sources. *Current Opinion in Food Science, 2*, 123–129. https://doi.org/10.1016/j.cofs.2014.12.009.
95. Wells, M. L., Potin, P., Craigie, J. S., Raven, J. A., Merchant, S. S., Helliwell, K. E., & Brawley, S. H., (2017). Algae as nutritional and functional food sources: revisiting our understanding. *Journal of Applied Phycology, 29*, 949–982. https://doi.org/10.1007/s10811-016-0974-5.
96. Chang, S. T., & Buswell, J. A., (1996). Mushroom nutraceuticals. *World Journal of Microbiology and Biotechnology, 12*(5), 473–476. https://doi.org/10.1007/BF00419460.
97. Erjavec, J., Kos, J., Ravnikar, M., Dreo, T., & Sabotič, J., (2012). Proteins of higher fungi-from forest to application. *Trends in Biotechnology, 30*, 259–273. https://doi.org/10.1016/j.tibtech.2012.01.004.
98. Kimatu, B. M., Zhao, L., Biao, Y., Ma, G., Yang, W., Pei, F., & Hu, Q., (2017). Antioxidant potential of edible mushroom (*Agaricus bisporus*) protein hydrolysates and their ultrafiltration fractions. *Food Chemistry, 230*, 58–67. https://doi.org/10.1016/j.foodchem.2017.03.030.
99. Ovando, Claudia Anahite, Carvalho, Julio Cesar de, VinÃcius de Melo Pereira, Gilberto, & Jacques, Philippe, (2018). Functional properties and health benefits of bioactive peptides derived from Spirulina: A review. *Food Reviews International, 34(1)*, 34–51. https://doi.org/10.1080/87559129.2016.1210632.
100. Patel, S., (2019). Immunomodulatory aspects of medicinal mushrooms. In: *Medicinal Mushrooms* (pp. 169–185). https://doi.org/10.1007/978-981-13-6382-5_5.
101. Shalinee Prasad, Rathore, H., Satyawati Sharma, & Anurag Yadav (2015). Medicinal mushrooms as a source of novel functional food. *International Journal of Food Science, Nutrition and Dietetics*, 221–225. https://doi.org/10.19070/2326-3350-1500040.
102. Sánchez, C., (2017). Bio-actives from mushroom and their application. In: *Food Bioactives: Extraction and Biotechnology Applications* (pp. 23–57). https://doi.org/10.1007/978-3-31951639-4_2.
103. Zhou, J., Chen, M., Wu, S., Liao, X., Wang, J., Wu, Q., & Ding, Y., (2020). A review on mushroom-derived bioactive peptides: Preparation and biological activities. *Food Research International, 134*, 109230. https://doi.org/10.1016/j.foodres.2020.109230.

CHAPTER 8

News and Trends in the Development of Functional Foods: Probiotic Dairy and Non-Dairy Products

ELIANE MAURÍCIO FURTADO MARTINS,[1]
WELLINGTA CRISTINA ALMEIDA DO NASCIMENTO BENEVENUTO,[1]
AURÉLIA DORNELAS DE OLIVEIRA MARTINS,[1]
AUGUSTO ALOÍSIO BENEVENUTO JUNIOR,[1]
ISABELA CAMPELO DE QUEIROZ,[1] THAINÁ DE MELO CARLOS DIAS,[1]
DANIELA APARECIDA FERREIRA SOUZA,[1]
DANIELE DE ALMEIDA PAULA,[2] and MAURÍLIO LOPES MARTINS[1]

[1]Federal Institute of Southeast of Minas Gerais, Food Science and Technology Department (DCTA/IF Sudeste MG), Rio Pomba, MG, CEP – 36180-000, Brazil

[2]Federal University of Viçosa, Food Technology Department, Viçosa, MG, CEP – 36570-000, Brazil

ABSTRACT

In recent years, there has been an increase in consumer search for healthy eating that promotes health and well-being. This demand has led to the creation of new market niches, consisting of foods with functional appeal. In this context, probiotics stand out, which are live microorganisms that, if consumed in adequate doses, confer health benefits. *Lactobacillus* and *Bifidobacterium* genera are more widespread in the market and more used in food industry. With the understanding of consumers about the benefits promoted by functional foods containing probiotics, its use expanded rapidly, with the dairy matrix being the most studied for the transport of these microorganisms. Dairy products such as cheese, fermented milk, yogurt, ice cream, infant formula and powdered milk are examples of these products. However, there is a demand for non-dairy products as a matrix of probiotic bacteria, in order to meet the portion of the population with lactose intolerance and allergy to dairy products, in addition to hypercholesterolemic, vegetarian, and those who do not consume dairy products due to habits cultural. Therefore, non-dairy foods such as fruits and vegetables, chocolates, baked, and meat products are also being studied and gaining notoriety as an alternative matrix to the incorporation of probiotics. The maintenance of these microorganisms during the production process, the product's useful life and during the passage through the human GIT is a challenge for the industries. Several intrinsic and extrinsic factors can affect the viability of probiotic cells. Thus, technological alternatives such as microencapsulation have gained prominence for protecting cells from adverse conditions and, consequently, increasing the viability of microorganisms. The incorporation of probiotics in dairy and non-dairy foods points to a promising future, arousing the interest of researchers and industry in the development of new healthy products.

8.1 INTRODUCTION

In recent years, there has been an increase in the search by consumers for a healthy diet that promotes health and well-being, less processed, without preservatives and that helps in the prevention of chronic non-transmissible diseases. To meet this demand, researchers and entrepreneurs have been looking for alternatives for the development of differentiated food products, which contributes to the creation of new market niches, consisting of foods with functional appeal.

This scenario is in line with functional and nutraceutical foods. Functional foods are those that have basic nutritional functions in addition to producing beneficial health effects, and nutraceuticals are bioactive compounds present in foods that play an important role in food, and can be consumed in an isolated form from food, in high doses, such as supplements.

In the scope of functional foods, probiotics stand out, which are live microorganisms that, if consumed in adequate doses, are beneficial to health [39, 54]. There are many benefits attributed to the consumption of probiotic bacteria such as regulation of the intestinal microbiota, reduction of intestinal pathogens, immunostimulation, increased bioavailability of nutrients, elimination of carcinogenic substances, reduction of the incidence of colon tumors, among others.

In order to guarantee consumer health benefits, it is extremely important that as many viable cells as possible are present at the time of ingestion.

According to the international literature, at least 10^6–10^7 CFU/g of probiotic microorganisms should be ingested daily [68].

The probiotic microorganisms most used in food and which are present on the market are those belonging to the genera *Lactobacillus* and *Bifidobacterium* [29, 83]. These two genera are predominant inhabitants of the human intestine, with *Lactobacillus* in the small intestine and *Bifidobacterium*, in the large intestine [135].

Fermented dairy products comprise the highest percentage of probiotic-carrying foods available on supermarket shelves, which are considered good matrices for these microorganisms. However, with the understanding of consumers about the benefits promoted by functional foods containing probiotics, there is a growing demand for other carriers such as ice cream, infant formula, and milk powder. In addition to dairy products, non-dairy foods such as fruits and vegetables products, chocolates, bakery, and meat products have also been studied and gaining notoriety.

8.2 PROBIOTIC IN DAIRY PRODUCTS

The most studied matrix for carrying probiotic microorganisms is milk and dairy products [121] and a variety of dairy products have been formulated with the addition of different probiotic bacteria, according to Table 8.1.

TABLE 8.1 Probiotic Bacteria Species and Strains in Dairy Products

Probiotic	Products	References
Lactobacillus reuteri	Infant formula	[122]
Lactobacillus casei Zhang	Minas frescal cheese	[28]
Lactobacillus plantarum	Yogurt	[143]
Lactobacillus casei	Yogurt	[15, 24, 33]
Lactobacillus plantarum	Milk fermented	[92]
Lactobacillus acidophilus	Milk fermented	[10]
Lactobacillus casei	Milk fermented	[34, 92, 136]
Lactobacillus rhamnosus GG	Pasta filata cheese	[27]
Lactobacillus plantarum	Feta cheese	[103]
Bifidobacterium lactis BB-12	Yogurt	[109]
Bifidobacterium animalis subsp. lactis BB-12	Ice cream sweetened with various polyols	[59]
Weissella cibaria D30 and *Lactobacillus rhamnosus* GG	Cottage cheese	[62]

TABLE 8.1 *(Continued)*

Probiotic	**Products**	**References**
Lactobacillus bulgaricus	Ice cream (milk powder)	[90]
Lactobacillus rhamnosus	Cream cheese	[94]
Lactobacillus casei DG, *L. paracasei* F19, *Lactobacillus paracasei* B21060, *Lactobacillus. rhamnosus* GG, *Lactobacillus rhamnosus* IMC 501 plus; *Lactobacillus paracasei ssp. paracasei* IMC 502	Mozzarella" and "scamorza	[113]
Bacillus coagulans MTCC 5856, *Bacillus coagulans* GBI-30 6086, *Bacillus subtilis* PXN 21, *Bacillus subtilis* PB6 and *B. flexus* HK1	"Requeijão cremoso" processed cheese	[125]
Lactobacillus casei-01	Prato cheese	[139]
Lactobacillus acidophilus	Yogurt	[32, 89]
Lactobacillus rhamnosus B 442	Ice cream (milk with 3.2% fat content)	[73, 102]
Lactobacillus acidophilus	Ice cream	[2]
Lactobacillus rhamnosus HN001 and *Lactobacillus paracasei LBC82*	Goat milk ice cream produced with cajá pulp	[30]
Bifidobacterium animalis ssp. *lactis*	Yogurt	[32]
Lactobacillus rhamnosus	Milk fermented	[100]
Lactobacillus acidophilus La-05	Buffalo milk ricotta cheese	[119]
Lactobacillus rhamnosus GG	Infant formula	[120]

Because to the benefits of probiotic strains to promote health, their use has expanded rapidly in products [143]. In addition to probiotic bacteria provide health benefits to consumers, their use in foods provides different patterns of flavor and texture; each mixture of microorganisms used can result in a specific product [79].

Fermented dairy products have bioactive compounds and metabolites derived from lactic acid bacteria (LAB) produced during fermentation. Due to their special characteristics, fermented dairy products are an excellent matrix for the incorporation of ingredients and/or nutrients that give the final product essential nutritional properties to the human being interested in a healthy diet [46].

8.2.1 CHEESE

The dairy industries have sought to diversify the product offering in order to meet the demands of the consumer market. For this purpose, the development of cheeses containing probiotic bacteria has shown to be a promising alternative.

The viability of probiotic microorganisms, as well as other microorganisms, is affected by factors intrinsic to food such as pH, oxygen availability, water activity, nutrient availability, by processing conditions such as binomial applied time and temperature, period, and storage conditions, making it a challenge to maintain adequate doses and to guarantee the functionality of probiotic products [113].

In this context, considering the intrinsic characteristics, cheeses have shown to be a very promising alternative for carrying probiotic microorganisms, which even stand out over yogurt and fermented milks, such as higher pH values, buffering capacity, solid matrix, which guarantees greater protection during the passage to the gastrointestinal (GI) tract and higher fat content [26, 61].

Figure 8.1 is displayed the characteristics that contribute to preserve the viability of the probiotic in cheeses.

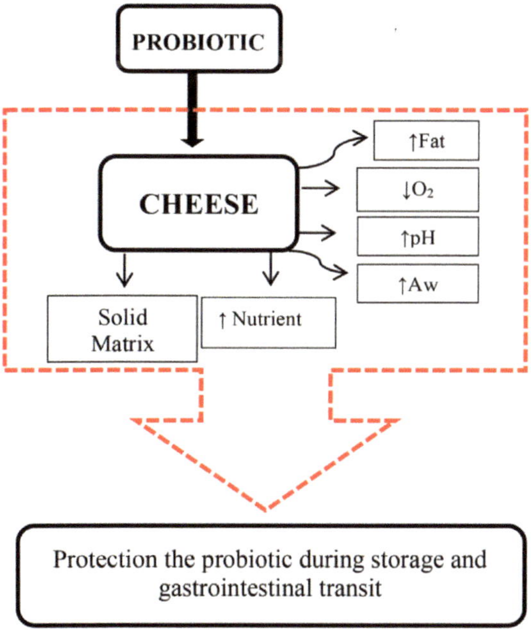

FIGURE 8.1 Characteristics of probiotics cheese.

For the production of probiotic cheeses, two alternatives can basically be adopted: promoting microbial growth after manufacture, during the maturation of the products, or promoting the maintenance of the inoculated microbial load in high concentration. In addition, it is necessary to evaluate, in addition to the final microbial load, which allows the product to be classified as a functional food, the changes caused in the cheeses, which define its acceptability by the consumer [50].

Regarding the form of inoculation, probiotic bacteria can be used as starter cultures or in conjunction with traditional starter cultures, in various combinations of starters and probiotics [99].

Some varieties of cheese, such as ricotta, have characteristics such as high humidity and pH, in addition to reduced salt concentration, which can provide protection to probiotic microorganisms during GI transit. However, their texture and acceptability characteristics may be negatively affected by the cultures. Added probiotics, making it necessary to study alternatives that allow combining the benefits conferred by the matrix and the maintenance of the sensory characteristics essential for the acceptance of cheese by the consumer [119].

Studies seek to evaluate the factors that most affect the development of probiotic cultures added in cheese production.

Reale et al. [119] studied the possibility of large-scale production of two much-appreciated Italian cheese specialties, mozzarella, and scamorza, incorporating commercial probiotic additives by direct inoculum into the vat according to the factory's standard procedure, and found that the filler step reduces the viability of these microorganisms, but that there is a possibility of their recovery during maturation for a minimum period of 30 days. They confirmed that an easy procedure such as direct-to-vat inoculation of lyophilized adjunct probiotic cultures at a concentration of 10^6 CFU/mL could be a proficient method for the production of functional scamorza, guaranteeing with a daily consumption of 100 g of this scamorza cheese a quantity of probiotic bacteria equal to 10^8 CFU/100 g. Furthermore, the probiotic adjunct did not modify the sensory features of the final product.

Cuffia et al. [26] evaluated the use of two probiotic, *Lactobacillus rhamnosus* GG and *Lactobacillus acidophilus* LA5 (either individually or together) to make pasta filata soft cheeses and assessed the effect of the storage temperature (4 and 12°C) on the pH evolution, the viability of both probiotic strains, and the influence on the sensory characteristics of the product. The study showed the storage temperature of probiotic cheeses can play a critical role, and this parameter must be considered to maintain cheese quality. The probiotic cheeses maintained at 4°C did not show significant differences in pH, presented a very good overall quality, and maintained the viability of both probiotics at levels higher than 7 log CFU/g after 29 days of storage while Cheeses stored at 12°C, presented lower pH values, significant post acidification with both probiotic strains toward the end of the ripening period, an increase in bitter, acid taste, granularity, and a decrease in overall quality of cheeses.

As shown in Table 8.1, studies with the addition of probiotic bacteria, of different strains and in different types of cheese have been successfully conducted in several countries.

Lactobacillus casei-01 counts in the Prato cheese samples were higher than 8 log CFU/g in a study conducted for Vasconcelos et al. [139], and the addition of the probiotic culture did not influence the starter culture counts, with similar *Lactococcus lactis* counts in conventional and probiotic cheeses. The search results showed that probiotic Prato cheese attenuates cigarette smoke-induced injuries in mice.

Prezz et al. [112] also evaluated the inoculation of *Lactobacillus rhamnosus* as a probiotic culture in Minas Frescal without negative effect on physicochemical characteristics evaluated.

In research carried out in Brazil [126], with Requeijão cremoso, the possibility of using five probiotic strains of the genus *Bacillus* was evaluated, which were inoculated at different stages of manufacture of this melted cheese. The main challenge for the use of probiotic cultures in this type of cheese is heating to 90°C by which the product is submitted during its manufacture. This high temperature causes the elimination of probiotic bacteria from the genera *Lactobacillus* and *Bifidobacterium.*

The results obtained in this study were very interesting, indicating that the evaluated strains supported the processing conditions of the Requeijão cremoso, without generating a negative impact on the proteolytic characteristics and on the fatty acid profile of the product.

In addition to nutritional interest, probiotic microorganisms can be added to cheeses for the purpose of bio-preservation, improving the microbiological safety of products by inhibiting the development of undesirable microorganisms.

Moghanjougi et al. [86] evaluated the possibility of using *Bifidobacterium animalis* subsp. *lactis* Bb-12 and *L. acidophilus* La-5 encapsulated to produce antifungal agents, seeking out increasing the shelf life of white cheese Feta and demonstrating significantly less results of yeasts during storage.

In Minas Frescal cheese, the addition of *Lactobacillus rhamnosus* reduced the *Listeria monocytogenes* count from the 7th day of storage [112].

It is likely that the inhibitory effect promoted by probiotic bacteria is related to the production of organic acids, bacteriocins or hydrogen peroxide [67], which are natural antimicrobial agents.

8.2.2 FERMENTED MILK AND YOGURT

Probiotics have been used for decades to promote human health benefits and were first known in the forms of yoghurt and fermented milk [115]. Dairy products and especially fermented milk are ideal carriers for the delivery of probiotic bacteria in the gastrointestinal tract (GIT) [109].

Dimitrellou et al. [34] investigated the probiotic *Lactobacillus casei* ATCC 393 encapsulated in alginates using the extrusion technique. The authors evaluated the survival of the microorganism in simulated GI conditions during the production of fermented milk and storage for four weeks at a temperature of 4°C. In the simulated GI conditions and after storage for 28 days in fermented milk, it was observed that encapsulated probiotic bacteria showed greater viability when compared to non-encapsulated microorganisms. Additionally, fermented milk showed a better aroma due to compounds produced by *L. casei*. The authors conclude that economically the use of encapsulated probiotic microorganisms is a sustainable process for the production of fermented milk.

The study focusing on the selection of probiotic lactobacilli as the main initial culture for the production of fermented milk with probiotics without commercial application concluded that five out 44 isolates showed the highest cholesterol removal capability; one strain (*Lactobacillus plantarum* DP3) showed the broadest inhibitory range against pathogens and *L. plantarum* DP3 and *L. casei* DP21 reduced the fermentation time to 10 h. The microbial count of the final products (> 10^8 CFU/mL) guaranteed its probiotic activity after 24 storage days [92].

Additionally, probiotics could have beneficial effects against carcinogenesis [43]. Research was carried out with the probiotic strain *Lactobacillus casei* CRL431 (PFM) to evaluate the immunomodulation exerted by fermented milk when administered to mice in the metastatic stage of breast cancer. The mice that received PFM were compared with the mice that received only milk. It was observed that the administration of PFM reduced the metastasis in the lungs and increased the survival of the mice.

The results of the research show that the modulation of immune cells in the lungs by PFM can be a strategy to fight cells that cause tumor in metastatic sites [136].

Dairy products have the capacity to carry probiotics, being an alternative to reduce the potentially pathogenic microbiota in the oral cavity [121]. Pahumunto et al. [100] evaluated the effect of maltitol and other sugars in milk fermented by *Lactobacillus rhamnosus*-SD11 on the growth and acid production of Streptococcus mutans. The authors found that fermented milk containing *L. rhamnosus*-SD11 with maltitol reduced the growth of *S. mutans*. Children, 123, were selected who randomly consumed the fermented milk with the probiotic or the control for 4 weeks once a day. The results showed that maltitol showed less acid production than simple sugars. Comparing the probiotic group with the control by means of a clinical trial, it was observed that there was a significant reduction in total salivary streptococci and *S. mutans*, while there was a significant increase in salivary lactobacilli after the consumption of probiotic fermented milk.

According to Bhalla et al. [13], dairy products such as cheese, yogurt, ice cream, and fermented milk in general when administered with probiotics tend to reduce buccal pH due to the buffering capacity of these foods. In addition, calcium, and calcium lactate that are present in milk can have anticariogenic properties, regardless these products are added with probiotics.

Yogurt can be produced with different combinations of cultures, including starter cultures and probiotic microorganisms [79]. Yogurt produced with probiotic cultures is a consumer trend and a challenge for the industry in relation to the development of functional foods [143].

Symbiotic yogurt was produced with high counts of probiotic *Lactobacillus acidophilus* and flaxseed using the response surface methodology. The flaxseed concentration varied from 0 to 4% w/w and the shelf life from 1 to 28 days. The authors found that probiotic yogurt added with flaxseed had a higher *L. acidophilus* count (up to 8.82 CFU/mL) compared to the control sample (6.87 CFU/mL). The results indicated that the addition of 4% of flaxseed to probiotic yogurt can result in a functional food with a desirability of 76.8%. In addition, the product maintained adequate properties for about 13 days under cold storage [89].

Delgado-Fernández et al. [32] evaluated the viability of the probiotic microorganisms *Lactobacillus acidophilus* and *Bifidobacterium animalis* ssp. *lactis* in yogurt produced also with the starter culture of *Streptococcus thermophilus* and *Lactobacillus delbrueckii* ssp. *bulgaricus*. In addition to the probiotic, symbiotic yogurts were added lactulose and the lactulose-derived oligosaccharide (OsLu). The addition of lactulose to yogurts significantly increased the count of *L. delbrueckii* ssp. *bulgaricus*, *L. acidophilus* and *B. animalis* ssp. *lactis* during fermentation when compared to the control product.

Wu et al. [143] found that yogurt produced with the probiotic microorganism *L. plantarum* had a higher content of nutrients and better sensory and texture characteristics, being suitable for healthy consumption.

Studies have reported that the addition of encapsulated or free cultures tends to reduce post-acidification during storage in addition to changing the product's texture properties [43]. Pinto et al. [109] evaluated the Greek-style yogurt without lactose as a new matrix to serve spray-dried microcapsules containing the probiotic microorganism *Bifidobacterium lactis* BB-12. For the formulation of the microcapsules, three different materials were used, gum Arabic, inulin, and maltodextrin. In all formulations, the viability of the probiotic was greater than 8 log CFU g^{-1} throughout 30 days of storage at 4°C. The addition of microcapsules did not affect the viability of the initial cultures and increased the pH, firmness, and adhesiveness of the product. After 30 days of storage, the viability of the probiotic microorganism was above 6.5 log CFU g^{-1}, indicating that Greek-style yogurt without lactose may be a good matrix to carry *B. lactis* BB-12.

8.2.3 ICE CREAM

Frozen dairy products are characterized by containing solid dairy compounds that may or may not include milk fat, and are consumed in a frozen solid state. Within the frozen dairy dessert's category, the most consumed product is ice cream [49].

Due to the worldwide popularity and its potential as a nutrients carrier, ice cream has aroused the food industry and the academia's interest in using it as a functional food. For Ferreira et al. [42] functional food is a product that contains nutrients that provide health benefits in addition to basic nutrition.

These foods are a trend of the future, since current science allows the development of processed products that offer health benefits [53].

According to Abdelazez et al. [1], frozen dairy products, such as yogurt and ice cream, maybe the persuasive carriers of probiotics. Probiotic ice cream is a trend in the dairy industry.

The sensory properties or the general quality of food should not be altered by the incorporation of probiotic cells [20].

For Cruz et al. [25], the incorporation of probiotic bacteria in a frozen dairy dessert formulation must not affect the general characteristics of the product. Therefore, it is essential that the physical-chemical parameters responsible for the quality of this product, such as fusion rate and organoleptic characteristics, present equal or even better characteristics when compared to a conventional ice cream.

The carrying of probiotics has been challenging, since microorganisms must survive technologically and physiologically harmful conditions. In addition, bacteria resistant to technological stresses are not always resistant to the environment of the digestive tract (low stomach pH, bile, digestive juices) and vice versa [48].

The determinants of the survival of microorganisms in ice cream include freezing, components of the mix, presence of oxygen and pH [73]. Cold stress is an important issue in the manufacture of probiotic ice creams, since the growth and viability of these microorganisms are influenced by the environmental temperature [41].

According to Tripathi and Giri [135], probiotic bacteria can survive in frozen products for a longer period of time. However, the freezing process can exert various effects on bacteria, depending on their phenotypic traits and the course of the process [73]. Ice crystals forming inside and outside cells during this process can cause mechanical damage to bacterial cells [102].

In addition to the freezing process, the initial injury caused by the increasing incorporation of oxygen, in addition to the homogenization stage during the ice cream processing, negatively interferes with the probiotic viability [41].

Some strategies can be used to minimize the loss in viability of probiotics added to food products. The use of microencapsulation, products that protect the microbial cell, food inputs that favor growth, packaging of materials that form an oxygen barrier, antioxidant agents, and modification of the storage atmosphere allowed microorganisms to survive better in various processes and formulations [135].

The encapsulation process is one of the strategies that favor the protection of microorganisms against technological and physiological agents [48]. Alginate is the most common encapsulating agent to be used due to its mild gelling conditions. Emulsions aiming at encapsulating LAB are generally water-in-oil emulsions, and alginate, pectins, or proteins are mainly used as water phases [20].

Protective agents commonly used to enhance probiotic viability during processing, storage, and transit through the GIT include dairy-based proteins, hydrocolloids (gelatin, gum Arabic) and polyalcohols. Carbohydrates have been the most widely used protective compounds during dehydration, storage, and exposure to the GIT [20].

Recent studies have been carried out with different probiotic species and strains tested in ice cream. For Muhardina et al. [90], the addition of probiotics to ice cream is one of the methods used to improve the product's benefit and functional value. The supplementation of foods with probiotic microorganisms makes it essential to adopt adequate measures to maintain their viability during the processing, packaging, and storage of these products [22].

Calligaris et al. [17] evaluated the potential use of structured emulsions with monoglyceride (anhydrous milk fat or sunflower oil), replacing sour cream, as a delivery system for *L. rhamnosus* in ice cream. The results obtained highlighted that the emulsions used protected *L. rhamnosus* cells against tensions and stresses during the ice cream processing and storage.

The bioactive efficiency of probiotics in the human body is usually compromised due to their low stability during storage and digestion processes [141].

Farias et al. [40] investigated the viability and evaluated the survival of *L. rhamnosus* and *L. casei* in yellow mombin ice cream. Tests were carried out comparing the cultures resistance at low temperature, efficiency in the encapsulated form with calcium alginate-chitosan and cell survivability in a simulated GI environment. The ice cream was stored at –18°C for 150 days. The results demonstrated that encapsulated *L. rhamnosus* ASCC 290 showed higher resistance to low temperature, while *L. casei* ATCC 334 had a higher survival rate during encapsulation and in the GI environment. It was observed that the best option for the preparation of functional yellow mombin ice cream, considering all the cellular losses suffered, is the use of *L. rhamnosus* ASCC 290 in free form or encapsulated of *L. casei* ATCC 334.

In vitro and *in vivo* studies are performed to test the probiotics survival during passage through the GIT. Afzaal et al. [2] evaluated the survival of free and microencapsulated *L. acidophilus* (sodium alginate and carrageenan) over a period of 120 days at –20°C. It was also assessed the survival of free and encapsulated probiotic bacteria under GIT conditions. The study results showed that encapsulation significantly improved ($p < 0.05$) the probiotic cells survival in ice cream compared to free cells, and also in *in vitro* conditions. The encapsulation guaranteed the recommended probiotics level (106 to 108 CFU/g) carried in food, offering health benefits.

A study of Akalin et al. [4] used different fibers in the manufacture of ice cream enriched with *L. acidophilus* and *B. lactis*. The authors evaluated the viability of probiotics over a period of 180 days in 6 ice cream formulations one control sample and the others with apple, orange, oat, and bamboo and wheat fibers. At the end of the study, it was concluded that wheat fiber had the potential to enhance rheological and textural characteristics, maintain sensory properties and probiotic viability. On day 120 of storage, the viability of *B. lactis* was higher in ice cream enriched with wheat fiber when compared to other fibers and the control sample.

8.2.4 INFANT FORMULA AND MILK POWDER

Colonization of the intestine by beneficial microorganisms early in life is essential to establish the barrier of the intestinal mucosa, increase the immune system and prevent infections caused by enteric pathogenic microorganisms. Factors such as mode of delivery, breastfeeding, prematurity, and use of antibiotics influence the beginning of human life. Studies suggest that the use of probiotics in food can prevent disease [12]. In an attempt to establish a microbiota in babies fed formula similar to breastfed ones, infant formula manufacturers are increasingly incorporating probiotics into their products [65].

An observational study was carried out to evaluate the effects of *Lactobacillus reuteri* DSM 17938 on the composition of the microorganisms in the GIT of babies. Fecal samples from 30 hospitalized children who received the microorganism and 30 who did not receive the probiotic were analyzed. The groups were different in composition and quantity of intestinal microbial strains. Babies who received the probiotic had a lower gram-negative total anaerobic count ($p = 0.03$) and a higher gram-positive total anaerobic count ($p = 0.02$). Enterobacteriaceae and enterococci were significantly higher ($p = 0.04$) in the group that did not receive the probiotic and lactobacilli and bifidobacteria did not differ between groups. Children who did not consume probiotics had greater colonization by diarrheal *E. coli* ($p = 0.04$). The results showed the importance of probiotics in intestinal health in pediatric patients. The administration of *L. reuteri* in infancy can reduce colonization of pathogens and improve intestinal health [122].

The prevalence of colic in babies is high and has a significant impact on the lives of children and their families. Study evaluated the effects of probiotic drops in the treatment of colic in children. The 72 children were randomly divided into two groups, the probiotic receiving group (PRR) and the placebo group (PCR). The results showed that there was a significant increase in weight in the PRR group ($p < 0.0001$) and there was no significant growth in the PCR group (p-value 0.437). Regarding fecal consistency, it was observed that there was no significant difference between the groups [3].

Probiotics have been proposed to be beneficial for the treatment and prevention of food allergy [130], once that for children, food allergy is considered a public health problem [120].

Santos et al. [120], in their studies, selected 18 clinical trials, which predominated babies and children of preschool age, and observed that the most used strain was *Lactobacillus rhamnosus* GG, alone or in combination. The most used carriers were capsules and infant formulas, and the intervention period ranged from four weeks to 24 months. Of the 46 evaluations, 27 demonstrated the benefits of using *Lactobacillus rhamnosus* GG. The authors observed that the use of probiotics helps to promote immunomodulation and reduces clinical symptoms.

There are not many powdered milk products on the market that contain probiotics and prebiotics. Although milk is recognized as a nutritionally rich and diverse food, the addition of probiotics is an alternative for improving an individual's intestinal health [16].

Teanpaisan et al. [131] evaluated the count of *Lactobacillus paracasei* SD1 in powdered milk by spray drying and observed that the survival of the microorganism varies according to the processing temperature. The authors concluded that spray drying is a process with potential use in large-scale production, in addition to facilitating the transport and storage of strains of the microorganism. Skimmed-milk powder can be used in different applications such as instant desserts and confectionery products, so it is possible that the culture contained in the powder can be used in a wide range of functional food applications.

Bradford et al. [16] produced powdered milk containing free and immobilized cells of *Lactobacillus plantarum*. The treatments had high viability of the microorganism before and after spray drying. After exposure to simulated gastric and intestinal conditions, counts of the microorganism greater than 8 log UFC/g were found in samples of powdered milk containing free and immobilized cells of *Lactobacillus plantarum*, which is more than recommended for probiotic products.

8.3 PROBIOTICS IN NON-DAIRY PRODUCTS

Probiotic cultures are traditionally carried in dairy products [9, 52], but there is an increasing number of individuals with intolerance and allergy to dairy products, hypercholesterolemic, vegetarians, who need diets with fat restriction and who do not consume dairy products for habits or cultural reasons. These dietary restrictions justify the search for new non-dairy matrices as a vehicle for probiotic bacteria, in order to serve this portion of the population.

The literature highlights that the addition of probiotic cultures in non-dairy products represents a challenge, especially with regard to the survival of these microorganisms [34]. Thus, the development of new non-dairy products must take into account the viability of the microorganism during storage and its survival to the GIT when carried in the product.

8.3.1 FRUITS AND VEGETABLES AND DERIVED PRODUCTS

Fruits and vegetables have been presented as an alternative to insert these bacteria in the diet since they have nutrients that help their growth [83]. These foods have micronutrients and fibers, carbohydrates, and are sources of C and B vitamins, pro-vitamin A, phytosterols, minerals, and phytochemicals, which are essential to the diet and microbial metabolism. Therefore, they reveal a potential to act as a vehicle for probiotics due to their intrinsic characteristics, such as, the presence of natural prebiotics that promote growth and act as protectors of probiotic microorganisms during product life and passage through the GIT [56, 82]. Moreover, fermented fruits and vegetables containing prebiotic compounds are sources of probiotics [56].

The most used species in probiotic products of plant origin are *L. rhamnosus*, *L. acidophilus*, *L. casei*, *L. plantarum* and *B. lactis* [83].

According to Dimitrellou et al. [34] consumers retain their interest and the global market of probiotic foods and beverages is still growing. So, it is believed that these foods have a promising future [9, 81, 117, 144], since there is a great interest from industry and researchers in evaluating new matrices in order to expand the market for fermented or fortified probiotic products [7, 117, 142]. Studies also show that besides foods based on fruits and vegetables being carriers of probiotic bacteria, they have good sensory characteristics [45].

Minimally processed products, juices, fermented and non-fermented fruits and vegetables, pickles, dehydrated fruits, ice cream, chocolate, vegetable appetizer, tea, coffee, fruit puree, among others, have been studied as carriers of probiotics (Table 8.2).

TABLE 8.2 Recent Studies on Plant Matrices as a Substrate for Probiotic Bacteria

Product	Probiotic Employed	Viability (log CFU/g or mL)	References
Minimally processed fruit salad	*L. rhamnosus* HN001	8.49–5 days	[81]
Tomato juice	*L. acidophilus, L. plantarum, L. casei*	> 8.0–3 days	[63]
Juçara and mango mixed juice	*L. rhamnosus* GG	> 8.0–30 days	[88]
Pickles	*L. casei*	8.0–63 days	[38]
Coconut fermented beverage	*L. plantarum* DW12	8.5–3 days	[60]
Cubes of osmotically dehydrated apples	*L. plantarum*	7.0–6 days	[37]
Strawberry ice cream and yacon flour	*L. acidophilus*	> 7.0–150 days	[104]
Chocolate	*L. paracasei, L. acidophilus*	> 6.6–90 days	[71]
Apple *Snacks*	*L. paracasei*	> 6.0–28 days	[5]
Juçara and pineapple mixed juice	*L. rhamnosus* GG	> 7.2–28 days	[18]
Vegetable appetizer	*L. plantarum LP299v* or *L. rhamnosus GG*	> 7.42 and 8.84, respectively	[19]
Tea and coffee	*B. coagulans* MTCC 5856	> 8.0–24 months	[78]
Mixed fermented vegetable juice (purple cabbage, tomato, and carrot)	*L. plantarum*	7.34	[144]

TABLE 8.2 *(Continued)*

Product	Probiotic Employed	Viability (log CFU/g or mL)	References
Mixed guava and beet beverages based on peanut extract and soy extract	*L. rhamnosus GG*	> 7.3–42 days	[87]
Fruit powder	*L. plantarum*	6.12–36 days	[140]
Mixed jussara and mango juice	*L. rhamnosus* GG	> 5.0–90 days	[111]
Carrot juice, apple puree, rice cream	*L. rhamnosus LMG S-30426*	> 8.7–1–2 days	[14]

L. acidophilus TISTR 1338 and *L. casei* TISTR 390 were added to grape juice jelly in the study of Ampornpat and Leenanon [6]. They found a reduction in counts from 9.23 log CFU/g and 8.97 log CFU/g (initial time) to 6.80 log CFU/g and 7.73 log CFU/g respectively, after nine days of storage at 5°C. The survival of bacteria against different harmful factors during product processing and storage depends on the characteristics of each species [135]. Sporulated bacteria can resist high temperatures, including pasteurization and low pH. Therefore, the use of spore-forming probiotic bacteria for the development of probiotic fruit jelly may be an option.

Panda et al. [101] studied prickly pear juice (*Opuntia* sp.) as a substrate for *Lactobacillus fermentum* ATCC 9338 and the product was well accepted by the panellists. Yang et al. [144] also observed in their work that the blended vegetable beverages were approved by the panelists, demonstrating their potential for commercial production.

According to Vivek et al. [140], probiotic fruit juice powder also can be a suitable alternative for dairy-based probiotic powders. They verified *L. plantarum* viability of 6.12 log CFU/g at 36 days of storage of the juice. Then, the authors concluded that the juice powder could present a potential application for the industry.

Khodaei et al. [69] studied *L. plantarum*, *L. casei*, and *Saccharomyces boulardii* incorporated in gelatin and low methoxyl pectin (LMP) films. Their viability was monitored during storage at different temperatures, and it was concluded that the material is suitable for safeguard and deliver LAB (>10^6 CFU/g) during storage time.

Probiotics have the potential to be used for allergies and treatment of various diseases. Lactobacillus probiotics survive the simulated GIT and can inhibiting foodborne microorganisms [117].

Probiotics and their bacteriocin application to fruit and vegetable products provide an important possibility to chemical compounds to control foodborne pathogens and spoilage microorganisms [56].

In this field, the work of George-Okafor et al. [47] showed that probiotic cultures are used in food products as a bio-preservative. They verified an increase in shelf-life of the home-processed tomato paste, suggesting the possible use of cell-free supernatants contained effective biomolecules of *L. plantarum* Cs and *L. acidophilus* ATCC 314 as bio-preservatives in tomato processing for > 25 days.

Probiotics are often reported to modify the gut microbiota structure of host [110]. So, *in vitro,* and *in vivo* studies are very important. *In vitro* studies are widely used and, when compared to *in vivo* testing, they are faster, safer methods and do not have the same ethical restrictions as *in vivo* testing [18, 145], being an efficient methodology to evaluate the probiotics transport along the TGI (Table 8.3).

These studies employ many enzymes such as α amylase, pepsin, lipase, pancreatin, and bovine bile from different sources, in order to simulate the oral, gastric, enteric I and enteric II phases of the *in vitro* assay [84].

The viability of *L. plantarum* LP299v or *L. rhamnosus* GG in vegetable appetizer were evaluated by Campos et al. [19] and our group found that the appetizer is apt to be considered probiotic.

Marcial-Coba et al. [80] studied *B. coagulans* BC4 spores embedded in dried date paste and the gastrointestinal resistance *in vitro* of the bacteria. They found that the product is a suitable vehicle for the delivery of *B. coagulans* spores, since the physical properties of this matrix are not conducive to spore germination and consequently do not affect the stability of the probiotic cells during storage.

A pectin coated dehydrated apple snack containing ≥ 9 log colony forming units/20 g portion of *L. paracasei* was developed by Valerio et al. [137]. The probiotic survived during fruit processing and simulated GIT digestion and the apple surface presented a good visual and nutritional quality which could be maintained for 30 days of storage.

Litchi juice was also used for fermentation of *L. casei*, and it was an effective method that increased the contents of total phenolic, total flavone, and exopolysaccharide [142]. The authors developed an animal experiment, and the results revealed that intake of fermented juice could improve the gut health of mice.

So, the studies demonstrate that fruits and vegetable bases are a promising alternative for add probiotic bacteria being one new potential option for the consumers.

TABLE 8.3 Studies Involving In Vitro Tests of Gastrointestinal Resistance (GIT) of Probiotic Microorganisms Carried by Fruits, Vegetables, and Derivatives

Matrix	**Probiotic**			**References**
	Specie	**GIT Resistance (log CFU/g or mL)**	**Time (Days)**	
Osmotically dehydrated apples	*L. plantarum*	7.0	6	[37]
Jabuticaba juice	*L. rhamnosus* GG	6.0	28	[97]
Tomato and feijoa juices	*Lactobacillus plantarum*	8.33 and 5.78, respectively	After fermentation	[138]
Mango juice	*L. acidophilus, L. pla*ntarum and *L. rhamnosus* GG	> 5.0	28	[44]
Mixed fermented pineapple and jussara juice	*L. rhamnosus* GG	5.6	28	[18]
Vegetable appetizer	*Lactobacillus plantarum LP299v* or *Lactobacillus rhamnosus GG*	8.67 and 9.53, respectively	90	[19]
Orange, mango, and carrot paste	*B. coagulans* BC4 cells and spores	> 8.0	45	[80]
Mixed guava and beet beverages based on peanut extract (MBPP) or soybean extract (MBSP)	*L. rhamnosus* GG	8.34 and 7.50 in BMPP and BMSP, respectively	45	[87]
Mango and carrot mixed juices	*L. plantarum*	> 7.0	35	[98]
Litchi juice	*L. casei*	9.27	After fermentation	[142]

8.3.2 *CHOCOLATE*

Another type of non-dairy and non-fermented food that has been studied as a functional food and specifically as a matrix for probiotic bacteria is chocolate. This product, in addition to being consumed by all age groups worldwide, has shown promising results as a probiotic food [72, 93]. Chocolate has an interesting antioxidant capacity and this capacity increases as the concentration of cocoa is higher, as it has a greater amount of phenolic compounds, proanthocyanidins, and flavan-3-ols [74] and this antioxidant capacity can protect probiotic bacteria against oxidative stress. Cocoa butter is also considered a material that can protect probiotics from the action of the GIT [106] and is present in chocolate.

Kemsawasd et al. [64] incorporated *L. casei* 01 and *L. acidophilus* LA5 in white chocolate, milk chocolate and dark chocolate and stored them at 4 and 25°C for 60 days. As a form of protection for the cultures, the researchers added them to skimmed milk and maltodextrin and carried out spray drying. They found that *L. casei* 01 had better resistance in all types of chocolate and that the cells survived less at 25°C. At 4°C the probiotic survival rates were greater than 6 log CFU/g. The best survival of probiotics in dark chocolate was verified, followed by milk chocolate and finally white chocolate. The authors justify this result due to the greater presence of cocoa in dark chocolate, which contains higher levels of phenolic compounds and, consequently, greater antioxidant activity.

Laličić-Petronijević et al. [76] evaluated the synergistic effect of three probiotic strains microencapsulated with proteins and polysaccharides on quality parameters of milk, semisweet, and dark chocolate at 4 and 20°C for 360 days. The tested bacteria were *L. acidophilus* LH5; *Streptococcus thermophilus* ST3 and *Bifidobacterium breve* BR2 (DUOLAC MIX L3). It was evidenced that *L. acidophilus* LH5 remained at a concentration of 8 log CFU/g regardless of the type of chocolate and the temperature during the 360 days. *S. thermophilus* ST3 had no significant reduction in milk chocolate, but it did in the other two types of chocolate. *B. breve* BR2 had a significant reduction at 4°C in milk chocolate after 90 days and after 30 days in the other two types of chocolate, reaching 3 log CFU/g, showing its greater fragility in relation to this matrix in relation to other bacteria. As it was a mixture of probiotic bacteria and their combined concentration was 8 log CFU/g at the end of the storage period, it can be considered that it can bring health benefits. The authors reported the success of the experiment due to the insertion of probiotic bacteria at a lower temperature (32°C) and highlighted the possibility of selling the refrigerated product and at room temperature. All chocolates containing probiotics showed excellent general sensory quality throughout the storage period, and furthermore, their rheological properties were not affected.

Another study that evaluated microencapsulation as a form of protection against probiotic bacteria was that of Nambiar et al. [91]. The authors developed milk chocolate containing *L. plantarum* HM47 microencapsulated with soft coconut water, *Moringa oleifera* gum and maltodextrin, storing it at 25°C for 180 days and after this period the cells remained viable in concentration above 8 log CFU/g. The bacteria do not alter the pH of the chocolate, indicating low metabolic activity, due to the microencapsulation and low water activity of this matrix, according to the authors. The product was very well accepted sensorially and also did not cause toxicity in rats. An increase in lactobacilli was also observed in the microbiota intestinal of mice that consumed probiotic chocolate compared to those that consumed only probiotic bacteria without chocolate, indicating that this matrix was able to provide additional protection to the probiotic to the action of the GIT.

The incorporation of *L. acidophilus* NCFM® and *B. lactis* HN019 and milk and dark chocolate was the object of study by Laličić-Petronijević et al. [76]. *L. acidophilus* NCFM® had better viability than *B. lactis* HN019 in both types of chocolate, mainly at a temperature of 4°C, with values above 8 log CFU/g after 180 days at this temperature, whereas *B. lactis* HN019 had its concentration above 7 log CFU/g for up to 90 days, which was reduced to below 6 log CFU/g after 150 days, no longer providing adequate quantity for consumption. The rheological properties of chocolates did not change much with the presence of probiotics, only the appearance of granulation, but this did not interfere with the sensory properties of the products.

Mirković et al. [85] developed dark chocolate containing *L. plantarum* 564 (potential probiotic) and *L. plantarum* 299v (commercial probiotic) and evaluated their influence on volatile compounds and sensory characteristics for 360 days storage at room temperature. The bacteria were washed and then resuspended in skimmed milk, being encapsulated by spray drying, and their addition was carried out at a temperature of about 30°C. The bacteria reached a concentration of 8 log CFU/g after 60 days of storage and 6 log CFU/g after 180 days, demonstrating an ideal shelf life of the product up to 6 months. The authors observed no difference in the volatile profile of chocolates, without affecting the sensory characteristics of the product.

Succi et al. [129] evaluated survival during simulated GI transit of some commercial probiotic formulations in dark chocolate (80% cocoa) after 90 days of storage. The bacteria in question were *L. rhamnosus* GG; *L. paracasei* F19. Chocolate offered bacteria good protection from simulated GI stress conditions.

These studies are encouraging for companies and institutions to invest in the development of probiotic chocolates, enabling a wide variety of people to consume these bacteria that bring health benefits, combined with the pleasure of consuming this type of food.

8.3.3 PROBIOTICS IN BAKERY PRODUCTS

Due to its high daily consumption around de the world, bread is considered an interesting non-dairy food vehicle for probiotics. It is the major type of bakery products and a staple food in most parts of the world. A typical, traditional bread-making is largely wheat-flour based, involving dough mixing, proving (i.e., fermentation) and baking.

Bread is identified as a potential food that can be enriched with probiotics due to the presence of a non-digestible carbohydrates like oligosaccharide that have been suggested to promote growth of probiotic bacteria, moreover, high-quality sourdough breads can be found by inoculation LAB [147]. Thus, the study of probiotics in bakery products have increased in recent years [123, 126, 147].

Soukoulis et al. [128] studied a probiotic pan bread constructed by the application of film-forming solutions (sodium alginate or sodium alginate and whey protein concentrate) containing *Lactobacillus rhamnosus GG*. The bread crust surface containing probiotic did not differ from the control bread. The *L. rhamnosus GG* viability was reduced during the first 24 h of storage, and viable count losses were low during the 2–3 days of storage, and growth was observed upon the 4–7 days of storage. Based on their calculations, an individual 30 e 40 g bread slice can deliver approx. 7.57–8.98 log CFU/portion before *in-vitro* digestion a 6.55–6.91 log CFU/portion after *in-vitro* digestion, meeting the World Health Organization (WHO) recommended. However, Soares et al. [126] reported that given the complexity and high cost of using edible films to prepare probiotic bread may not be economically viable for the industry.

The incorporation of probiotics in bakery products is challenging due to their viability and sensitivity to the high temperatures during baking. The methodology combining microencapsulation and edible coatings has been successfully.

In 2016, *Lactobacillus acidophilus* was encapsulated in the Seyedain-Ardabili, Sharifan, and Tarzi study [123]. *Lactobacillus acidophilus* LA-5 and *L. casei* 431 were encapsulated (with calcium alginate and Hi-maize resistant starch), coated (chitosan) and inoculated (1 g of microencapsulated bacteria per 100 g of final product) into the two types of bread (hamburger buns and white pan bread). The hamburger buns were baked for 15 min and white pan bread baked for 25 min, both at 180°C. The initial cells count before and after encapsulation was approx. 10^{11} CFU/g. With an encapsulated, viable cells survived the baking process and both types of bread met the standard for probiotic products. Using the chitosan coating, a significant increase in probiotic survival was observed. Among the probiotics, *L. casei* 431 was more resistant to high temperature than *L. acidophilus* LA-5, so this study showed that bacteria in unfavorable conditions are dependent on species. The type of bread, also, affected the bacteria survival, which was higher in hamburger bun, probably due to the shorter baking time, than in white pan bread. This study, additionally, indicated that the production of symbiotic bread using microencapsulation is possible and increase the viability and thermal resistance of probiotic bacteria in breads [123].

Zhang et al. [147] studied the better impact of different bread sizes, baking temperature and subsequent storage on the survival of probiotic bacteria (*Lactobacillus plantarum P8*). The viability of bacteria was evaluated for both bread crust and crumb. They showed approx. 4–5 log reduction of viable counts of *L. plantarum* in both the crust and crumb of 5, 30, or 60 g bread after baking at 175, 205, or 235°C for 8 min, while the initial viable count (N0) in dough was 8.8 ± 0.1 log CFU/g. Different bread sizes had little influence on survival of probiotics during 8 min baking, every way, higher residual viability can be obtained in less baking time. The bacteria in the bread crust were found more stable than those in the crumb despite the higher temperature in the crust during baking under certain conditions. The lower moisture content and the dense and glassy microstructure of the crust may have a positive effect on the thermo-stability of bacteria in the crust during baking. After 4-day storage, the population of probiotic bacteria restored to an amount higher than 10^6 CFU/g in the crumb and 10^8 CFU/g in the crust. These authors suggest that future research should also focus on strategies, such as micro-encapsulation or optimization of processing parameters, to retain high viability of probiotics after baking.

Thus, in the same year, Zhang et al. [148] investigated the survival of dried probiotics subjected to isothermal heating and bread baking by encapsulating. In this study, *Lactobacillus plantarum* P8 was freeze-dried in four different matrices (reconstituted skim milk: RSM, gum Arabic: GA, maltodextrin, and inulin) as protectants. The probiotic powders were added to bread by distributing it on the surface of the dough, and a control group was made without adding this probiotic. The RSM matrix showed the highest protective effect on cells during baking at either 100°C or 175°C, followed by the inulin matrix. However, no protective effect was observed for gum Arabic and maltodextrin matrices during baking.

Another way that can help overcome various technical challenges and expand the possibility of applying probiotic microorganisms in products which are submitted to extremely high temperatures is the use of sporulated bacteria with potential probiotic property.

In the Soares et al. [125] study, the bread dough was separately inoculated with *Bacillus* strains (*Bacillus subtilis* PXN 21, *Bacillus coagulans* GBI30 6086 and *Bacillus coagulans* MTCC 5856) and were baked at 180°C for 20 min. The loaves were packaged in plastic bags and stored at room temperature (approximately 25°C) for seven days, and the counts of the strains with probiotic properties were determined throughout the product's shelf-life. The counts of *Bacillus* strains with probiotic properties showed reductions below 1 log cycle in the bread samples. The counts of *B. subtilis* PXN 21 at the end of seven days (4.4 log CFU/g) was lower than that initially observed (5.1 log CFU/g), indicating less resistance or possible germination during the storage. On the other hand, the counts of *B. coagulans* strains remained practically stable at the same time (7 log CFU/g). These findings indicate that the reduction in the counts of *Bacillus* strains with probiotic properties should be considered and "corrected" to allow the desired dose of these microorganisms to be achieved in commercial products.

8.3.4 MEAT PRODUCTS

Meat is the source of protein, group B vitamins and minerals important for health. However, they are not sources of carbohydrates and dietary fiber and face negative criticism due to the presence of saturated fats and cholesterol. However, with meat processing, it is possible to introduce these missing nutrients, obtain greater nutritional balance in meat products and produce functional meat products [95].

Studies focused on functional meat products have fermented meat products as an excellent source of microorganisms with probiotic characteristics

[146]. According to Ojha et al. [96], consumer demands for high quality meat products including fermented meats.

The bioprotective action on fermented meat products is possible with the use of probiotics and LAB [11, 96].

Target products include kinds of dry sausage that are processed by fermentation without heating or only mildly heated. In fermented sausages, sugar is fermented by bacteria in an anaerobic environment to produce acid, which lowers pH. The ingredients found in a fermented sausage include meat, fat, salt, spices, sugar, and nitrite. De Vuyst et al. [31] alert about the potential negative impact of the fermented sausages environment with high content of curing salt and its low pH and water activity on bacteria viability.

Studies with probiotic bacteria in meat are usually carried out in two stages. In a first stage, Kołozyn-Krajewska, and Dolatowski [70] argue about the need to verify whether the probiotic bacterial strains that can be used in the manufacture of dry fermented meat products should be able to survive under the conditions found in fermented products; in addition, they must dominate other microorganisms found in the finished product. Probiotic contribution in food biopreservation act extending shelf life of food and their ability to inhibit spoilage and foodborne pathogens [114]. So, they can act both as a probiotic and bioprotective culture.

Reviews of probiotics in meat products such as De Vuyst et al. [31]; Khan et al. [66]; Kołozyn-Krajewskaa and Dolatowski [70]; Rouhi et al. [118], presented many studies about the potential use of probiotic bacteria in meat products.

Recent studies verified interesting results. Quantities of 8.0 log CFU/g of *L. paracasei* LPC02 was inoculated in dry sausage, and at the end of the process, the average count in was 7.59 ± 0.37 log CFU/g, showing that it was able to dominate the endogenous microbiota of the meat batter [23].

Example of probiotic culture like bioprotective was observed by Trabelsi et al. [134], which incorporated probiotic strain in raw minced beef meat during refrigerated storage. The results showed that the incorporation of the probiotic strain can inhibit the proliferation of spoilage microorganisms, such as *Listeria monocytogenes* and *Salmonella* spp., delay the lipid oxidation, improve texture parameters, extend the shelf life, and then can be used as a biopreservative agent for extending the safety and quality.

In a second stage, is required human clinical studies documenting health promoting effects. In contrast to the successes obtained in the dairy industry in meat industry, human studies using probioticsprobiotics in fermented meats have been very scarce. Jahreis et al. [55] checked the effect of the daily consumption probiotic sausage containing *L. paracasei* in healthy volunteers for several weeks showed only moderately successful in fecal samples. A statistically significant increase in the numbers of bacteria was observed for some of the volunteers.

Lactobacillus rhamnosus dominated fermented sausages during the ripening process and temporarily colonized the GIT of healthy volunteers, confirming that this strain could be delivered as a potential probiotic [57].

A German producer launched a salami product containing three intestinal LAB strains (*L. acidophilus*, *L. casei*, *Bifidobacterium* spp.). This product is claimed to have health benefits and is thought to be the first probiotic-like salami product to be marketed. Japanese producers too began to market a new range of meat spread products fermented with probiotic LAB *L. rhamnosus* FERM P-15120 [8]. Despite commercial examples, the commercial application of probiotic microorganisms in dry fermented meat products is not yet common.

Use of different probiotic administration strategies for use in meat products to ensure the viability of microorganisms mainly for cooked products like frankfurter sausages, have microorganism immobilization techniques such as encapsulation [21]. Pérez-Chabela et al. [107] observed that thermotolerant LAB encapsulated with acacia gum and sprayed dry were inoculated in cooked meat batters has enhanced initial count.

Probiotic delivery strategies for use in meat products like encapsulation techniques are promising. The potential use of probiotic bacteria in meat products is a reality, although human studies using probiotic fermented meats have been very scarce. Now, dominated by dairy products, but in future probiotic meat products will become an important part of the meat processing industry.

8.4 TECHNOLOGICAL PROBLEMS AND CHALLENGES FOR ADDING AND MAINTAINING PROBIOTICS IN DAIRY AND NON-DAIRY FOODS: TECHNOLOGIES USED TO MAINTAIN THE VIABILITY OF PROBIOTICS DURING STORAGE

The maintenance of probiotic microorganisms during production processing, the product shelf life, and during passage through the human GIT is a challenge for industries. Factors such as temperature, pH, salt content, added of preservatives and oxygen can affect the viability of probiotic cells. Thus, technology alternatives can be used aiming to reduce the damage on probiotics, among them, the microencapsulation has gained importance.

The microencapsulation technique is based on the incorporation of probiotic cells into an encapsulating matrix or membrane forming a physical barrier between the microorganism and the external environment [124]. Thus, the main objective of microencapsulation is to protect the probiotic cells from adverse conditions and, consequently, increase the viability of microorganisms. The microencapsulated probiotics are release in appropriate locations (e.g., small intestine), promoting adhesion and colonization in the intestinal epithelium.

Therefore, modern, and innovative methods have been developed over the last decades to creating a wide variety of probiotic microcapsules [105]. Although many industries prefer economic processes thus the balance between cost and benefit must be considered. Some technologies require specific devices or materials affecting production costs. The following are the most important technologies used for probiotics encapsulation:

1. **Spray-Dryer:** The basic principle of the spray drying process is the atomization of a polymer suspension containing probiotic cells. Thus, the suspension with appropriate viscosity or consistency is sprayed into contact with a hot stream and instantly producing powdered microcapsules. Heat and mass transfer occur simultaneously from air to atomized drops and vice versa, respectively [116].
2. **Freezing Dryer:** The fundamental principle in freeze-drying consists in the removal of the water present in the material by sublimation. The microencapsulation of probiotics using the freeze-drying stands out when compared to other techniques because drying using low temperatures increases the survival rate of probiotic cells [77]. Besides, freeze-dried products reconstituted very easily and have a long shelf life [58].
3. **Extrusion:** The principle of this technique consists in the extrusion of a liquid mixture containing the polymers of the wall material and probiotic cells through in the form of droplets into a hardening solution or setting bath [35]. For the microencapsulation of probiotics by extrusion, various polymers can be used but alginate, carrageenan, and pectin are the most frequent. The hardening solution contains the divalent salts, generally calcium and potassium, because the polymers used form gels when they in contact with minerals solutions.

4. **Complex Coacervation:** It is a liquid-liquid phase separation process, which occurs spontaneously from mixed solutions of polymers with opposite charges. In complex coacervation, at a specified pH, occur the electrostatic interaction between polymers with opposite charges [36]. This causes a formation and aggregation of nearly neutral complexes to reduce the free energy of the system, becoming insoluble. Subsequently, there is a phase separation process; an insoluble polymer-rich phase that contains the coacervate complex stabilized, and a solvent-rich phase [132]. During the encapsulation, the coacervated complex is deposited around the active ingredient (core) leading to sedimentation of encapsulated core [133].
5. **Emulsification Techniques:** The principle of this technique is based on the formation of an emulsion with a continuous phase (generally vegetable oil) and a dispersed phase where probiotic and hydrocolloid will be present. The mixture containing the continuous phase and the dispersed phase is homogenized. After the homogenization, occur the formation of the water-in-oil (W/O) emulsion and, thus, resulting in aqueous/polymeric droplets containing the probiotic cells. After the formation of the emulsion, the polymer present in the aqueous droplets can be gelled. In this case, the emulsification technique can be divided into internal or external ionic gelation process [127]. The coating material used must have some, for example, easy workability during encapsulation; solubility in solvents suitable for use in food; low viscosity at high concentrations; protection of the core against adverse conditions, food-grade status; emulsifying and stabilizing capacity; non-reactivity with the core; they must easily release solvents or other materials used during the process, absence of taste or odor, and low cost [51]. Among the coating material, the following stand out gelatin, whey protein, and various plant proteins, Arabic gum, chitosan, pectin, alginates, xanthan gum, carrageenan, and carboxymethyl cellulose.

Therefore, the use of microencapsulated probiotics is promising and represents an alternative to increase the supply of probiotic products, meet consumer demand and, also allow a wider application of probiotics in the market. However, studies should be executed to select the encapsulation technique, the efficient coating material, and appropriate conditions for the process. Consequently, allowing the formation of microcapsules suitable for applications in food.

8.5 FINAL CONSIDERATIONS

Fortification of foods with probiotics is a tendency. Research on probiotic products with heterogeneous food matrices is necessary in order to meet consumer demand. Thus, it is believed that the functional food industry has a promising future and should focus on research and development of biofortified products.

The carrier vehicle is very important for probiotic survival in order to maintain its positive effects. An industry challenge is to ensure that the viability of probiotics is maintained throughout the shelf life of products at an appropriate cost-benefit in order to provide benefits to consumers in general.

In addition, to be used for the health of the host, probiotics have an advantage once they can be used as bioprotective cultures in food in order to avoid spoilage. Our research group works on enriching foods with probiotic bacteria such as *L. rhamnosus*, *L. plantarum*, *L. acidophilus*, *L. casei*, and *Bacillus coagulans* and *in vitro* and *in vivo* resistance to TGI, being the results promising.

KEYWORDS

- **carrageenan**
- **gastrointestinal resistance**
- **low methoxyl pectin**
- **oligosaccharide**
- **placebo group**
- **probiotic receiving group**

REFERENCES

1. Abdelazez, A., Muhammad, Z., Zhang, Q. X., Zhu, Z. T., Abdelmotaal, H., Sami, R., & Meng, X. C., (2017). Production of a functional frozen yogurt fortified with *Bifidobacterium* spp. *Biomed Res. Int., 4*, 1–10.
2. Afzaal, M., Khan, A. U., Saeed, F., Arshad, M. S., Khan, M. A., Saeed, M., Maan, A. A., et al., (2020). Survival and stability of free and encapsulated probiotic bacteria under simulated gastrointestinal conditions and in ice cream. *Food Sci Nutr., 8*, 1649–1656.
3. Ahmadipour, S., Fallahi, A., & Rahmani, P., (2020). Probiotics for infantile colic. *Clin. Nutr. Exp., 31*, 1–7.

4. Akalin, A. S., Kesenkas, H., Dinkci, N., Unal, G., Ozer, E., & Kinik, O., (2018). Enrichment of probiotic ice cream with different dietary fibers: Structural characteristics and culture viability. *J. Dairy Sci., 101*, 1, 37–46.
5. Akman, P. K., Uysal, E., Ozkaya, G. U., Tornuk, F., & Durak, M. Z., (2019). Development of probiotic carrier dried apples for consumption as snack food with the impregnation of *Lactobacillus paracasei*. *LWT-Food Sci and Technol., 103*, 60–68.
6. Ampornpat, W., & Leenanon, B., (2017). Survival of probiotic bacteria in fruit juice jelly products. *Proceedings of International Postgraduate Symposium on Food, Agriculture and Biotechnology* (pp. 63–70). Maha Sarakham, Thailand.
7. Argyri, A. A., Panagou, E. Z., & Tassou, C. C., (2016). Probiotics from the olive microbiota. In: Farnworth, E. R., & Champagne, C. P., (eds.), *Probiotics, Prebiotics, and Synbiotics, Bioactive Foods in Health Promotion* (pp. 371–389). Academic Press: Cambridge, Chapter 25.
8. Arihara, K., (2006). Strategies for designing novel functional meat products. *Meat Sci., 74*, 219–229.
9. Aspri, M., Papademas, P., & Tsaltas, D., (2020). Review on non-dairy probiotics and their use in non-dairy based products. *Fermentation, 6*, 1–20.
10. Azevedo, P. O. D. S. D., Aliakbarian, B., Casazza, A. A., LeBlanc, J. G., Perego, P., & Oliveira, R. P. D. S., (2018). Production of fermented skim milk supplemented with different grape pomace extracts: Effect on viability and acidification performance of probiotic cultures. *PharmaNutrition, 6*, 64–68.
11. Belgacem, Z. B., Abriouel, H., Omar, N. B., Lucas, R., Martínez-Canamero, M., Gálvez, A., & Manai, M., (2010). Antimicrobial activity, safety aspects, and some technological properties of bacteriocinogenic *Enterococcus fecium* from artisanal Tunisian fermented meat. *Food Control., 21*, 462–470.
12. Bertelsen, R. J., Jensen, E. T., & Ringel-Kulka, T., (2016). Use of probiotics and prebiotics in infant feeding. *Best Pract. Res. Clin. Gastroenterol., 30*, 39–48.
13. Bhalla, M., Ingle, N. A., Kaur, N., & Yadav, P., (2015). Mutans streptococci estimation in saliva before and after consumption of probiotic curd among school children. *J. Int. Soc. Prevent Communit. Dent., 5*(1), 31–34.
14. Bocchi, S., Sagheddu, V., Elli, M., Lim, C. Y., & Morelli, L., (2020). the synergistic interaction between probiotics and food affects their beneficial features. *Advances in Nutrition and Food Science, 2020*(02), 1–12.
15. Bosnea, L. A., Kopsahelis, N., Kokkali, V., Terpou, A., & Kanellaki, M., (2017). Production of a novel probiotic yogurt by incorporation of *L. casei* enriched fresh apple pieces, dried raisins, and wheat grains. *Food Bioprod Process, 102*, 62–71.
16. Bradford, R., Reyes, V., Bonilla, F., Bueno, F., Dzandu, B., Liu, C., Chouljenko, A., & Sathivel, S., (2019). Development of milk powder containing *Lactobacillus plantarum* NCIMB 8826 immobilized with prebiotic hi-maize starch and survival under simulated gastric and intestinal conditions. *Food Production, Processing and Nutrition*, 1–10.
17. Calligaris, S., Marino, M., Maifreni, M., & Innocente, N., (2018). Potential application of monoglyceride structured emulsions as delivery systems of probiotic bacteria in reduced saturated fat ice cream. *LWT-Food Sci and Technol., 96*, 329–334.
18. Campos, P. A., Martins, E. M. F., Martins, M. L., Martins, A. D. O., Leite, Jr. B. R. C., Silva, R. R., & Trevizano, L. M., (2019b). *In vitro* resistance of *Lactobacillus plantarum* LP299v or *Lactobacillus rhamnosus* GG carried by vegetable appetizer. *LWT-Food Sci and Technol., 116*.

19. Campos, R. C. A. B., Martins, E. M. F., Pires, B. A., Peluzio, M. C. G., Campos, A. N. R., Ramos, A. M., Leite, Jr. B. R. C., et al., (2019a). *In vitro* and *in vivo* resistance of *Lactobacillus rhamnosus* GG carried by a mixed pineapple (*Ananas comosus* L. Merril) and jussara (*Euterpe edulis* Martius) juice to the gastrointestinal tract. *Food Res. Int., 116*, 1247–1257.
20. Cassani, L., Gomez-Zavaglia, A., & Simal-Gandara, J., (2020). Technological strategies ensuring the safe arrival of beneficial microorganisms to the gut: From food processing and storage to their passage through the gastrointestinal tract. *Food Res. Int., 129*, 108852.
21. Cavalheiro, C. P., Ruiz-Capillas, C., Herrero, A. M., Jiménez-Colmenero, F., Menezes, C. R., & Fries, L. L. M., (2015). Application of probiotic delivery systems in meat products. *Trends Food Sci. Tech., 46*(1), 120–131.
22. Chugh, B., & Kamal-Eldin, A., (2020). Bioactive compounds produced by probiotics in food products. *Curr. Opin. Food Sci., 32*, 76–82.
23. Coelho, S. R., Lima, I. A., Martins, M. L., Júnior, A. A. B., Robledo, D. A. T., Filho, R.A.T., Ramos, A. L. S., & Ramos, E. M., (2019). Application of *Lactobacillus paracasei* LPC02 and lactulose as a potential symbiotic system in the manufacture of dry-fermented sausage. *LWT-Food Sci. and Technol., 102*, 254–259.
24. Costa, G. M., Paula, M. M., Barão, C. E., Klososki, S. J., Bonafé, E. G., Visentainer, J. V., Cruz, A. G., & Pimentel, T. C., (2019). Yoghurt added with *Lactobacillus casei* and sweetened with natural sweeteners and/or prebiotics: Implications on quality parameters and probiotic survival. *Int. Dairy J., 97*, 139–148.
25. Cruz, A. G., Antunes, A. E. C., Sousa, A. L. O. P., Faria, J. A. F., & Saad, S. M. I., (2009). Ice-cream as a probiotic food carrier. *Food Res. Int., 42*, 1233–1239.
26. Cuffia, F., George, G., Renzulli, P., Reinheimer, J., Meinardi, C., & Burns, P., (2017). Technological challenges in the production of a probiotic pasta filata soft cheese. *LWT-Food Sci and Technol., 81*, 111–117.
27. Cuffia, F., Pavón, Y., George, G., Reinheimer, J., & Burns, P., (2019). Effect of storage temperature on the chemical, microbiological, and sensory characteristics of pasta filata soft cheese containing probiotic lactobacilli. *Food Sci. Technol. Int., 7*, 588–596.
28. Dantas, A. B., Jesus, V. F., Silva, R., Almada, C. N., Esmerino, E. A., Cappato, L. P., Silva, M. C., et al., (2016). Manufacture of probiotic minas frescal cheese with *Lactobacillus casei* zhang. *J. Dairy Sci., 99*, 18–30.
29. Davis, C., (2014). Enumeration of probiotic strains: Review of culture-dependent and alternative techniques to quantify viable bacteria. *J. Microbiol. Methods, 103*, 9–17.
30. De Paula, C. M., Dos, S. K. M. O., Oliveira, L. S., Oliveira, J. D. S., Buriti, F. C. A., & Saad, S. M. I., (2020). Fat substitution by inulin in goat milk ice cream produced with cajá (*Spondias mombin*) pulp and probiotic cultures: Influence on composition, texture, and acceptability among consumers of two Brazilian regions. *Emir. J. Food Agr., 32*(2), 140–149.
31. De Vuyst, L., Falony, G., & Leroy, F., (2008). Probiotics in fermented sausages. *Meat Sci., 80*, 75–78.
32. Delgado-Fernández, P., Hernández-Hernández, O., Olano, A., Moreno, F. J., & Corzo, N., (2020). Probiotic viability in yoghurts containing oligosaccharides derived from lactulose (OsLu) during fermentation and cold storage. *Int. Dairy J. [Online], 102*. https://doi.org/10.1016/j.idairyj.2019.104621.

33. Demirci, T., Aktas, K., Sözeri, D., Öztürk, H. I., & Akın, N., (2017). Rice bran improve probiotic viability in yoghurt and provide added antioxidative benefits. *J. Funct. Foods, 36*, 396–403.
34. Dimitrellou, D., Kandylisa, P., Levićc, S., Petrovićc, T., Ivanovićd, S., Nedovićc, V., & Kourkoutas, Y., (2019). Encapsulation of *Lactobacillus casei* ATCC 393 in alginate capsules for probiotic fermented milk production. *LWT-Food Sci. Technol., 116*, 108501.
35. Đorđević, V., Balanč, B., Belščak-Cvitanović, A., Lević, S., Trifković, K., Kalušević, A., Kostić, I., et al., (2015). Trends in encapsulation technologies for delivery of food bioactive compounds. *Food Eng. Rev., 7*, 452–490.
36. Eghbal, N., Yarmand, M. S., Mousavi, M., Degraeve, P., Oulahal, N., & Gharsallaoui, A., (2016). Complex coacervation for the development of composite edible films based on LM pectin and sodium caseinate. *Carbohydr. Polym., 151*, 947–956.
37. Emser, K., Barbosa, J., Teixeira, P., & Morais, A. M. M. B. D., (2017). *Lactobacillus plantarum* survival during the osmotic dehydration and storage of probiotic cut apple. *J. Funct. Foods, 38*, 519–528.
38. Fan, S., Breidt, F., Price, R., & Díaz, I. P., (2017). Survival and growth of probiotic lactic acid bacteria in refrigerated pickle products. *J. Food Sci., 82*, 167–173.
39. FAO, (2001). *Food and Agriculture Organization of United Nations (FAO)/World Health Organization (WHO)*. Evaluation of Health and Lactic Acid Bacteria. Report of a Joint FAO/WHO Expert Consultation, Córdoba, Argentina.
40. Farias, T. G. S. D., Ladislau, H. F. L., Stamford, T. C. M., Medeiros, J. A. C., Soares, B. L. M., Arnaud, T. M. S., & Stamford, T. L. M., (2019). Viabilities of *Lactobacillus rhamnosus* ASCC 290 and *Lactobacillus casei* ATCC 334 (in free form or encapsulated with calcium alginate-chitosan) in yellow mombin ice cream. *LWT-Food Sci Technol., 100*, 391–396.
41. Ferraz, J. L., Cruz, A. G., Cadena, R. S., Freitas, M. Q., Pinto, U. M., Carvalho, C. C., Faria, J. A. F., & Bolini, H. M. A., (2012). Sensory acceptance and survival of probiotic bacteria in ice cream produced with different overrun levels. *J Food Sci., 77*, 1, S24–S28.
42. Ferreira, J. M. M., Azevedo, B. M., Luccas, V., & Bolini, H. M. A., (2017). Sensory profile and consumer acceptability of prebiotic white chocolate with sucrose substitutes and the addition of goji berry (*Lycium barbarum*). *J. Food Sci., 82*(3), 818–824.
43. Frakolaki, G., Giannou, V., Kekos, D., & Tzia, C., (2020). A review of the microencapsulation techniques for the incorporation of probiotic bacteria in functional foods. *Crit. Rev. Food Sci. Nutr.* [Online]. https://doi.org/10.1080/10408398.2020.1761773 (accessed on 19 June 2021).
44. Furtado, L. L., Martins, M. L., Ramos, A. M., Silva, R. R., Leite, Jr. B. R. C., & Martins, E. M. F., (2019). Viability of probiotic bacteria in tropical mango juice and the resistance of the strains to gastrointestinal conditions simulated *in vitro*. *Semina: Ciênc. Agrár., 40*(1), 149–162.
45. Galgano, F., Condelli, N., Caruso, M. C., Colangelo, M. A., & Favati, F., (2015). Probiotics and prebiotics in fruits and vegetables: Technological and sensory aspects. In: Rai, V. R., & Bai, J. A., (eds.), *Beneficial Microbes in Fermented and Functional Foods* (pp. 189–206). CRC Press: Boca Raton, London, Chapter 10.
46. García-Burgos, M., Moreno-Fernández, J., Alférez, M. J. M., Díaz-Castro, J., & López-Aliaga, I., (2020). New perspectives in fermented dairy products and their health relevance. *J. Funct. Foods, [Online], 72*. https://doi.org/10.1016/j.jff.2020.104059.

47. George-Okafor, U., Ozoani, U., Tasie, F., & Mba-Omeje, K., (2020). *The Efficacy of Cell-Free Supernatants from Lactobacillus Plantarum Cs and Lactobacillus Acidophilus ATCC 314 for the Preservation of Home-Processed Tomato-Paste*. Scientific African.
48. Ghibaudo, F., Gerbino, E., Dall,' O. V. C., & Gómez-Zavaglia, A., (2017). Pectin-iron capsules: Novel system to stabilize and deliver lactic acid bacteria. *J. Funct. Foods, 39*, 299–305.
49. Goff, D. H., & Hartel, R. W., (2013). *Ice Cream* (7th edn., p. 462). digital; New York: Springer.
50. González-Olivares, L. G., López-Cuellar, Z. L., Añorve-Morga, J., Franco-Fernández, M. J., Castañeda-Ovando, A., & Contreras-López, E., (2014). Viability and proteolytic capacity of *Lactobacillus bulgaricus* and *Lactobacillus rhamnosus* GG during cheese ripening. *J. Biosci and Medic., 2*, 7–12.
51. Goud, K., & Park, H. J., (2005). Recent developments in microencapsulation of food ingredients. *Dry. Technol., 23*, 1361–1394.
52. Granato, D., Nazzaro, F., Pimentel, T. C., Esmerino, E. A., & Cruz, A. G., (2018). Probiotic food development: An updated review based on technological advancement. *Reference Module in Food Science*, 1–7.
53. Guimarães, J. T., Balthazar, C. F., Silva, R., Rocha, R. S., Graça, J. S., Esmerino, E., Silva, M. C., et al., (2020). Impact of probiotics and prebiotics on food texture. *Curr. Opin. Food Sci., 33*, 38–44.
54. Hill, C., Guarner, F., Reid, G., Gibson, G. R., Merenstein, D. J., Pot, B., Morelli, L., et al., (2014). Expert consensus document: The international scientific association for probiotics and prebiotics consensus statement on the scope and appropriate use of the term probiotic. *Nat. Rev. Gastroenterol. Hepatol., 11*, 506–514.
55. Jahreis, G., Vogelsang, H., Kiessling, G., Schubert, R., Bunte, C., & Hammes, W. P., (2002). Influence of probiotic sausage (*Lactobacillus paracasei*) on blood lipids and immunological parameters of healthy volunteers. *Food Res. Int., 35*, 133–138.
56. James, A., & Wang, Y., (2019). Characterization, health benefits and applications of fruits and vegetable probiotics. *CYTA-J. Food., 17*(1), 770–780.
57. Jofré, A., Aymerich, T., & Margarita, G., (2015). Probiotic fermented sausages: Myth or reality? *Procedia Food Sci., 5*, 133–136.
58. Joye, I. J., & McClements, D. J., (2014). Biopolymer-based nanoparticles and microparticles: Fabrication, characterization, and application. *Curr. Opin. Colloid Interface Sci., 19*, 417–427.
59. Kalicka, D., Znamirowska, A., Pawlos, M., Buniowska, M., & Szajnar, K., (2019). Physical and sensory characteristics and probiotic survival in ice cream sweetened with various polyols. *Int. J. Dairy Technol. [Online], 72*, 3. https://doi.org/10.1111/1471-0307.12605 (accessed Jun. 10, 2020).
60. Kantachote, D., Ratanaburee, A., Hayisama, W., Sukhoom, A., & Nunkaew, T., (2017). The use of potential probiotic *Lactobacillus plantarum* DW12 for producing a novel functional beverage from mature coconut water. *J. Funct. Foods, 32*, 401–408.
61. Karimi, R., Mortazavian, A. M., & Cruz, A., (2011). Viability of probiotic microorganism in cheese during production and storage: A review. *Dairy Sci. Technol., 91*, 283–308.
62. Kariyawasam, K. M. G. M. M., Jeewanthi, R. K., Lee, N. K., & Paik, H. D., (2019). Characterization of cottage cheese using *Weissella cibaria* D30: Physicochemical, antioxidant, and antilisterial properties. *J Dairy Sci., 102*, 1–7.

63. Kaur, S., Kaur, H. P., & Grover, J., (2016). Fermentation of tomato juice by probiotic lactic acid bacteria. *Int. J. Adv. Pharm., Biol. Chem., 2*, 212–219.
64. Kemsawasd, V., Chaikham, P., & Rattanasena, P., (2016). Survival of immobilized probiotics in chocolate during storage and with an *in vitro* gastrointestinal model. *Food Biosci., 16*, 37–43.
65. Kent, R. M., & Doherty, S. B., (2014). Probiotic bacteria in infant formula and follow-up formula: Microencapsulation using milk and pea proteins to improve microbiological quality. *Food Res. Int., 64*, 567–576.
66. Khan, M. I., Arshad, M. S., Anjum, F. M., Sameen, A., Rehman, A., & Gill, W. T., (2011). Meat as a functional food with special reference to probiotic sausages. *Food Res. Int., 44*, 3125–3133.
67. Khaneghah, A. M., Abhari, K., Eş, I., Soares, M. B., Oliveira, R. B. A., Hosseini, H., Rezaei, M., et al., (2020). Interactions between probiotics and pathogenic microorganisms in hosts and foods: A review. *Trends in Food Sci. Technol., 95*, 205–218.
68. Khodaei, D., & Hamidi-Esfahani, Z., (2019). Influence of bioactive edible coatings loaded with *Lactobacillus plantarum* on physicochemical properties of fresh strawberries. *Postharvest Biol. Technol., 156*, 110944.
69. Khodaei, D., Hamidi-Esfahani, Z., & Lacroix, M., (2020). Gelatin and low methoxyl pectin films containing probiotics: Film characterization and cell viability. *Food Biosci.*
70. Kołozyn-Krajewska, D., & Dolatowski, Z. J., (2012). Probiotic meat products and human nutrition. *Process Biochem., 47*(12), 1761–1772.
71. Konar, N., Palabiyik, I., Toker, O. S., Polat, D. G., Kelleci, E., Pirouzian, H. R., Akcicek, A., & Sagdic, O., (2018). Conventional and sugar-free probiotic white chocolate: Effect of inulin DP on various quality properties and viability of probiotics. *J. Funct. Foods, 43*, 206–213.
72. Konar, N., Toker, O. S., Oba, S., & Sagdic, O., (2016). Improving functionality of chocolate: A review on probiotic, prebiotic, and/or synbiotic characteristics. *Trends Food Sci. Technol, 49*, 35–44.
73. Kozłowicz, K., Góral, M., Góral, D., Pankiewicz, U., & Bronowicka-Mielniczuk, U., (2019). Effect of ice cream storage on the physicochemical properties and survival of probiotic bacteria supplemented with zinc ions. *LWT-Food Sci. Technol., 116*, 108562.
74. Laličić-Petronijević, J., Komes, D., Gorjanović, S., Belščak-Cvitanović, A., Pezo, L., Pastor, F., Ostojić, S., et al., (2016). Content of total phenolics, flavan-3-ols and proanthocyanidins, oxidative stability and antioxidant capacity of chocolate during storage. *Food Technol. Biotechnol., 54*(1), 13–20.
75. Laličić-Petronijević, J., Popov-Raljic′, J., Lazic′, V., Pezo, L., & Nedovic′, V., (2017). Synergistic effect of three encapsulated strains of probiotic bacteria on quality parameters of chocolates with different composition. *J. Funct. Foods, 38*, 329–337.
76. Laličić-Petronijević, J., Popov-Raljic′, J., Obradovic′, D., Radulovic′, Z., Paunovic′, D., Petrušic′, M., & Pezo, L., (2015). Viability of probiotic strains *Lactobacillus acidophilus* NCFM® and Bifidobacterium lactis HN019 and their impact on sensory and rheological properties of milk and dark chocolates during storage for 180 days. *J. Funct. Foods, 15*, 541–550.
77. Li, K. L., Wang, B. W., Wang, W. J., Liu, G. D., Ge, W. H., Zhang, M. G., Yue, B., & Kong, M., (2019). Microencapsulation of *Lactobacillus casei* BNCC 134415 under lyophilization enhances cell viability during cold storage and pasteurization, and in simulated gastrointestinal fluids. *LWT-Food Sci and Technol., 116*, 108521.

78. Majeed, M., Majeed, S., Nagabhushanam, K., Arumugam, S., Beede, K., & Ali, F., (2019). Evaluation of probiotic *Bacillus coagulans* MTCC 5856 viability after tea and coffee brewing and its growth in GIT hostile environment. *Food Res. Int., 121*, 497–505.
79. Mani-López, E., Palou, E., & López-Malo, A., (2014). Probiotic viability and storage stability of yogurts and fermented milks prepared with several mixtures of lactic acid bacteria. *J. Dairy Sci., 97*, 2578–2590.
80. Marcial-Coba, M. S., Pjaca, A. S., Andersen, C. J., Knøchel, S., & Nielsen, D. S., (2019). Dried date paste as carrier of the proposed probiotic *Bacillus coagulans* BC4 and viability assessment during storage and simulated gastric passage. *LWT-Food Sci and Technol.*
81. Martins, E. M. F., Ramos, A. M., Martins, M. L., & Leite, Jr. B. R. C., (2016). Fruit salad as a new vehicle for probiotic bacteria. *Food Sci. Technol., 36*(3), 540–548.
82. Martins, E. M. F., Ramos, A. M., Martins, M. L., & Rodrigues, M. Z., (2015). Research and development of probiotic products from vegetable bases: A new alternative for consuming functional food. In: Rai, V. R., & Bai, J. A., (eds.), *Beneficial Microbes in Fermented and Functional Foods* (pp. 207–223). CRC Press: Boca Raton, London.
83. Martins, E. M. F., Ramos, A. M., Vanzela, E. S. L., Stringheta, P. C., Pinto, C. L. O., & Martin, S. J. M., (2013). Products of vegetable origin: A new alternative for the consumption of probiotic bacteria. *Food Res. Int., 51*, 764–770.
84. Minekus, M., Alminger, M., Alvito, P., Ballance, S., Bohn, T., Bourlieu, C., Carrière, F., et al., (2014)., A standardized static *in vitro* digestion method suitable for food: An international consensus. *Food Funct.*, 1113–1124.
85. Mirkovi´c, M., Seratli´c, S., Kilcawley, K., Mannion, D., Mirkovi´c, N., & Radulovi´c, Z., (2018). The sensory quality and volatile profile of dark chocolate enriched with encapsulated probiotic *Lactobacillus plantarum* bacteria. *Sensors, 18*, 1–16.
86. Moghanjougi, Z. M., Bari, M. R., Khaledabad, M. A., Almasi, H., & Amiri, S., (2020). Bio-preservation of white brined cheese (Feta) by using probiotic bacteria immobilized in bacterial cellulose: Optimization by response surface method and characterization. *LWT-Food Sci Technol., 117*, 108603.
87. Montanari, S. R., Leite, Jr. B. R. C., Martins, M. L., Ramos, A. M., Binoti, M. L., Campos, R. C. A. B., Campos, A. N. R., & Martins, E. M. F., (2020). *In vitro* gastrointestinal digestion of a peanut, soybean, guava, and beet beverage supplemented with *Lactobacillus rhamnosus* GG. *Food Biosci., 36*, 100623.
88. Moreira, R. M., Martins, M. L., Leite, Jr. B. R. C., Martins, E. M. F., Ramos, A. M., Cristianini, M., Campos, A. N. R., et al., (2017). Development of a juçara and Ubá mango juice mixture with added *Lactobacillus rhamnosus* GG processed by high pressure. *LWT-Food Sci. and Technol., 77*, 259–268.
89. Mousavi, M., Heshmati, A., Garmakhany, A. D., Vahidinia, A., & Taheri, M., (2019). Optimization of the viability of *Lactobacillus acidophilus* and physicochemical, textural, and sensorial characteristics of flaxseed-enriched stirred probiotic yogurt by using response surface methodology. *LWT-Food Sci. Technol., 102*, 80–88.
90. Muhardina, V., Sari, P. M., Aisyah, Y., & Haryani, S., (2019). Physical characteristic of probiotic ice cream substituted by encapsulated lactic acid bacteria (LAB) with variety of aging time. *Journal of Physics: Conference Series., 1232 012042.*
91. Nambiar, R. B., Sellamuthu, P. S., & Perumal, A. B., (2018). Development of milk chocolate supplemented with microencapsulated *Lactobacillus plantarum* HM47 and to determine the safety in a Swiss albino mice model. *Food Control., 94*, 300–306.

92. Nami, Y., Haghshenas, B., Vaseghi, R., Jalaly, H. M., Lotfi, H., Eslami, S., & Hejazi, M. A., (2018). Novel autochthonous lactobacilli with probiotic aptitudes as a main starter culture for probiotic fermented milk. *LWT-Food Sci Technol., 98*, 85–93.
93. Neffe-Skocińska, K., Rzepkowska, A., Szydłowska, A., & Kołożyn-Krajewska, D., (2018). Trends and possibilities of the use of probiotics in food production. In: Holban, A. M., & Grumezescu, A. M., (eds.), *Alternative and Replacement Foods* (Vol. 17, pp. 65–94). Academic Press: Cambridge, Chapter 3.
94. Ningtyas, D. W., Bhandari, B., Bansal, N., & Prakash, S., (2019). The viability of probiotic *Lactobacillus rhamnosus* (non-encapsulated and encapsulated) in functional reduced-fat cream cheese and its textural properties during storage. *Food Control, 100*, 8–16.
95. Nollet, M. L., & Toldrá, F., (2006). *Advanced Technologies for Meat Processing (p. 483)*. CRC Press: Boca Raton.
96. Ojha, K. S., Kerry, J. P., Duffy, G., Beresford, T., & Tiwari, B. K., (2015). Technological advances for enhancing quality and safety of fermented meat products. *Trends Food Sci. Tech., 44*, 105–116.
97. Oliveira, D. C. D., Martins, E. M. F., Martins, M. L., Martins, G. B., Binoti, M. L., Campos, A. N. R., Ramos, A. M., et al., (2017). Blanching effect on the bioactive compounds and on the viability of *Lactobacillus rhamnosus* GG before and after *in vitro* simulation of the digestive system in jabuticaba juice. *Semina: Ciênc. Agrár., 38*(3), 1277–1294.
98. Oliveira, P. M., Leite, Jr. B. R. C., Martins, E. M. F., Martins, M. L., Vieira, É. N. R., Barros, F. A. R., Cristianini, M., et al., (2020). Mango and carrot mixed juice: A new matrix for the vehicle of probiotic lactobacilli. *J. Food Sci. Technol.*
99. Ong, J., & Shah, N. P., (2008). Influence of Probiotic *Lactobacillus acidophilus* and *L. helveticus* on proteolysis, organic acid profiles, and ACE-inhibitory activity of cheddar cheeses ripened at 4, 8, and 12°C. *J. Food Sci., 73*, 111–120.
100. Pahumunto, N., Piwat, S., Chanvitan, S., Ongwande, W., Uraipan, S., & Teanpaisan, R., (2020). Fermented milk containing a potential probiotic *Lactobacillus rhamnosus* SD11 with maltitol reduces *Streptococcus mutans*: A double-blind, randomized, controlled study. *J. Dent. Sci.* [Online]. https://doi.org/10.1016/j.jds.2020.03.003 (accessed on 21 June 2021).
101. Panda, S. K., Behera, S. K., Qaku, X. W., Sekar, S., Ndinteh, D. T., Nanjundaswamy, H. M., Ray, R. C., & Kayitesi, E., (2017). Quality enhancement of prickly pears (*Opuntia* sp.) juice through probiotic fermentation using *Lactobacillus fermentum*-ATCC 9338. *LWT-Food Sci and Technol., 75*, 453–459.
102. Pankiewicz, U., Góral, M., Kozlowicz, K., & Góral, D., (2020). Application of pulsed electric field in production of ice cream enriched with probiotic bacteria (*L. rhamnosus* B 442) containing intracellular calcium ions. *J. Food Eng., 275*, 109876.
103. Papadopoulou, O. S., Argyri, A. A., Varzakis, E. E., Tassou, C. C., & Chorianopoulos, N. G., (2018). Greek functional Feta cheese: Enhancing quality and safety using a *Lactobacillus plantarum* strain with probiotic potential. *Food Microbiol., 74*, 21–33.
104. Parussolo, G., Busatto, R. T., Schmitt, J., Pauletto, R., Schons, P. F., & Ries, E. F., (2017). Synbiotic ice cream containing yacon flour and *Lactobacillus acidophylus* NCFM. *LWT-Food Sci. and Technol., 82*, 192–198.

105. Pech-Canul, A. C., Ortega, D., García-Triana, A., González-Silva, N., & Solis-Oviedo, R. L., (2020). A brief review of edible coating materials for the microencapsulation of probiotics. *Coatings, 10*, 197.
106. Pedroso, D. L., Dogenski, M., Thomazini, M., Heinemann, R. J. B., & Favaro-Trindade, C. S., (2013). Microencapsulation of bifidobacterium animal is subsp. lactis and *Lactobacillus acidophilus* in cocoa butter using spray chilling technology. *Braz. J. Microbiol., 44*(3), 777–783.
107. Pérez-Chabela, M. L., Lara-Labastida, R., Rodriguez-Huezo, E., & Totosaus, A., (2013). Effect of spray drying encapsulation of thermotolerant lactic acid bacteria on meat batters properties. *Food Bioproc. Tech., 6*, 1505–1515.
108. Pinto, D., Castro, I., Vicente, A., Bourbon, A. I., & Cerqueira, M. A., (2014). Functional bakery products: An overview and future perspectives. In: Zhou, W., Hui, Y. H., Leyn, I., Pagani, M. A., Rosell, C. M., Selman, J. D., & Therdthai, N., (eds.), *Bakery Products Science and Technology* (pp. 431–449). John Wiley & Sons, Ltd: Hoboken, Chapter 25.
109. Pinto, S. S., Fritzen-Freire, C. B., Dias, C. O., & Amboni, R. D. M. C., (2019). A potential technological application of probiotic microcapsules in lactose-free Greek-style yoghurt. *Int Dairy J., 97*, 131–138.
110. Piper, H. G., Coughlin, L. A., Hussain, S., Nguyen, V., Channabasappa, N., & Koh, A. Y., (2020). The impact of *lactobacillus* probiotics on the gut microbiota in children with short bowel syndrome. *J. Am. Coll. Surg., 251*, 112–118.
111. Prates, F. C., Leite, Jr. B. R. C., Martins, E. M. F., Cristianini, M., Silva, R. R., Campos, A. N. R., Gandra, S. O. S., et al., (2020). Development of a mixed jussara and mango juice with added *Lactobacillus rhamnosus* GG submitted to sub-lethal acid and baric stresses. *J. Food Sci. Technol.*
112. Prezzi, L. E., Lee, S. H. I., Nunes, V. M. R., Corassin, C. H., Pimentel, T. C., Rocha, R. S., Ramos, G. L. P., et al., (2020). Effect of *Lactobacillus rhamnosus* on growth of *Listeria monocytogenes* and *Staphylococcus aureus* in a probiotic minas frescal cheese. *Food Microbiol., 92*, 103557.
113. Reale, A., Renzo, T. D., & Coppola, R., (2019). Factors affecting viability of selected probiotics during cheese-making of pasta filata dairy products obtained by direct-to-vat inoculation system. *LWT-Food Sci and Technol., 116*, 108476.
114. Reis, J. A., Paula, A. T., Casarotti, S. N., & Penna, A. L. B., (2012). Lactic acid bacteria antimicrobial compounds: Characteristics and applications. *Food Eng. Rev., 4*, 124–140.
115. Reis, S. A. D., Conceição, L. L. D., Siqueira, N. P., Rosa, D. D., Silva, L. D., & Peluzio, M. D., (2017). Review of the mechanisms of probiotic actions in the prevention of colorectal cancer. *Nutr Res., 37*, 1–19.
116. Riveros, B., Ferrer, J., & Bórquez, R., (2009). Spray drying of a vaginal probiotic strain of *Lactobacillus acidophilus*. *Dry. Technol., 27*, 123–132.
117. Roobab, U., Batool, Z., Manzoor, M. F., Shabbir, M. A., Khan, M. R., & Aadil, R. M., (2020). Visualization of food probiotics with technologies to improve their formulation for health benefits and future perspectives. *Curr. Opin. Food Sci.*
118. Rouhi, M., Sohrabvandi, S., & Mortazavian, A. M., (2013). Probiotic fermented sausage: Viability of probiotic microorganisms and sensory characteristics. *Crit. Rev. Food Sci. Nutr., 53*, 331–348.

119. Sameer, B., Ganguly, S., Khetra, Y., & Sabikhi, L., (2020). Development and characterization of probiotic buffalo milk ricotta cheese. *LWT-Food Sci Technol., 121*, 108944.
120. Santos, S. C., Konstantyner, T., & Cocco, R. R., (2020). Effects of probiotics in the treatment of food hypersensitivity in children: A systematic review. *Allergol Immunopathol., 48*(1), 95–104.
121. Sarmento, E. G., Cesar, D. E., Martins, M. J., Góis, E. G. O., Martins, E. M. F. M., Campos, A. N. R., Del'Duca, A., & Martins, A. D. O., (2019). Effect of probiotic bacteria in composition of children's saliva. *Food Res Int., 116*, 1282–1288.
122. Savino, F., Fornasero, S., Ceratto, S., De Marco, A., Mandras, N., Roana, J., Tullio, V., & Amisano, G., (2015). Probiotics and gut health in infants: A preliminary case-control observational study about early treatment with *Lactobacillus reuteri* DSM 17938. *Clin. Chim. Acta, 451*(Part A), 82–87.
123. Seyedain-Ardabili, M., Sharifan, A., & Tarzi, B. G., (2016). Synbiotic bread with encapsulated probiotics. *Food Technol. Biotechnol., 54*(1), 52–59.
124. Shori, A. B., (2017). Microencapsulation improved probiotics survival during gastric transit. *HAYATI J. Biosci., 24*, 1–5.
125. Soares, M. B., Almada, C. N., Almada, C. N., Martinez, R. C. R., Pereira, E. P. R., Balthazar, C. F., Cruz, A. G., et al., (2019a). Behavior of different *Bacillus* strains with claimed probiotic properties throughout processed cheese ("requeijão cremoso") manufacturing and storage. *Int. J. Food Microbiol., 307*, 1–9.
126. Soares, M. B., Martinez, R. C. R., Pereira, E. P. R., Balthazar, C. F., Cruz, A. G., Ranadheera, C. S., & Sant'Ana, A. S., (2019b). The resistance of *Bacillus, Bifidobacterium*, and *Lactobacillus* strains with claimed probiotic properties in different food matrices exposed to simulated gastrointestinal tract conditions. *Food Res. Int., 125*, 108542.
127. Song, H., Yu, W., Gao, M., Liu, X., & Ma, X., (2013). Microencapsulated probiotics using emulsification technique coupled with internal or external gelation process. *Carbohydr. Polym., 96*, 181–189.
128. Soukoulis, C., Yonekura, L., Gan, H., Behboudi-Jobbehdar, S., Parmenter, C., & Fisk, I., (2014). Probiotic edible films as a new strategy for developing functional bakery products: The case of pan bread. *Food Hydrocoll., 39*, 231–242.
129. Succi, M., Tremonte, P., Pannella, G., Tipaldi, L., Cozzolino, A., Coppola, R., & Sorrentino, E., (2017). Survival of commercial probiotic strains in dark chocolate with high cocoa and phenols content during the storage and in a static *in vitro* digestion model. *J. Funct. Foods, 35*, 60–67.
130. Tan-Lim, C. S. C., & Esteban-Ipac, N. A. R., (2018). Probiotics as treatment for food allergies among pediatric patients: A meta-analysis, *World Allergy Organ. J., 6*, 11–25.
131. Teanpaisan, R., Chooruk, A., Wannun, A., Wichienchot, S., & Piwat, S., (2012). Survival rates of human-derived probiotic *Lactobacillus paracasei* SD1in milk powder using spray drying. *Songklanakarin J. Sci. Technol., 34*, 241–245.
132. Tiebackx, F. W., (1911). Gleichzeitige Ausflockung zweier Kolloide. *Zeitschrift für Chemie und Industrie der Kolloide., 8*, 198–201.
133. Timilsena, Y. P., Akanbi, T. O., Khalid, N., Adhikari, B., & Barrow, C. J., (2019). Complex coacervation: Principles, mechanisms, and applications in microencapsulation. *Int. J. Biol. Macromol., 121*, 1276–1286.

134. Trabelsi, I., Slima, S. B., Ktari, N., Triki, M., Abdehedi, R., Abaza, W., Moussa, H., et al., (2019). Incorporation of probiotic strain in raw minced beef meat: Study of textural modification, lipid and protein oxidation and color parameters during refrigerated storage. *Meat Sci., 154*, 29–36.
135. Tripathi, M. K., & Giri, S. K., (2014). Probiotic functional foods: Survival of probiotics during processing and storage. *J. Funct. Foods., 9*, 225–241.
136. Utza, V. E. M., Perdigóna, G., & Leblanca, A. D. M. D., (2019). Oral administration of milk fermented by *Lactobacillus casei* CRL431 was able to decrease metastasis from breast cancer in a murine model by modulating immune response locally in the lungs. *J. Funct. Foods, 54*, 263–270.
137. Valerio, F., Volpe, M. G., Santagata, G., Boscaino, F., Barbarisi, C., Di Biase, M., Bavaro, A. R., et al., (2020). The viability of probiotic *Lactobacillus paracasei* IMPC2.1 coating on apple slices during dehydration and simulated gastro-intestinal digestion. *Food Biosci., 34*, 100533.
138. Valero-Cases, E., Roy, N., Frutos, M. J., & Anderson, R., (2017). Influence of the fruit juice carriers on the ability of *Lactobacillus plantarum* DSM20205 to improve *in vitro* intestinal barrier integrity and its probiotic properties. *J. Agric. Food Chem.*
139. Vasconcelos, F. M., Silva, H. L. A., Poso, S. M. V., Barroso, V. M., Lanzetti, M., Rocha, R. S., Graça, J. S., et al., (2019). Probiotic Prato cheese attenuates cigarette smoke-induced injuries in mice. *Food Res. Int., 123*, 697–703.
140. Vivek, K., Mishra, S., & Pradhan, R. C., (2020). Characterization of spray dried probiotic sohiong fruit powder with *Lactobacillus plantarum. LWT-Food Sci and Technol.*
141. Wang, L., Canção, M., Zhao, Z., Chen, X., Cai, J., Cao, Y., & Xiao, J., (2020). *Lactobacillus acidophilus* loaded Pickering double emulsion with enhanced viability and colon-adhesion efficiency. *LWT-Food Sci. Technol., 121*, 108928.
142. Wen, J., Ma, L., Xu, Y., Wu, J., Yu, Y., Peng, J., Tang, D., et al., (2020). Effects of probiotic litchi juice on immunomodulatory function and gut microbiota in mice. *Food Res. Int., 137*, 109433.
143. Wu, Z., Wu, J., Cao, P., Jin, Y., Pan, D., Zeng, X., & Guo, Y., (2017). Characterization of probiotic bacteria involved in fermented milk processing enriched with folic acid. *J. Dairy Sci., 100*, 4223–4229.
144. Yang, F., Wang, Y., & Zhao, H., (2020). Quality enhancement of fermented vegetable juice by probiotic through fermented yam juice using *Saccharomyces cerevisiae. Food Sci. Technol., 40*(1), 26–35.
145. You, L., Zhao, M., Regenstein, J. M., & Ren, J., (2010). Changes in the antioxidant activity of loach (*Misgurnus anguillicaudatus*) protein hydrolysates during a simulated gastrointestinal digestion. *Food Chem., 120*, 810–816.
146. Zdolec, N., (2017). *Fermented Meat Products: Health Aspects.* CRC Press: Boca Raton.
147. Zhang, L., Chen, X. D., Boom, R. M., & Schutyser, M. A. I., (2018b). Survival of encapsulated *Lactobacillus plantarum* during isothermal heating and bread baking. *LWT-Food Sci. and Technol., 93*, 396–404.
148. Zhang, L., Taal, M. A., Boom, R. M., Chen, X. D., & Schutyser, M. A. I., (2018a). Effect of baking conditions and storage on the viability of *Lactobacillus plantarum* supplemented to bread. *LWT-Food Sci. and Technol.*, 87318–87325.

CHAPTER 9

Microencapsulation: An Alternative for the Application of Probiotic Cells in the Food and Nutraceuticals Industries

DANIELE DE ALMEIDA PAULA,[1] CARINI APARECIDA LELIS,[2] and NATALY DE ALMEIDA COSTA[3]

[1]*Federal Institute of São Paulo (IFSP), Campus Avaré - Av. Professor Celso Ferreira da Silva, 1333, Jardim Europa, CEP 18707-150, SP, Brazil*

[2]*Federal University of São Carlos (UFSCar), Campus Lagoa do Sino–Rodovia Lauri Simões de Barros, Aracaçu, Buri, CEP 18290-000, SP, Brazil*

[3]*Department of Food Technology, Federal University of Viçosa (UFV), P.H. Rolfs Avenue, Campus, Viçosa – 36570-900, MG, Brazil*

ABSTRACT

In recent years, the use of probiotic microorganisms has increased considerably. Probiotics are live microorganisms that which when administered in adequate amounts, confer health benefits on the host. Because of the numerous benefits and consumer concern about the importance of healthy diets, a wide variety the products containing probiotics microorganisms are available, represented by food and pharmaceutical formulations. However, maintaining the adequate viability of these microorganisms during processing, shelf life, storage conditions, and during passage through GI conditions is a challenge for industries. Thus, technological alternatives can be used to minimize the damage on probiotics, among them, stands out the microencapsulation. The objective of encapsulation is to create a physical barrier between the microorganism and the external environment, increasing cell viability during processing, storage, and passage through the GIT. Thus, viability will be maintained and the cells released in appropriate locations (for example, small intestine) for adhesion and colonization of the intestinal epithelium to occur.

9.1 INTRODUCTION

In recent years, the use of probiotic microorganisms has increased considerably. The Food and Agriculture Organization (FAO) and World Health Organization (WHO) defined probiotics as: "live microorganisms that which when administered in adequate amounts confer health benefits on the host." The more important benefits are effects on the gastrointestinal tract (GIT); on the urogenital system; on the immune system modulation, and even on the cardiovascular system. All health benefits are dependent on the strain used, the frequency, and daily dosage.

Because of the numerous benefits and consumer concern about the importance of healthy diets, a wide variety the products containing probiotics microorganisms are available, represented by food and pharmaceutical formulations. The various options aim to attend different publics, such as lactose intolerant, allergic to milk proteins, vegans, vegetarians, and others. Also, the diversity of products allows the consumption of probiotics at various moments of the day and, so, favoring the frequency of ingestion, fundamental factors the provide benefits to the body.

However, maintaining the adequate viability of these microorganisms during processing, shelf life, storage conditions, and during passage through GI conditions is a challenge for industries. Factors such as temperature, pH, oxygen, salt content, and added preservatives can affect the survival of the probiotic's cells. Thus, technological alternatives can be used to minimize the damage on probiotics, among them, stands out the microencapsulation.

The microencapsulation technique consists of packing the material of interest using polymeric coatings and creating an important barrier between the microorganisms and the extern environment. In the industry, the microencapsulation has solved many limitations, and it is used to protect sensible compounds of external agents, to release of these compounds under controlled conditions, to mask undesirable flavors or odors, and others. Among the materials that can be encapsulated, stand out: acids, bases, oils, vitamins, salts, flavors, dyes, enzymes, and microorganisms.

Therefore, the microencapsulation of probiotic bacteria has been used to protect these organisms from adverse conditions and, so, increase cell viability. The release of the cells occurs at appropriate locations (for example, small intestine), promoting the adhesion and colonization of the intestinal epithelium. Currently, modern and innovative methods of microencapsulation have been developed and, consequently, creating a wide variety of probiotic microcapsules. Thus, it represents an alternative for the food and pharmaceutical industry to increase the offer of probiotic products, attend a greater number of consumers and, also allow a varied application of probiotics in the market.

9.2 PROBIOTICS

Probiotics are live microorganisms that which when administered in adequate amounts, confer health benefits on the host, act as therapeutic agents [1]. The microorganisms used as probiotics are *Lactobacillus*: *L. acidophilus, L. bulgaricus, L. casei, L. delbrueckii ssp. bulgaricus, L. fermentum, L. johnsonii, L. rispatus, L. salivarius, L. bifidus, L. reuteri, L. plantarum, L. helveticus, L. casei subsp. rhamnosus, L. gallinarum, L. brevis, L. gasseri, L. cellobiosus, L. vitulinus, L. collinoides, L. cremoris, L. ruminis, L. dextranicum, L. lactis, L. rhamnosus, L. curvatus, L. faecium* and *L. paracasei*; *Bifidobacteria*: *B. bifidum, B. adolescentis, B. brevis, B. longum, B. animalis, B. infantis, B. thermophilum, B. breve, B. essencis, and B. lactis; Bacillus-B. coagulans, B. lactis, B. licheniformis, B. subtili*s and *B. subtilis*; *Pediococcus*: *P. acidilactici, P. pentosaceus* and *P. halophilus*; *Lactococcus-L. lactis subspp. lactis* and *cremoris*; *Leuconostoc*: *L. mesenteroides subsp. dextranium, paramesenteroides*, or *lactis; Streptococcus: S. diacetilactis, S. cremoris, S. lactis, salivarius subsp. thermophilus, S. faeciu* and *S. equinu*s, *Weissella*: *W. cibaria* and *confusa*; Yeast: *Saccharomyces cerevisia*e and *S. cerevisiae var. boulardii*; Fungi: *Aspergillus oryzae* and *Scytalidium acidophilum* [2].

Probiotics belong to the different genres, species, and even phylum, and have been associated with numerous beneficial effects. High numbers of microorganisms are considered probiotics and, currently, research has discovered new species. The development and/or improvement of analyses also reveal and emphasize the beneficial effects derived by the consumption of probiotic microorganisms, as promoters of human health. *Lactobacillus* and *Bifidobacterium* species are the most commonly used as probiotics and are normally presented in the GIT in a commensal relationship. Thus, the intestinal microbiota of healthy individuals may represent an important source of these species. Several studies have been demonstrated the isolation of probiotics from the oral and vaginal cavity, human milk and other species, colostrum, and various fermented products, indicating the diversity of sources for obtaining probiotics strains [3–5].

However, the selection of microorganisms considered as probiotic candidates, isolated from these sources, requires a systematic realization of tests to ensure food security and the functional technological aspects. Moreover, the taxonomic identification is very important to avoid the choice of strains antibiotic-resistant, producers of harmful metabolic, infectious, or virulent. Also, microorganisms must have certain characteristics, such as survives physiological stress, acid pH of the stomach, and the presence of bile salts, to be able to adhere to the intestinal epithelium and colonize the colon and ability to compete against pathogenic bacteria. About the functional technological aspects, the probiotics generally need to be suited to the food matrix to maintain their viability, it is necessary to maintain an adequate level of viable cells during the processing and storage of the product and during passage through the human GIT and, besides, they cannot modify the sensory characteristics of food.

Although probiotics have demonstrated their benefits for years only in the last few decades, they have been widely used both in foods and in pharmaceutical formulations. Among the beneficial effects caused by consumption of probiotic microorganisms, there are: prevent the colonization of pathogens; to prevention for different gastrointestinal (GI) disorders (e.g., lactose intolerance, viral, and bacterial gastroenteritis, and inflammatory bowel disease (IBD)); antimicrobial activities due to the production of short-chain fatty acids, production of hydrogen peroxide and bacteriocins, and pH modulation. Moreover, probiotic bacteria can produce important nutrients from foods originating in the diet, for example, vitamin B_{12}, vitamin K, pyridoxine (vitamin B_6), folate, and thiamine, riboflavin, and menaquinone (vitamin K2). *Lactobacillus* and *Bifidobacterium* have been associated which the production of conjugated linoleic acid (CLA) and conjugated α-linolenic acid (CLNA). These compounds have anti-inflammatory, antioxidant, anti-allergic, anticarcinogenic, and anti-obesity properties. Simultaneously, probiotic bacteria can: ferment proteins by producing peptides, amino acids, lactones, indoles, and phenols that collaborate in the energy balance and have antioxidant and anti-inflammatory effects; convert indigestible carbohydrates into smaller sugars with the production of different compounds such as propionate, acetate, and butyrate; reduce the symptoms of lactose intolerance due to the presence of β-galactosidase and modification of the intestinal microbiota [6]; make the anaerobic intestinal environment and, therefore, unsuitable for pathogenic bacteria; produce exopolysaccharides (EPS) that form biofilms which select beneficial bacteria and have anti-tumor activity in human cells, antioxidant, anti-inflammatory effects, inhibit mutagens and cholesterol reduction [7–9].

However, for probiotics to have beneficial effects on the host, a sufficient number of live cells is necessary during consumption and GI administration. Additionally, the probiotics cells must promote the adhesion and colonization in specific locations. Probiotic microorganisms promote specific benefits and, therefore, it may be necessary to use different strains to obtain all the expected results.

9.3 APPLICATION OF PROBIOTICS

The consumption of probiotic microorganisms has increased considerably in recent years due to the numerous beneficial effects. Currently, the commercialization of probiotics products can occur as pharmaceutical formulations and functional foods.

9.3.1 PHARMACEUTICAL FORMULATIONS

The probiotics are available in pharmaceutical formulations, in the form of powder, capsule, tablets, eye drops, etc., based on the nutraceutical appeal. For the manufacture of probiotic pharmaceutical formulations, the following steps are carried out: (i) production of biomass under *pH control*; (ii) cell recovery and concentration; (iii) drying of the concentrated; (iv) commercialization in capsule, powers, compressed, etc. These products require dosing and storage under conditions that security the viability of the microorganisms.

Regarding the powder formulation, these are presented in the form of dry and fine solid particles. Thus, these formulations form suspensions when dispersed in the water, easing its administration. The capsules consist of a hard or soft soluble container or "shell" with variable shape, capacity, and properties. Generally, the capsules are unit dosage containers containing the probiotic cells inside. Tablets are solid dosage forms containing the appropriate dose of cells and may be obtained by compressing a volume of particle or extruding, molding, and lyophilizing. The oral administration of tablets can be done through chewing, swallowed, or remaining in the mouth until the cells are released [10]. Medicated chewing gums also represent pharmaceutical formulations for the delivery of different probiotics and principally indicated for children. These products usually contain a gelatinous base and the active substance is released by chewing [11]. Other pharmaceutical formulas are bioadhesive gels, eye drops, and lozenges.

During all fabrication process is extremely essential to control the processing conditions, to check the dosage and viability of the probiotic's cells. In conclusion, pharmaceutical formulations are excellent for the delivery of probiotic cells presenting great stability, security, and efficiency.

9.3.2 FUNCTIONAL FOOD

The lactic acid bacteria (LAB) are widely used in fermented products. In the past, these bacteria were identified to cause food spoilage, altering the components, and decreasing the acceptability. Ancient people consumed fermented milk and cheeses obtained by the spontaneous action of these microorganisms. At present, these microorganisms are used to promote health benefits and their use by the food industry has grown. Food matrices have factors that benefit the growth and the survival of probiotic microorganisms, such as water, sugars, fat, proteins, and available peptides. Thus, the consumer benefits from nutrition and also the benefits related to the consumption of probiotics.

The recommended number of viable cells to be ingested to obtain health benefits associated with probiotics is around 9 log $CFU.g^{-1}$ of the product [12]. For food applications, a single microorganism, or a combination of them can be used.

Dairy products have the largest probiotic food market share, especially yogurts and fermented milk. Different combinations between initial and probiotics lactic cultures permit the fabrication of products with desirable technological, sensory, nutritional characteristics, and health benefits. The *nutritional* composition of *milk* is highly complex, containing many compounds, such as proteins, minerals, fat, sugars. Therefore, it represents a food matrix adequate for the development and viability of probiotic microorganisms. Moreover, foods such as dairy products are known to improve the survival of probiotic microorganisms in gastric juice, probably due to the buffering power.

Studies report that the addition of milk or milk proteins significantly increases the pH of gastric juice and, consequently, improving the survival of some species of *Lactobacillus* and *Bifidobacterium* [13]. Recently, probiotic cheeses have been successfully developed for the market. Cheese as a food matrix carrying probiotic microorganisms has enormous potential because it is a solid matrix, with high fat content, proteins, higher pH values compared to fermented milk and have a buffer effect against acidic stomach conditions, creating an environment more favorable for the survival of probiotic microorganisms [14].

Ice cream, beyond composition, is a product widely appreciated product representing a good vehicle for probiotic microorganisms. Besides, it has relatively high pH values when compared to fermented products favoring the survival of probiotic cultures [15]. Dairy desserts also represent an interesting option for incorporating probiotic microorganisms. Due to the widespread consumption of these products by all age groups, they can contribute to reaching a wide consumer market.

Recent research reports that chocolate with a high cocoa content may confer to consumer health benefits, principally due to the presence of antioxidant compounds [16, 17]. Thus, because it is a product appreciated worldwide, the incorporation of probiotic microorganisms represents a strategy for the food industries. Succi et al. demonstrated that chocolate with a high cocoa content (80%) and phenols are an excellent environment to carrier probiotic bacteria such as *Lactobacillus paracasei* and *Lactobacillus rhamnosus* [18]. Silva et al. also observed that *Lactobacillus acidophilus* and *Bifidobacterium animalis* subsp. *lactis* were successfully incorporated in dark chocolate, remaining viable for 120 days at 25°C and maintaining survival after *in vitro* simulation of GI conditions [19].

Currently, the food industry has developed new carrier matrices for probiotic microorganisms because of factors such as lactose intolerance, milk protein allergy, the search for products with low fat and no cholesterol, the veganism, and vegetarianism.

Food matrices with potential for application of probiotic microorganisms include juices, meats, cereals, fresh vegetables, and fruits. Probiotic fruit juices are already commercialized and represent an interesting food matrix because of the presence of nutrients such as vitamins, minerals, fibers, and antioxidants. Additionally, they have no starter cultures that compete with probiotic cells for nutrients, and they are widely consumed, without restricting the population allergic to milk protein or lactose intolerant [20]. Besides, vegetable products such as cucumber, sauerkraut, and olives have shown good results as matrices for probiotic bacteria [21, 22]. Fermented meat products have demonstrated to be a viable option for obtaining a probiotic product. The probiotics microorganisms can be naturally present in meat or added as *starter* cultures. Studies have shown satisfactory results compared to the growth of probiotic bacteria added to cereal-based products [23, 24]. Cereal grains are rich in proteins, carbohydrates, minerals, vitamins, and fibers that contribute to the metabolism of the probiotic microorganisms.

Ribeiro et al. found that the mixed drink composed of banana, strawberry, and juçara has the potential to be a vehicle for *Lactobacillus plantarum* and *Lactobacillus casei*. The results obtained indicated high cell viability for 90 days of cold storage, besides, *L. plantarum* presented viable populations above 6 log $CFU.g^{-1}$ after *in vitro* simulation of GI conditions [25]. Santos et al. found that fermented cocoa juice added with sucralose maintained the viability of *Lactobacillus casei* for 42 days (7.05 ± 0.04 log $CFU.mL^{-1}$) [26]. Akman et al. impregnated *Lactobacillus paracasei* in apple slices and verified that the viability probiotic remained above 6 and 7 log $CFU.g^{-1}$ after vacuum and oven drying, respectively [27]. Shigematsu et al. coated minimally processed carrots with sodium alginate added of Lactobacillus acidophilus (7.36 log $CFU.g^{-1}$), followed by immersion in calcium chloride and verified that the cell viability of *L. acidophilus* remained at 7 log $CFU.g^{-1}$ [28]. Rubio et al. when evaluating different strains of *Lactobacillus* during the manufacture of fermented sausages, observed that *L. rhamnosus* was able to grow up to 8 log $CFU.g^{-1}$, dominating the endogenous population of LAB throughout the process of maturation [29].

9.4 LIMITATIONS ON THE APPLICATIONS OF PROBIOTICS

As mentioned before, several beneficial effects have been reported due to the consumption of probiotic microorganisms. In recent years, there has been a significant increase in the variety of pharmaceutical formulas and probiotic foods. New research is constant, aiming to improve cell viability, develop new products, and, consequently, meet the highest consumer demand. The ingestion of probiotic microorganisms by food represents a more natural way when compared to pharmaceutical formulations. However, if we have several food matrices and the health benefits are numerous, why don't we have an infinite variety of probiotic foods available on the market?

Although studies report that numerous food matrices are promising alternative as a vehicle for probiotic microorganisms, some limitations limit the incorporation. Therefore, the food industry, together with researchers, is trying to find alternatives to overcome these limitations and offer a wider variety of probiotic products at an affordable price. The main limitations in the development of probiotic products will be presented in this section.

9.4.1 CHARACTERISTICS OF THE FOOD MATRIX AND PROCESSING

There are different types of microorganisms with probiotic potential and, therefore, present different characteristics and nutritional requirements. However, some factors are common to all the types of microorganisms used as probiotics, for example, the insertion in a nutrient-rich medium is extremely important for cellular survival. These nutrients may benefit one specifics probiotic, depending on the nutritional needs of each species. Additionally, to nutrients, other dietary properties also affect the viability of probiotic cells, such as water availability, pH, salt content, buffering capacity, presence of antimicrobials, among others. Also, probiotic microorganisms are anaerobic or microaerophilic, therefore, the presence of oxygen is toxic to these microorganisms. Moreover, it should be considered that the addition of probiotic microorganisms can cause changes in the sensory characteristics of foods. For example, some probiotic strains can produce enzymes that act on proteins, lipids, and carbohydrates producing compounds that modify the flavor, aroma, color, and texture.

The different stages of technological processing can expose probiotic microorganisms to adverse conditions, such as high temperatures, freezing, dehydration, presence of gases (CO_2, O_2), among other factors, and, consequently, significantly reduce the cell viability. Furthermore, in probiotic foods, the fermentation temperature can affect cell viability. The optimum temperature for growth of most probiotics is between 37°C and 43°C. Thus, temperatures above 45°C should be avoided if the probiotic microorganisms participate in the fermentation process, as many LABs. Besides, a vacuum fermentation system is recommended, because the oxygen content formed during the process is harmful to probiotic bacteria.

The storage temperature also influences the viability of probiotics being recommended temperatures between 4°C and 5°C. At freezing temperatures, the formation of ice crystals can cause damage to cells. Also, the crystallization of water during freezing causes the cryogenic concentration of solutes, which can induce osmotic damage and reduce cellular metabolic activities. During thawing, cells are exposed to osmotic stress, hydrogen ions, oxygen, organic acids, etc.

Probiotic cultures usually come in the form of a dry powder, easily stored, and with a long shelf life. For this, bacterial dehydration processes are used, including freeze-drying and spray drying. In spray drying, high temperatures, around 200°C, are used despite being a fast method, microorganisms are exposed to heat, osmotic, and dehydration stress, among others, which can reduce their viability. In the lyophilization process, in addition to the freezing, the dehydration step can modification the protein structure and the physical state of the lipid membranes, reducing the metabolic activity of the microorganisms [30].

During the storage of probiotic product factors such as the thickness of the packaging material, the permeability to light and oxygen, the presence of gases (CO_2, O_2, water vapor), among others, can affect the probiotic viability. Therefore, when developing pharmaceutical formulations or probiotic functional foods, the selection of appropriate processing steps is an essential factor for the survival of the microorganism.

9.4.2 SURVIVAL DURING PASSAGE THROUGH THE GASTROINTESTINAL TRACT (GIT)

To promote the various health benefits, probiotic microorganisms must be alive and consumed in adequate quantities. However, the use of probiotic microorganisms is associated with some obstacles, such as the low survival rate during exposure to adverse conditions of the human GIT and the small residence time in the intestine [13, 133] . In the stomach and small intestine, for example, the environment is unfavorable for colonization and bacterial proliferation, due to the low pH of gastric juice, the presence of bile and pancreatic juice, and, also, the peristaltic movements.

After ingesting the probiotic cells, they are exposed to the acidic conditions present in the stomach, the hydrochloric acid (HCl) has pH 0.9. However, the presence of the food can increase this value to pH 3 [31]. Probiotic cells are typically very sensitive to the highly acidic conditions of the human stomach. After passing through the stomach, probiotic cells are released into the small intestine, the decrease in probiotic viability caused by bile salts in the intestine occurs due to cellular disorganization caused by changes in the membrane. Subsequently, the cells reach the large intestine and, in the colon, the microorganisms find favorable conditions for their proliferation due to the absence of intestinal secretions, the slow peristalsis, and the abundant nutritional supply. The colon contains the highest microbial density and the population is known as microflora or intestinal microbiota [32].

To provide beneficial effects to the host, the probiotic must survive passage through the GIT and reach the intestine in adequate quantities. However, the loss of viability of probiotic microorganisms during passage through the GIT has led to the search for new strategies for maintaining viability. In this context, microencapsulation is the most prominent method for the protection of probiotics.

9.5 MICROENCAPSULATION

Microencapsulation is an effective technique to protect probiotic cells against adverse conditions and keep them in adequate amounts to confer health benefits on the host. Todd defined microencapsulation as the technology of packaging with a thin polymeric coating applied in solid, liquid, and gaseous materials [131]. Thus, forming particles known as capsules that release their contents at controlled rates over prolonged periods and under specific conditions. Arshady in 1993 [33] describes microcapsules as extremely small packages, composed of a polymer as wall material and an active material called core. Currently, the coated material in addition to being called a core is referred to as active material, internal phase, load, and the coating are called a wall material, membrane, shell, matrix, or external phase [34].

The particles formed may have regular or irregular shapes, and classified as mononuclear, polynuclear, or matrix. Mononuclear microcapsules contain the core bypassed by a defined and continuous film of the wall material. The multinucleated have many cores enclosed in the shell material. In the matrix, the microencapsulated material is uniformly distributed in the shell material [35].

One of the main factors that influence the stability of the encapsulated compounds is the type of coating used. Therefore, these materials must have specific characteristics, for example, easy control; low hygroscopicity; low viscosity at high concentrations; ability to disperse or emulsify and stabilize the core material; complete release of the solvent or other materials used during the encapsulation; maximum protection of the core under adverse conditions (light, pH, and oxygen); solubility in commonly used solvents; no taste or odor and low cost [36].

In the industry, the microencapsulation has decreased limitations and is used to protect compounds of external agents; release them in a controlled means, to mask undesirable flavors or odors. Materials that can be encapsulated consist of: acids, bases, oils, vitamins, salts, flavors, dyes, enzymes, and microorganisms.

9.6 MICROENCAPSULATION OF PROBIOTICS AND RELEVANCE INDUSTRIAL

As mentioned before, due to the varied applications, the probiotics market is one of the most promising sectors in the industry. Though, the inclusion of probiotics in products presents challenges and, thus, the microencapsulation represents an alternative to face the limitations related to the applications of probiotics. Microencapsulation is considered one of the more efficient technology to protect probiotic cells from adverse conditions.

The objective of encapsulation is to create a physical barrier between the microorganism and the external environment, increasing cell viability during processing, storage, and passage through the GIT. Thus, viability will be maintained and the cells released in appropriate locations (for example, small intestine) for adhesion and colonization of the intestinal epithelium to occur.

The lactic acid bacterium was first immobilized in 1975 on Berl saddles, and *Lactobacillus lactis* was encapsulated in alginate gel beads years later [37]. Currently, the encapsulation of probiotic cells is advancing and permitting the development of innovative systems for probiotic products. About the food industry, the use of microencapsulation in probiotics has increased because of the new demands of probiotics products. Numerous food systems containing probiotic encapsulation have been introduced and accepted by consumers [38]. In this context, cellular microencapsulation has gained interest and, thus, increasing the varied application of probiotics in the market.

9.6.1 TECHNIQUES FOR MICROENCAPSULATION OF PROBIOTICS

Probiotics cells are affected by different factors and, thus, the encapsulation techniques used need to be carried out in mild conditions, for example, low temperature, controlled agitation, low oxygen, moderate pH. Furthermore, not be used solvents that are toxic for microorganisms, non-GRAS (no generally recognized as safe) encapsulating agents, or that affect the sensory characteristics of the product. The size of the particles obtained must be adequate to protect probiotic cells and not modify the sensory characteristics of the product. The perfect characteristics for a microencapsulated probiotic are in the form of a dry powder, easily stored, and with a long shelf life, or a moist gel with long-term stability [39].

Modern and innovative methods of microencapsulation have been developed in recent decades, permitting the creation of a wide variety of probiotic microcapsules. However, it is important to note that industries probably prefer economic processes. Thus, the balance between cost and benefit must be considered. Then, the most important technologies to encapsulate probiotic cells will be highlighted.

9.6.1.1 SPRAY DRYING

Historically, the spray drying process has been extensively used in the food industry. However, there was a rapid expansion of this technique for other sectors such as in pharmaceuticals, cosmetics, and textiles [40].

The main mechanisms involved in the microencapsulation using spray drying include the solubilization of a specific polymer together with the probiotic microorganisms and, then, the sprayed of the solution in contact whit a hot air stream, instantly producing powdered microcapsule [41]. Thus, it is a conversion process, in one step, fluid materials in solid or semi-solid particles [42]. In this technique, heat, and mass transfer occur simultaneously from hot air to atomized drops and vice versa, respectively [43].

The microencapsulation method using spray drying involves the components presented in Figure 9.1. The process is carried out according to the following sequence of operations: Initially, the solution containing the probiotic microorganism and the polymer is prepared; posteriorly, this solution is homogenized and the atomized inside the drying chamber. During the passage through the drying chamber, heat, and mass transfer from the hot air to the atomized droplets occur and vice versa. Finally, the dry material is separated and collected by a cyclone. The process can produce micron- or nano-scale particles, in a short time [40].

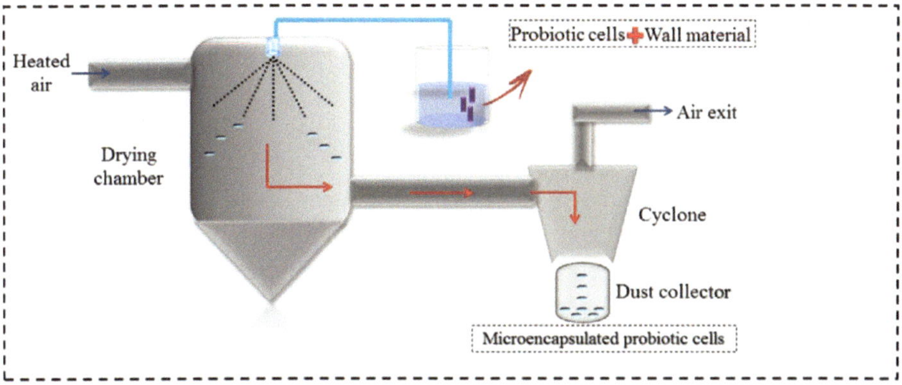

FIGURE 9.1 Microencapsulation of probiotics using spray drying.

Spray drying is flexible and produces particles with a moisture content between 4 and 7%, causing better stability during storage [44]. Moreover, spray drying microencapsulation has low operating costs and high production rates, justifying the preference of the industries [45].

Vanden Braber et al. by microencapsulating *Kluyveromyces marxianus* and optimizing the spray drying process, observed that the air outlet temperature of 68°C allowed higher encapsulation efficiency (EE) and *probiotic viability* of 8.38 $CFU.g^{-1}$. Besides, microencapsulated microorganisms using chitosan showed higher tolerance under simulated GI conditions compared to free cells and microencapsulated cells using whey protein concentrate [46]. Rosolen et al. used the spray drying technique for microencapsulation *Lactococcus lactis subsp lactis R7.* The probiotic cells in the microcapsules presented high viability (13.0 log $CFU.g^{-1}$) and remained stable for 6 months (> 8.0 log $CFU.g^{-1}$). After simulation *in vitro* of human GI conditions, the cells remained viable and, thus, surviving the effects of gastric and intestinal juices [47].

However, maintaining the high cell viability represents a challenge for microencapsulation thus, inlet air temperature, outlet air temperature, and type of the polymer are conditions that can negatively impact the survival of microorganisms [48]. Outlet air temperature above 85–90°C can injury macromolecules, such as proteins, DNA, and RNA, ribosomes, and membranes, being lethal to the probiotic microorganisms [49].

Thus, the factors optimization including the type of polymers and operational drying conditions can minimize damage [50]. Besides, the use of prebiotics can improve the survival of probiotics during processing steps [47, 51, 52]. Generally, the water-soluble polymers are used as wall material in the spray drying techniques, for example, whey proteins, maltodextrin, β-cyclodextrin, and gum Arabic [53].

9.6.1.2 FREEZE-DRYING

For decades, the freeze-drying technique was used for the manufacture of probiotic powders. However, currently, the technique has been associated with encapsulation and used for the production of microencapsulated probiotics [54]. The main application of this technique is the drying of thermo-sensitive compounds aiming to improve the conservation of their functional properties [55].

Freeze drying (Figure 9.2) is a *low-temperature* dehydration process that involves the removal of water by sublimation [56]. The process depends on heat and mass transfer and can be divided into three main steps: freezing, primary drying (sublimation), and secondary drying (desorption) [57, 58].

Initially, the material is frozen and exposed under vacuum conditions, thus, occur the sublimation process and the water content of dry material are about 15% [59]. Subsequently, secondary drying occurs by desorption, and the defrosted water absorbed by the material is removed. The residual moisture content of the freeze-dried material is around 1–3% [60] and then it is triturated, resulting in microparticles [61]. Among the steps mentioned, freezing is the most complex stage of freeze-drying due to the development of the structure that previously determines the properties of the dry material [62]. The materials commonly used as coatings are chitosan, gelatin, carrageenan, gum Arabic, gum guar, soy protein, disaccharides, and others [61, 63].

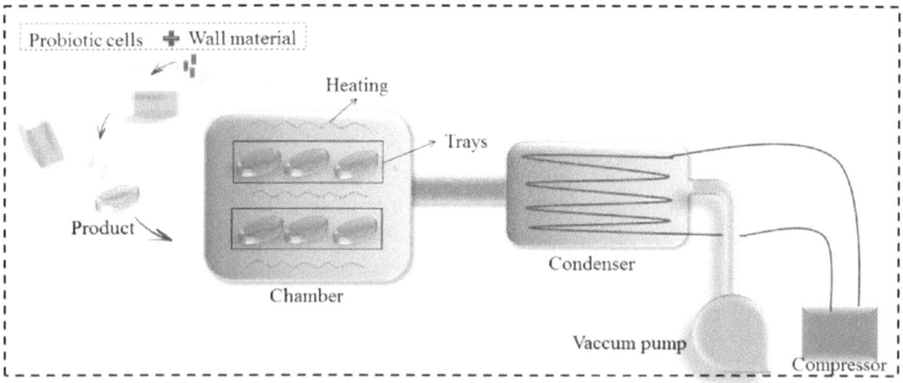

FIGURE 9.2 Encapsulation process by freeze-drying.

The microencapsulation of probiotics by freeze-drying stands out because drying at lower temperatures results in higher survival rates of the probiotic cells [64, 65]. Moreover, the freeze-drying material usually reconstitutes easily and has a long service life [66]. Besides, the selection of the wall material used is extremely important because it can act as a cryoprotectant and contribute to the stabilization of the encapsulated compound [61]. Some cryoprotectants such as sucrose, still, and mannitol are used to minimize the effects of freezing, stabilize the final particle size and prevent aggregation of particles during the process [67].

Li et al. performed the microencapsulation of *Lactobacillus casei* by freeze-drying technique. The authors used combinations of whey protein, gellan gum, and cellulose acetate phthalate as wall materials on the microencapsulation. The results obtained showed that microencapsulation protected the probiotic cells and maintained high viability after *in vitro* simulation of GI conditions and when exposed to heating [65]. Maleki et al. applied the technique to microencapsulate *Lactobacillus rhamnosus.* The formulation of the wall materials containing 57.22% whey protein, 25.00% crystalline nanocellulose, and 17.78% inulin presented the highest EE (89.60%). Furthermore, the formulation significantly improved the survival of probiotic bacteria during passage through simulated GI conditions. Thus, it can be a *promising alternative* for the production of probiotic microcapsules for use in food and pharmaceutical products [68].

However, the freeze-drying technique has some disadvantages, for example, the need for an extended drying period, resulting in higher energy consumption; drying time may be different according to the sample thickness and the deficiency of standardization of the particle size [69]; the formation of crystals during the freezing can damage the cell membrane and cause stress due to high osmolarity [49].

9.6.1.3 EXTRUSION

The extrusion is commonly applied for microencapsulation of probiotic microorganisms. The technique consists in the extrusion of a mixture containing the polymer and compound that will be encapsulated through an orifice and subsequent formation of droplets at the discharge point of the nozzle. Extrusion technique can be classified into simple drip, electrostatic extrusion, coaxial airflow, jet or vibratory nozzle, jet cutting, and rotary disc atomization. In the simple drip, low speed is applied and droplets with large diameters, around 2 mm, are formed, and production is insufficient for industrial application [70]. Electrostatic extrusion and coaxial airflow allow the production of smaller particles, from 50 μm to 200 μm, with uniform size distribution controlled by variation of the applied potential [49, 71]. Jet cutting, rotating disk atomization, and vibrating jet or nozzle are ideal techniques for mass production [72].

The extrusion technique is relatively simple and consists of the preparation of a solution containing the hydrocolloid and the probiotic microorganism (Figure 9.3). Then the material is forced through a nozzle, for example, a needle of the syringe or spray machine, and the cells suspension is then dripped into a hardening solution resulting in the formation of gelled drops [61].

For the microencapsulation of probiotics by extrusion, some polymers are frequently used as encapsulating material, for example, alginate, starch, agar, carrageenan, gelatin, and pectin [73, 74]. These polymers form a gel in contact with solutions containing minerals, particularly calcium and potassium [75]. Among polymers, sodium alginate is the most frequently used [39, 74] and has a wide range of applications in the food and pharmaceutical industry [39, 76]. Sodium alginate forms a network structure similar to an "egg-box," forming covalent crosslinking between alginate molecules capable of trapping materials such as probiotic cells [77].

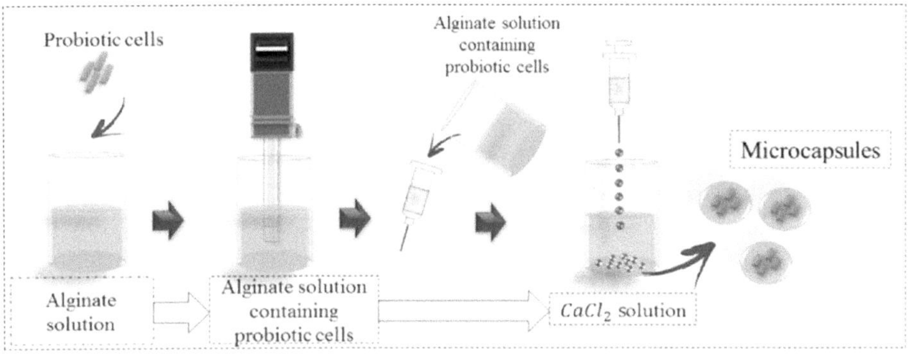

FIGURE 9.3 Probiotic encapsulation by extrusion using alginate and calcium chloride.

The extrusion technique is frequently used because of its low cost and the high viability of probiotic cells [78]. Seth et al. evaluated the effect of microencapsulation using the extrusion technique on cell viability during the spray drying of sweetened yogurt. The results obtained showed a 2-log increase in the survival of the encapsulated cells. Besides, it was found that the concentration of sodium alginate significantly affected the size of the microcapsules, the EE, and the survival of the bacteria [79]. Dimitrellou et al. used the technique to microencapsulate *Lactobacillus casei* in an alginate capsule for the manufacture of fermented milk. After *in vitro* simulation of GI conditions and storage for 28 days of fermented milk, the encapsulation increased significantly the viability of the probiotic microorganism [80].

However, the extrusion technique produces large particles of a large size (500 to 1000 nm) that could influence the sensory characteristics of the product [41, 81]. Another factor is the porosity of the microspheres that allow the exposition of the encapsulated material, principally under acidic conditions [78]. The addition of more polymer layers during the microcapsule production may be an alternative to decrease the porosity [81].

9.6.1.4 COMPLEX COACERVATION

Historically, the word "coacervate" is derived from the Latin "acervus" meaning aggregation, and the prefix "co" means the colloidal particles. Thus, in colloidal chemistry, the term "Coacervation" is used to denote the process of phase separation by modification of the environment (pH, ionic force, temperature, solubility). This process is characterized by the separation of a colloidal dispersion into two immiscible liquid phases: One phase has a high colloid concentration (coacervate phase) and another diluted with low amounts of colloid (equilibrium phase) [82].

In complex coacervation, at a specified pH, the electrostatic interaction between polymers of opposite charge results in the formation of soluble complexes. These complexes aggregate to decrease the free energy of the system, becoming insoluble and, subsequently, there is a separation of phases. In this process, the driving force is electrostatic interaction [83]. However, under conditions in which the electrostatic force is suppressed (for example, high concentration of salt), hydrogen bonds and hydrophobic interactions, among others, can contribute to the formation of coacervates [84]. During the *encapsulation procedures* based *on complex coacervation,* the coacervate polymer is deposited around the active ingredient (core), occurring the sedimentation of the encapsulated cores (Figure 9.4).

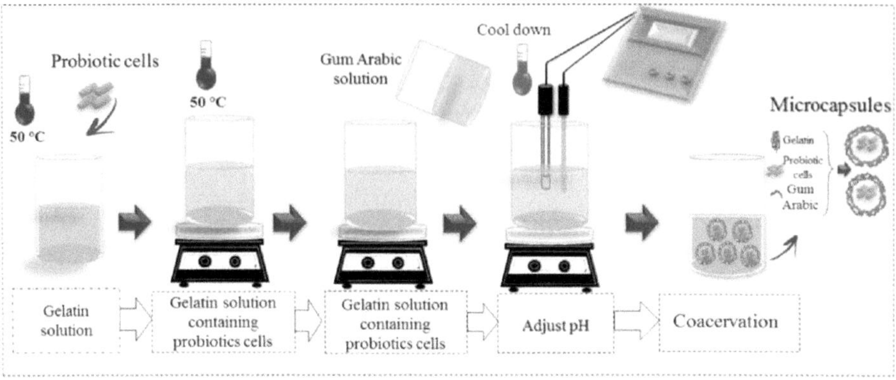

FIGURE 9.4 Encapsulation process based on complex coacervation.

The *coacervation* process is affected by *multiple* factors, including, the electrical charge of polymers, polymer concentration, ionic strength, temperature, pH, etc., [85]. These parameters are usually optimized aiming the highest yield and functionality of the complex coacervates. The analysis of zeta potential, absorbance, and yield of the dehydrated coacervates are used to optimize. The complex coacervation technique has varied advantages compared to other techniques such as versatility, ease of operation, mild conditions, and *low cost* and environmental impact [86]. Besides, the microcapsules produced have an excellent controlled release, modulated by changes in ionic strength, pH, and temperature [18].

Complex coacervation is an old method that has been used in different industrial applications such as pharmaceutical, chemical, cosmetic, and food industries. The polymers generally used for encapsulation by complex coacervation are gelatin, whey protein, Arabic gum, chitosan, pectin, alginates, xanthan gum, carrageenan, and carboxymethyl cellulose. *One disadvantage* of the complex coacervation technique is the low mechanical and thermal resistance because of the ionic nature of the interactions, it is necessary to strengthen the structure, for example, to crosslink the polymeric chains [87].

Da Silva et al. produced microcapsules containing *B. lactis* by complex coacervation using gelatin and gum Arabic as coating materials. Microencapsulated probiotics-maintained viability after *in vitro* simulation GI conditions and, also, the complex coacervation method was efficient in maintaining the viability of probiotics during storage at temperatures of –18°C for 120 days, 7°C for 120 days and 25°C for 90 days [86]. Paula et al. by microencapsulating probiotic cells of *Lactobacillus plantarum* through emulsification followed by complex coacervation using gelatin and gum Arabic. The authors observed that after *in vitro* simulation of GI conditions, the viability of encapsulated cells was 80.4%, while for free cells, it was 25%. Additionally, cell viability was maintained during storage at 8°C and –18°C for 45 days. Thus, the results obtained show that the complex coacervation is an appropriate alternative to increase the viability of probiotics [88].

However, although the complex coacervation process is considered as promising for the encapsulation of probiotics, it is little used. Therefore, there is a need to explore more studies on coating materials, the variation of concentrations and association with drying techniques, are desirable to increase the protection of probiotic and allow a more effective application [86].

9.6.1.5 EMULSIFICATION TECHNIQUES

9.6.1.5.1 EMULSIONS

Emulsions are defined as heterogeneous systems and thermodynamically unstable composed of a mixture of two immiscible liquids, in which one liquid (the dispersed phase) is in the form of droplets dispersed in the other (continuous) phase [89]. In general, simple emulsions are categorized according to their dispersed phase. When the oil is the dispersed phase, the emulsion is of the oil-in-water (O/W) type, however, if the water is the dispersed phase, the emulsion is of the water-in-oil (W/O) type. In the food industry, products such as milk, sour cream, soups, yogurts, ice cream, butter, and margarine are examples of simple emulsions.

The emulsions do not form spontaneously and require a considerable involvement of energy, usually mechanical [90]. In food processing are used high-speed mixers, colloid mills, homogenizers, and ultrasonic mixers. Moreover, a third component or combination of active and surface agents, often called an emulsifier or emulsifying agent is used [91]. The most used emulsifiers are lecithin; mono and di-glycerides of fatty acids, proteins, phospholipids, and, in certain cases, polysaccharides (hydrocolloids).

Typically, each method has associated advantages and disadvantages. The main disadvantage of the emulsification technique is difficulty in standardizing the microcapsule size distribution. Additionally, it also presents difficulties in implementation and the requirement of vigorously agitated which can be potentially injurious to cells, and the inability to sterilize vegetable oil if it is necessary strict aseptic conditions.

The principle of this technique is based on the dispersion of the solution of polymer and probiotic cells (dispersed phase), in a large volume of oil, usually vegetable oil (continuous phase). The mixture is homogenized continuously resulting in aqueous/polymer droplets containing the probiotic cells, specifically, the formation of an emulsion water-in-oil (W/O) type. However, after formation, the water-soluble polymer (present in the aqueous droplets) can be insolubilized causing gelation of the aqueous/polymer droplets. In this case, the emulsification technique can be divided into internal or external ionic gelation.

9.6.1.5.2 EMULSIFICATION/INTERNAL IONIC GELATION

In this method, occur the addition of a solution containing insoluble calcium salt (usually calcium carbonate) in a polymeric solution (for example, sodium alginate) containing probiotic cells, with subsequent dispersion of this mixture in an oily phase containing surfactant for the formation of emulsion W/O (Figure 9.5). Subsequently, for the gelation process to occur, it is necessary to add an acid solution, that is, reduce the pH and thus release calcium ions (present inside the emulsion droplets), allowing its complexation with the alginate carboxylic groups [92]. In this case, Ca^{2+} is crosslinked with sodium alginate from the inside out of the droplet. The microcapsules are recuperated by filtration or centrifugation.

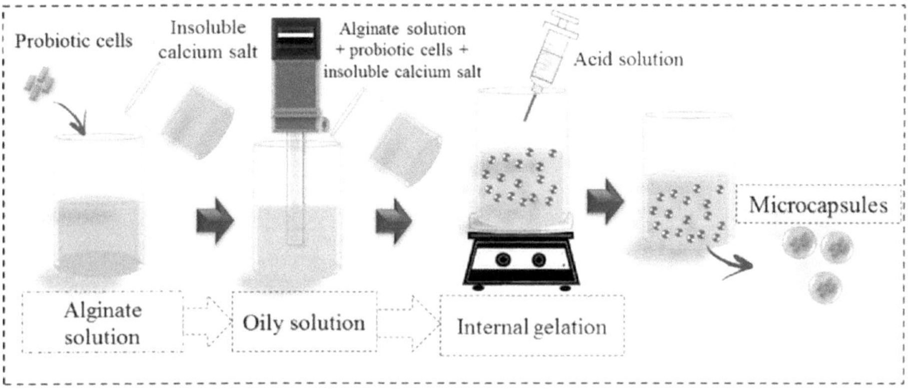

FIGURE 9.5 Encapsulation of probiotics by internal ionic gelation.

Many studies report that microcapsules prepared by internal ionic gelation have greater size standardization and better EE [93].

9.6.1.5.3 EMULSIFICATION/EXTERNAL IONIC GELATION

In external ionic gelation, the polymeric solution containing the probiotic cells is initially dispersed in an oily phase containing surfactant for the formation of emulsion W/O. Subsequently, a solution containing insoluble calcium salt is incorporated, usually calcium carbonate, with the ion diffusion into the aqueous alginate droplets. Consequently, gel formation occurs from the surface into the droplets (Figure 9.6) [94]. According to King [134], it is possible to manipulate the strength of the gel through changes in processing conditions, such as pH, calcium concentration, concentration, and source of alginate (algae species), etc.

9.7 INCREASED RESISTANCE OF MICROENCAPSULATED PROBIOTICS

In some cases, the use of the encapsulation technique alone does not maintain the satisfactory viability of microorganisms. Thus, technological resources can be used, such as the addition of cryoprotectants, the use of prebiotics, the formation of covalent bonds called crosslinking, and the increase of polymer layers around the microcapsules.

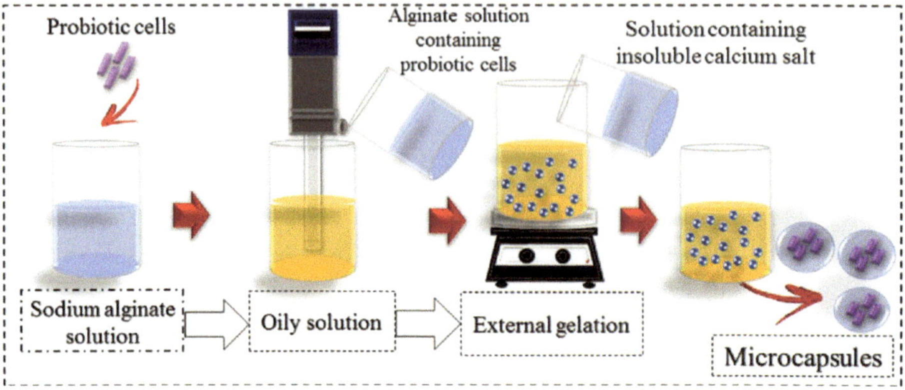

FIGURE 9.6 Encapsulation of probiotics by external ionic gelation.

The addition of cryoprotectants such as sugars (e.g., trehalose and sucrose) is used to reduce the osmotic difference between the interior of a cell and the exterior environment and can minimize the damage caused by cold. The addition of the prebiotics in the formulation of the microparticles results in a symbiotic combination. Currently identified prebiotics are non-digestible carbohydrates, however, beneficial bacteria can ferment, stimulating the growth and/or activity of bacteria in the colon [135]. The most frequently studied examples are fructans and fructooligosaccharides of the inulin-type. Raddatz et al. using different prebiotics in combination with the microencapsulation of *Lactobacillus acidophilus* LA-5 found greater cell viability after *in vitro* simulation of GI simulation and under storage conditions of 25°C to –18°C [95].

The formation of covalent bonds that hold portions of several polymer chains together is called. The crosslinking can be used to improve the thermal properties of the microcapsules [87]. However, most of the chemical crosslinking such as formaldehyde, glutaraldehyde, glyoxal, diisocyanate, epichlorohydrin have their applications limited due to toxicity and difficulty of complete removal [96]. Consequently, enzymatic crosslinking represents an alternative [97] and transglutaminase is an enzyme widely used to improve the rheological and physical properties of microcapsules [98]. Da Silva et al. evaluated the use of transglutaminase to improve the resistance of microcapsules obtained by complex coacervation. After *in vitro* simulation of GI conditions, it was found that microencapsulation together with crosslinking showed good results. Besides, after storage, probiotic viability was maintained for up to 60 days in freezing temperature, with counts of up to 9.59 log $CFU.g^{-1}$. The results obtained are innovative and present a promising alternative for the protection of probiotics [99].

The increase of polymer layers around the microcapsules provides extra protection for cells from adverse conditions [39, 100]. According to Grigoriev and Miller for the formation of polymer layers, electrostatic deposition layer by layer can be used. A wide variety of materials have been explored to protect the capsules, including alginate, pectin, starch, and chitosan [101]. Etchepare et al. show that *L. acidophilus* microencapsulated in calcium alginate particles coated with multilayers showed greater survival under simulated GI conditions, heat treatment, and during storage [102].

9.8 BIOPOLYMERS USED IN MICROENCAPSULATION

In the microencapsulation of probiotics, the microcapsules must be able to maintain their structure even under adverse conditions, releasing your content at controlled rates and/or under specific conditions. Generally, the microcapsules release their content because of pH changes, chelating agents, and enzymatic action. Thus, as mentioned before, the coating material must have some characteristics, among them, easy handling; low hygroscopicity; low viscosity at high concentrations; disperse/emulsify and stabilize the core material; good film-forming properties; complete release of the solvent; protection of the core against adverse conditions (for example, light, pH, and oxygen); solubility in commonly used solvents; absence of taste or odor and low cost [36]. The polymers most widely used for this purpose are then highlighted:

1. **Gelatin:** It is a naturally derived polymer that can be obtained through acidic (Type A) or alkaline (Type B) hydrolysis of collagen. The gelatin represents the main commercial option as wall material due to its excellent water solubility, emulsification, and thickening capacity and high crosslinking activity. Also, the polypeptide structure of the molecule facilitates interactions with other polymers of opposite electrical charge, making it an important wall material [103]. However, gelatin solutions, even in low concentrations, have high viscosity and can cause problems of aggregation and agglutination during the preparation of the microcapsule process [136].
2. **Arabic Gum:** It is an exudate of acacia trees, of which there are species distributed over tropical and subtropical regions. In general, the composition of gum Arabic consists basically of two fractions.

One of the fractions, representing 70% of the gum, is composed of a polysaccharide chain with little or no protein. The other fraction contains molecules of higher molecular mass with proteins as part of their structure [104]. This composition confers an efficient surface property, besides having cold solubility due to the presence of loaded groups and peptide fragments [105]. About the electrical charge, gum Arabic is a weak polyelectrolyte that has a negative charge at a pH greater than 2.2, due to its carboxyl groups [106, 107]. Industrially it is widely used due to its surface activity, low viscosity, emulsifying capacity in aqueous solutions, as well as being non-toxic, odorless, and tasteless.

3. **Chitosan:** It is a biopolymer derived from chitin composed of N-***acetyl***-d-glucosamine units, linked by β (1,4)-glycosidic bonds. Chitosan can be obtained by deacetylation of chitin, which, which is the major constituent of exoskeletons of crabs, shrimp, and other crustaceans.

Chemically, chitosan contains groups of free amines in a neutral or alkaline environment, while due to the protonated amine groups (NH^{3+}) at pH > 6.3, the polymer becomes soluble in water and positively charged. Thus, the characteristics make chitosan suitable for controlled release technologies under specific conditions at the local target. Besides, chitosan is a biocompatible, biodegradable, inexpensive, and non-toxic polymer, making it attractive for applications in the medical, cosmetics, agriculture, food, and textiles [108].

4. **Sodium Alginate:** It consists of an anionic polysaccharide extracted from the cell wall of brown algae (*Laminaria spp.*), it is composed of β-1,4 glycosidic bonds formed between the β-D-mannuronate and α-L-guluronate residues [109]. Alginate molecules can form a structure called "egg-box" through crosslinking and exchanging sodium ions for divalent cation such as calcium [110]. Sodium alginate is one of the most used wall materials for microencapsulation [111] due to its biocompatibility, non-toxicity, low cost, and ability to form gel in the presence of calcium ions [112]. However, microcapsules are sensitive in acidic conditions and have a porous structure that affects their stability and efficiency [113].
5. **Maltodextrin:** It is composed of multiple glucose units linked by α-(1,4) glycosidic bonds and obtained from the partial hydrolysis of starch with acid or enzymes [114]. They are classified according to the degree of hydrolysis, expressed in dextrose equivalent (DE)

which is the percentage of reducing sugar calculated as dextrose on a dry-weight basis [115]. They are commonly used as wall material in microencapsulation technique because they have low density, low viscosity when used at high temperature, good solubility in water, do not alter the characteristics of the product, and have a low cost [116]. However, they have some limitations, such as low emulsifying capacity and retention of volatile compounds, and thus are used in mixtures with other wall materials [117].

6. **Gellan Gum:** It is an anionic polysaccharide with a linear structure of a repetitive tetrasaccharide sequence consisting of two β-D-glucose residues, one of β-D-glucuronate and one of α-L-rhamnose [74]. Obtained by microbial fermentation from the bacteria *Sphingomonas elodea* or *Pseudomonas elodea* [118]. Commercially, gellan gum is available in two forms, being high acyl (acylated) and low acyl (deacylated) and each type has individual properties [119]. In the presence of cations, high-acyl gelatin gum forms soft, flexible hydrogels after cooling to 65°C, while low-acyl gum forms rigid and brittle hydrogels after cooling to 40°C [120]. The deacylated form has been used successfully as a coating material for probiotic microencapsulation [74], and when mixed with other types of gum, such as xanthan gum, it has a high resistance to acidic conditions [121].
7. **Xanthan Gum:** Among the gums used in microencapsulation, xanthan gum is the most used. It consists of a high molecular weight extracellular polysaccharide that is produced by fermenting the bacteria *Xanthomonas campestris* [122]. Xanthan gum has some characteristics such as tasteless, odorless, has low viscosity, stability at high temperatures, good solubility in water, and can be widely used [123].
8. **Carrageenans Gum:** These are neutral polysaccharides extracted from red algae (*Rhodophyta*) and commonly used in the food industry. Red algae are capable of producing three distinct types of commercial carrageenan (κ-, ι-e λ-carrageenan) that differ in their structures and chemical properties [124]. They are extremely compatible with microbial cells, ensuring high viability during microencapsulation techniques, however, the gel formed presents physical instability in adverse conditions [41, 125].
9. **Milk Proteins:** These are divided into two groups, caseins (80%) and whey proteins (20%). Caseins are relatively hydrophobic, but they also have polar and charged residues, thus representing natural

and viable polymers for the encapsulation of compounds. The physic-chemical properties and structural enable its functionality, with emphasis on the binding of ions and small molecules, surfactant properties, emulsifying, and water-binding capacity. Moreover, if the ability to protect or compose its structure, allows it to control its bioavailability [126]. In addition to low cost, it is considered non-toxic, heat-stable, and GRAS. The main whey proteins are β-lactoglobulin, α-lactoalbumin, bovine serum albumin, lactoferrin, and immunoglobulins, which can vary in size, molar mass, and function [127]. Whey proteins are also recognized as GRAS, low cost, high nutritional value, have techno-functional properties (ability to form gels, foams, and emulsions), good sensory properties, and structural properties making them suitable for the transport of other molecules. However, they are globular proteins with a high level of structural organization, being susceptible to denaturation. Depending on the conditions of the environment, such as pH, strength, and temperature, these substances can exist in individual or agglomerated form, these characteristics being important for the development of distribution systems.

9.9 FINAL CONSIDERATIONS

The growing number of benefits presented by regular consumption of probiotics has attracted the attention of the pharmaceutical and food industry in recent years. The development of different probiotic formulations possibility the optimization of the delivery of microorganisms. The commercialization of a wide variety of probiotic foods is a market trend. The different options such as juices, chocolates, cereals, and yogurts, reach different audiences, with different dietary restrictions, providing an improvement in health from children to the elderly. Besides to *Lactobacillus* and *Bifidobacteria*, others have been reported as probiotic species and thus allowing the industry to use the microorganism with the greatest potential for processing and manufactured product. Besides, the use of more than one microbial species permits a wide range of benefits provided to human health by probiotics.

Although numerous pharmaceutical formulas and food products containing probiotic microorganisms are already available on the market, the new product development still represents a true challenge. Among all factors, the main problem is to maintain the viability of probiotic microorganisms.

Besides, the nutritional requirements, the processing steps, the characteristics of the food matrix, and the conditions during the passage through the GIT can affect the survival of the cells and, therefore, the health benefits of consumers.

In this way, the development of innovative technologies is an important area to optimize the manufacture of probiotic products. Among the technologies used, microencapsulation is considered one of the most efficient to protect probiotic cells from adverse conditions. The objective of microencapsulation is to create a physical barrier between the microorganism and the external environment, increasing cell viability during processing, storage, and passage through the GIT. Thus, viability will be maintained and cells released at appropriate sites (for example, small intestine) for adhesion and colonization of the intestinal epithelium.

However, encapsulation systems must be economically viable, relatively easy to handle, efficient, and safe. These characteristics depend on the technology and the polymers used as a coating. The search for biocompatible, non-toxic, biodegradable, and naturally sourced materials represents a preferred solution. Thus, making it possible to expand the application of probiotics in different foods, the development of different pharmaceutical formulas, and, consequently, reach a larger target audience.

KEYWORDS

- **conjugated linoleic acid**
- **dextrose equivalent**
- **exopolysaccharides**
- **gastrointestinal tract**
- **generally recognized as safe**
- **lactic acid bacteria**
- **oil-in-water**

REFERENCES

1. FAO/WHO, (2002). Joint FAO/WHO group report on drafting guidelines for the evaluation of probiotics in food, London, Ontario, Canada. *Guidelines for the Evaluation of Probiotics in Food: Report of a Joint FAO/WHO Working Group on Drafting Guidelines for the Evaluation of Probiotics in Food* (pp. 1–11). London, ON, Canada.

2. Bultosa, G., (2016). Functional foods: Dietary fibers, prebiotics, probiotics, and synbiotics. *Encyclopedia of Food Grains* (pp. 11–16). Elsevier,
3. Boricha, A. A., Shekh, S. L., Pithva, S. P., Ambalam, P. S., & Manuel, V. B. R., (2019). *In vitro* evaluation of probiotic properties of *Lactobacillus* species of food and human origin. *LWT-Food Science and Technology, 106*, 201–208. https://doi.org/10.1016/j.lwt.2019.02.021.
4. Nader-Macías, M. E. F., & Juárez, T. M. S., (2015). Profiles and technological requirements of urogenital probiotics. *Advanced Drug Delivery Reviews, 92*, 84–104. https://doi.org/10.1016/j.addr.2015.03.016.
5. Roobab, U., Batool, Z., Manzoor, M. F., Shabbir, M. A., Khan, M. R., & Aadil, R. M., (2020). Sources, formulations, advanced delivery and health benefits of probiotics. *Current Opinion in Food Science, 32*, 17–28. https://doi.org/10.1016/j.cofs.2020.01.003.
6. Suri, S., Kumar, V., Prasad, R., Tanwar, B., Goyal, A., Kaur, S., Gat, Y., Kumar, A., Kaur, J., & Singh, D., (2019). Considerations for development of lactose-free food. *Journal of Nutrition & Intermediary Metabolism, 15*, 27–34. https://doi.org/10.1016/j.jnim.2018.11.003.
7. Dhillon, P., & Singh, K., (2020). Therapeutic applications of probiotics in ulcerative colitis: An updated review. *Pharma Nutrition, 13*, 100194. https://doi.org/10.1016/j.phanu.2020.100194.
8. Novik, G., & Savich, V., (2020). Beneficial microbiota. Probiotics and pharmaceutical products in functional nutrition and medicine. *Microbes and Infection, 22*(1), 8–18. https://doi.org/10.1016/j.micinf.2019.06.004.
9. Preedy, V. R., & Watson, R. R., (2016). *Probiotics, Prebiotics, and Synbiotics: Bioactive Foods in Health Promotion* (1st edn.). Academic Press is an imprint of Elsevier.
10. Govender, M., Choonara, Y. E., Kumar, P., Toit, L. C., Vuuren, S., & Pillay, V., (2014). A review of the advancements in probiotic delivery: Conventional vs. non-conventional formulations for intestinal flora supplementation. *AAPS PharmSciTech, 15*(1), 29–43. https://doi.org/10.1208/s12249-013-0027-1.
11. Kaur, K., Nekkanti, S., Madiyal, M., & Choudhary, P., (2018). Effect of chewing gums containing probiotics and xylitol on oral health in children: A randomized controlled trial. *Journal of International Oral Health, 10*(5), 237–243. https://doi.org/10.4103/jioh.jioh_170_18.
12. FAO/WHO, (2018). *Joint FAO/WHO Food Standards Program Codex Committee on Nutrition and Foods for Special Dietary Uses: Discussion Paper on Harmonized Probiotic Guidelines for Use in Foods and Dietary Supplements* (pp. 26–30) Berlin, Germany.
13. Ross, R. P., Desmond, C., Fitzgerald, G. F., & Stanton, C., (2005). Overcoming the technological hurdles in the development of probiotic foods. *Journal of Applied Microbiology, 98*(6), 1410–1417. https://doi.org/10.1111/j.1365-2672.2005.02654.x.
14. Gomes, D. C. A., Buriti, F. C. A., Batista, D. S. C. H., Faria, J. A. F., & Saad, S. M. I., (2009). Probiotic cheese: Health benefits, technological and stability aspects. *Trends in Food Science & Technology, 20*, 344–354. https://doi.org/10.1016/j.tifs.2009.05.001.
15. Cruz, A. G., Antunes, A. E. C., Sousa, A. L. O. P., Faria, J. A. F., & Saad, S. M. I., (2009). Ice-cream as a probiotic food carrier. *Food Research International, 42*(9), 1233–1239. https://doi.org/10.1016/j.foodres.2009.03.020.
16. Kerimi, A., & Williamson, G., (2015). The cardiovascular benefits of dark chocolate. *Vascular Pharmacology, 71*, 11–15. https://doi.org/10.1016/j.vph.2015.05.011.

17. Seem, S. A., Yuan, Y. V., & Tou, J. C., (2019). Chocolate and chocolate constituents influence bone health and osteoporosis risk. *Nutrition, 65*, 74–84. https://doi.org/10.1016/j.nut.2019.02.011.
18. Bosnea, L. A., Moschakis, T., & Biliaderis, C. G., (2017). Microencapsulated cells of *Lactobacillus paracasei* subsp. *paracasei* in biopolymer complex coacervates and their function in a yogurt matrix. *Food & Function, 8*, 554–562. https://10.1039/C6FO01019A.
19. Silva, M. P., Tulini, F. L., Marinho, J. F. U., Mazzocato, M. C., De Martinis, E. C. P., Luccas, V., & Favaro-Trindade, C. S., (2017). Semisweet chocolate as a vehicle for the probiotics *Lactobacillus acidophilus* LA3 and *Bifidobacterium animalis* subsp. Lactis BLC1: Evaluation of chocolate stability and probiotic survival under *in vitro* simulated gastrointestinal conditions. *LWT, 77*, 640–647. https://doi.org/10.1016/j.lwt.2016.10.025.
20. Granato, D., Nazzaro, F., Pimentel, T. C., Esmerino, E. A., & Gomes, D. C. A., (2019). Probiotic food development: An updated review based on technological advancement. *Encyclopedia of Food Security and Sustainability* (pp. 422–428). Elsevier.
21. Argyri, A. A., Nisiotou, A. A., Mallouchos, A., Panagou, E. Z., & Tassou, C. C., (2014). Performance of two potential probiotic *Lactobacillus strains* from the olive microbiota as starters in the fermentation of heat-shocked green olives. *International Journal of Food Microbiology, 171*, 68–76. https://doi.org/10.1016/j.ijfoodmicro.2013.11.003.
22. Peres, C. M., Peres, C., Hernández-Mendoza, A., & Malcata, F. X., (2012). Review on fermented plant materials as carriers and sources of potentially probiotic lactic acid bacteria-with an emphasis on table olives. *Trends in Food Science & Technology, 26*(1), 31–42. https://doi.org/10.1016/j.tifs.2012.01.006.
23. Mantzouridou, F., Karousioti, A., & Kiosseoglou, V., (2013). Formulation optimization of a potentially prebiotic low-in-oil oat-based salad dressing to improve *Lactobacillus paracasei* subsp. *paracasei* survival and physicochemical characteristics. *LWT-Food Science and Technology, 53*(2), 560–568. https://doi.org/10.1016/j.lwt.2013.04.005.
24. Zubaidah, E., Nurcholis, M., Wulan, S. N., & Kusuma, A., (2012). Comparative study on synbiotic effect of fermented rice bran by probiotic lactic acid bacteria *Lactobacillus casei* and newly isolated *Lactobacillus plantarum* B2 in Wistar rats. *APCBEE Procedia, 2*, 170–177. https://doi.org/10.1016/j.apcbee.2012.06.031.
25. Ribeiro, A. P. O., Gomes, F. S., Santos, K. M. O., Matta, V. M., Freitas, D. S. D. G. C., Santiago, M. C. P. A., Conte, C., et al., (2020). Development of a probiotic non-fermented blend beverage with Juçara fruit: Effect of the matrix on probiotic viability and survival to the gastrointestinal tract. *LWT-Food Science and Technology, 118*, 108756. https://doi.org/10.1016/j.lwt.2019.108756.
26. Santos, F. A. L. D., Freitas, H. V., Rodrigues, S., Abreu, V. K. G., Lemos, T. D. O., Gomes, W. F., Narain, N., & Pereira, A. L. F., (2019). Production and stability of probiotic cocoa juice with sucralose as sugar substitute during refrigerated storage. *LWT-Food Science and Technology, 99*, 371–378. https://doi.org/10.1016/j.lwt.2018.10.007.
27. Akman, P. K., Uysal, E., Ozkaya, G. U., Tornuk, F., & Durak, M. Z., (2019). Development of probiotic carrier dried apples for consumption as a snack food with the impregnation of *Lactobacillus paracasei*. *LWT- Food Science and Technology, 103*, 60–68. https://doi.org/10.1016/j.lwt.2018.12.070.
28. Shigematsu, E., Dorta, C., Rodrigues, F. J., Cedran, M. F., Giannoni, J. A., Oshiiwa, M., & Mauro, M. A., (2018). Edible coating with probiotic as a quality factor for minimally

processed carrots. *Journal of Food Science and Technology, 55*(9), 3712–3720. https://doi.org/10.1007/s13197-018-3301-0.
29. Rubio, R., Jofré, A., Aymerich, T., Guàrdia, M. D., & Garriga, M., (2014). Nutritionally enhanced fermented sausages as a vehicle for potential probiotic lactobacilli delivery. *Meat Science, 96*(2), 937–942. https://doi.org/10.1016/j.meatsci.2013.09.008.
30. Cassani, L., Gomez-Zavaglia, A., & Simal-Gandara, J., (2020). Technological strategies ensuring the safe arrival of beneficial microorganisms to the gut: From food processing and storage to their passage through the gastrointestinal tract. *Food Research International, 129*, 108852. https://doi.org/10.1016/j.foodres.2019.108852.
31. Erkkila, S., & Petaja, E., (2000). Screening of commercial meat starter cultures at low pH and in the presence of bile salts for potential probiotic use. *Meat Science, 55*, 297–300. https://doi.org/10.1016/S0309-1740(99)00156-4.
32. Guarner, F., & Malagelada, J., (2003). Gut flora in health and disease. *The Lance, 361*(9356), 512–519. https://doi.org/10.1016/S0140-6736(03)12489-0.
33. Arshady, R., (1993). Microcapsules for food. *Journal of Microencapsulation, 10*, 413–435. https://doi.org/10.3109/02652049309015320.
34. Xiao, Q., Gu, X., & Tan, S., (2014). Drying process of sodium alginate films studied by two-dimensional correlation ATR-FTIR spectroscopy. *Food Chemistry, 164*, 179–184. https://doi.org/10.1016/j.foodchem.2014.05.044.
35. Ghosh, S. K., (2006). Functional coatings and microencapsulation: A general perspective. *Functional Coating*, 1–28.
36. Desai, K. G. H., & Park, H. J., (2005). Recent developments in microencapsulation of food ingredients. *Drying Technology: An International Journal, 23*(7), 1361–1394. https://doi.org/10.1081/DRT-200063478.
37. Linko, P., (1985). Immobilized lactic acid bacteria. In: Larson, A., (ed.), *Enzymes and Immobilized Cells in Biotechnology* (pp. 25–36). Benjamin Cummings: Meno Park, CA.
38. Yeung, T. W., Uçok, E. F., Tiani, K. A., McClements, D. J., & Sela, D. A., (2016). Microencapsulation in alginate and chitosan microgels to enhance viability of *Bifidobacterium longum* for oral delivery. *Frontiers in Microbiology, 7*(494), 1–11. https://doi.org/10.3389/fmicb.2016.00494.
39. Cook, M. T., Tzortzis, G., Charalampopoulos, D., & Khutoryanskiy, V. V., (2012). Microencapsulation of probiotics for gastrointestinal delivery. *Journal of Controlled Release, 162*(1), 56–67. https://doi.org/10.1016/j.jconrel.2012.06.003.
40. Singh, A., & Van, D. M. G., (2016). Spray drying formulation of amorphous solid dispersions. *Advanced Drug Delivery Reviews, 100*, 27–50. https://doi.org/10.1016/j.addr.2015.12.010.
41. Chen, M. J., & Chen, K. N., (2007). Applications of probiotic encapsulation in dairy products. In: Lakkis, J. M., (ed.), *Encapsulation and Controlled Release Technologies in Food Systems* (pp. 83–112). Blackwell Publishing: Ames, Iowa, USA. https://doi.org/10.1002/9780470277881.ch4.
42. Murugesan, R., & Orsat, V., (2012). Spray drying for the production of nutraceutical ingredients: A review. *Food Bioprocess Technol, 5*(1), 3–14. https://doi.org/10.1007/s11947-011-0638-z.
43. Riveros, B., Ferrer, J., & Bórquez, R., (2009). Spray drying of a vaginal probiotic strain of *Lactobacillus acidophilus*. *Drying Technology, 27*(1), 123–132. https://doi.org/10.1080/07373930802566002.

44. Ananta, E., Volkert, M., & Knorr, D., (2005). Cellular injuries and storage stability of spray-dried *Lactobacillus rhamnosus* GG. *International Dairy Journal, 15*(4), 399–409. https://doi.org/10.1016/j.idairyj.2004.08.004.
45. Bustamante, M., Villarroel, M., Rubilar, M., & Shene, C., (2015). *Lactobacillus acidophilus* La-05 encapsulated by spray drying: Effect of mucilage and protein from flaxseed (*Linum usitatissimum* L.). *LWT-Food Science and Technology, 62*(2), 1162–1168. https://doi.org/10.1016/j.lwt.2015.02.017.
46. Vanden, B. N. L., Díaz, V. L. I., Rossi, Y. E., Aminahuel, C. A., Mauri, A. N., Cavaglieri, L. R., & Montenegro, M. A., (2020). Effect of microencapsulation in whey protein and water-soluble chitosan derivative on the viability of the probiotic *Kluyveromyces marxianus* VM004 during storage and in simulated gastrointestinal conditions. *LWT-Food Science and Technology, 118*, 108844. https://doi.org/10.1016/j.lwt.2019.108844.
47. Rosolen, M. D., Bordini, F. W., De Oliveira, P. D., Conceição, F. R., Pohndorf, R. S., Fiorentini, Â. M., Griwang, W. P., & Pieniz, S., (2019). Symbiotic microencapsulation of *Lactococcus lactis* subsp. *lactis* R7 using whey and inulin by spray drying. *LWT-Food Science and Technology, 115*, 108411. https://doi.org/10.1016/j.lwt.2019.108411.
48. Ghandi, A., Powell, I. B., Howes, T., Chen, X. D., & Adhikari, B., (2012). Effect of shear rate and oxygen stresses on the survival of *Lactococcus lactis* during the atomization and drying stages of spray drying: A laboratory and pilot-scale study. *Journal of Food Engineering, 113*(2), 194–200. https://doi.org/10.1016/j.jfoodeng.2012.06.005.
49. Martín, M. J., Lara-Villoslada, F., Ruiz, M. A., & Morales, M. E., (2015). Microencapsulation of bacteria: A review of different technologies and their impact on the probiotic effects. *Innovative Food Science & Emerging Technologies, 27*, 15–25. https://doi.org/10.1016/j.ifset.2014.09.010.
50. Bao, C., Jiang, P., Chai, J., Jiang, Y., Li, D., Bao, W., Liu, B., et al., (2019). The delivery of sensitive food bioactive ingredients: Absorption mechanisms, influencing factors, encapsulation techniques, and evaluation models. *Food Research International, 120*, 130–140. https://doi.org/10.1016/j.foodres.2019.02.024.
51. Fritzen-Freire, C. B., Prudêncio, E. S., Amboni, R. D. M. C., Pinto, S. S., Negrão-Murakami, A. N., & Murakami, F. S., (2012). Microencapsulation of *Bifidobacteria* by spray drying in the presence of prebiotics. *Food Research International, 45*(1), 306–312. https://doi.org/10.1016/j.foodres.2011.09.020.
52. Rajam, R., & Anandharamakrishnan, C., (2015). Microencapsulation of *Lactobacillus plantarum* (MTCC 5422) with fructooligosaccharide as wall material by spray drying. *LWT-Food Science and Technology, 60*(2, Part 1), 773–780. https://doi.org/10.1016/j.lwt.2014.09.062.
53. Arslan-Tontul, S., & Erbas, M., (2017). Single and double-layered microencapsulation of probiotics by spray drying and spray chilling. *LWT-Food Science and Technology, 81*, 160–169. https://doi.org/10.1016/j.lwt.2017.03.060.
54. Zhang, C., Ada, K. S. L., Chen, X. D., & Quek, S. Y., (2020). Microencapsulation of fermented noni juice via micro-fluidic-jet spray drying: Evaluation of powder properties and functionalities. *Powder Technology, 361*, 995–1005. https://doi.org/10.1016/j.powtec.2019.10.098.
55. González-Ortega, R., Faieta, M., Di Mattia, C. D., Valbonetti, L., & Pittia, P., (2020). Microencapsulation of olive leaf extract by freeze-drying: Effect of carrier composition on process efficiency and technological properties of the powders. *Journal of Food Engineering, 285*, 110089. https://doi.org/10.1016/j.jfoodeng.2020.110089.

56. Ceballos, A. M., Giraldo, G. I., & Orrego, C. E., (2012). Effect of freezing rate on quality parameters of freeze-dried soursop fruit pulp. *Journal of Food Engineering, 111* (2), 360–365. https://doi.org/10.1016/j.jfoodeng.2012.02.010.
57. Porfire, A., Tomuta, I., Iurian, S., & Casian, T., (2019). Chapter 10-quality by design considerations for the development of lyophilized products. In: Beg, S., & Hasnain, M. S., (eds.), *Pharmaceutical Quality by Design* (pp. 193–207). Academic Press. https://doi.org/10.1016/B978-0-12-815799-2.00011-3.
58. Tang, X., & Pikal, M. J., (2004). Design of freeze-drying processes for pharmaceuticals: practical advice. *Pharm Res., 21*(2), 191–200. https://doi.org/10.1023/B:PHAM.0000016234.73023.75.
59. Berk, Z., (2013). Chapter 23-freeze drying (lyophilization) and freeze concentration. In: Berk, Z., (ed.), *Food Process Engineering and Technology* (2nd edn., pp. 567–581). Food Science and Technology; Academic Press: San Diego. https://doi.org/10.1016/B978-0-12-415923-5.00023-X.
60. Ratti, C., (2001). Hot air and freeze-drying of high-value foods: A review. *Journal of Food Engineering, 49*(4), 311–319. https://doi.org/10.1016/S0260-8774(00)00228-4.
61. Ozdal, T., Yolci-Omeroglu, P., & Tamer, E. C., (2020). Role of encapsulation in functional beverages. In: *Biotechnological Progress and Beverage Consumption* (pp. 195–232). Elsevier. https://doi.org/10.1016/B978-0-12-816678-9.00006-0.
62. Kasper, J. C., & Friess, W., (2011). The freezing step in lyophilization: Physico-chemical fundamentals, freezing methods, and consequences on process performance and quality attributes of biopharmaceuticals. *European Journal of Pharmaceutics and Biopharmaceutics, 78*(2), 248–263. https://doi.org/10.1016/j.ejpb.2011.03.010.
63. Bodade, R. G., & Bodade, A. G., (2020). Microencapsulation of bioactive compounds and enzymes for therapeutic applications. In: *Biopolymer-Based Formulations* (pp. 381–404). Elsevier. https://doi.org/10.1016/B978-0-12-816897-4.00017-5.
64. Dimitrellou, D., Kandylis, P., Petrović, T., Dimitrijević-Branković, S., Lević, S., Nedović, V., & Kourkoutas, Y., (2016). Survival of spray-dried microencapsulated *Lactobacillus casei* ATCC 393 in simulated gastrointestinal conditions and fermented milk. *LWT-Food Science and Technology, 71*, 169–174. https://doi.org/10.1016/j.lwt.2016.03.007.
65. Li, K., Wang, B., Wang, W., Liu, G., Ge, W., Zhang, M., Yue, B., & Kong, M., (2019). Microencapsulation of *Lactobacillus casei* BNCC 134415 under lyophilization enhances cell viability during cold storage and pasteurization, and in simulated gastrointestinal fluids. *LWT-Food Science and Technology, 116*, 108521. https://doi.org/10.1016/j.lwt.2019.108521.
66. Joye, I. J., & McClements, D. J., (2014). Biopolymer-based nanoparticles and microparticles: Fabrication, characterization, and application. *Current Opinion in Colloid & Interface Science, 19*(5), 417–427. https://doi.org/10.1016/j.cocis.2014.07.002.
67. Ezhilarasi, P. N., Karthik, P., Chhanwal, N., & Anandharamakrishnan, C., (2013). Nanoencapsulation techniques for food bioactive components: A review. *Food Bioprocess Technol., 6*(3), 628–647. https://doi.org/10.1007/s11947-012-0944-0.
68. Maleki, O., Khaledabad, M. A., Amiri, S., Asl, A. K., & Makouie, S., (2020). Microencapsulation of *Lactobacillus rhamnosus* ATCC 7469 in whey isolate-crystalline manocellulose-inulin composite enhanced gastrointestinal survivability. *LWT- Food Science and Technology, 126*, 1–20, https://doi.org/10.1016/j.lwt.2020.109224.

69. Anandharamakrishnan, C., Rielly, C. D., & Stapley, A. G. F., (2010). Spray-freeze-drying of whey proteins at sub-atmospheric pressures. *Dairy Science Technology, 90*(2), 321–334. https://doi.org/10.1051/dst/2010013.
70. Đorđević, V., Balanč, B., Belščak-Cvitanović, A., Lević, S., Trifković, K., Kalušević, A., Kostić, I., et al., (2015). Trends in encapsulation technologies for delivery of food bioactive compounds. *Food Engineering Reviews, 7*(4), 452–490. https://doi.org/10.1007/s12393-014-9106-7.
71. Nedovic, V., Kalusevic, A., Manojlovic, V., Levic, S., & Bugarski, B., (2011). An overview of encapsulation technologies for food applications. *Procedia Food Science, 1*, 1806–1815. https://doi.org/10.1016/j.profoo.2011.09.265.
72. Whelehan, M., & Marison, I. W., (2011). Microencapsulation using vibrating technology. *Journal of Microencapsulation, 28*(8), 669–688. https://doi.org/10.3109/02652048.2011.586068.
73. Estevinho, B. N., & Rocha, F., (2018). Application of biopolymers in microencapsulation processes. In: *Biopolymers for Food Design* (pp. 191–222). Elsevier. https://doi.org/10.1016/B978-0-12-811449-0.00007-4.
74. Pech-Canul, A. D. L. C., Ortega, D., García-Triana, A., González-Silva, N., & Solis-Oviedo, R. L., (2020). A brief review of edible coating materials for the microencapsulation of probiotics. *Coatings, 10*(3), 197. https://doi.org/10.3390/coatings10030197.
75. Champagne, C. P., & Kailasapathy, K., (2008). Encapsulation of probiotics. In: *Delivery and Controlled Release of Bioactives in Foods and Nutraceuticals* (pp. 344–369). Elsevier. https://doi.org/10.1533/9781845694210.3.344.
76. Huq, T., Fraschini, C., Khan, A., Riedl, B., Bouchard, J., & Lacroix, M., (2017). Alginate based nanocomposite for microencapsulation of probiotic: Effect of cellulose nanocrystal (CNC) and lecithin. *Carbohydrate Polymers, 168*, 61–69. https://doi.org/10.1016/j.carbpol.2017.03.032.
77. Fareez, I. M., Lim, S. M., Mishra, R. K., & Ramasamy, K., (2015). Chitosan coated alginate-xanthan gum bead enhanced pH and thermotolerance of *Lactobacillus plantarum* LAB12. *International Journal of Biological Macromolecules, 72*, 1419–1428. https://doi.org/10.1016/j.ijbiomac.2014.10.054.
78. Gouin, S., (2004). Microencapsulation: industrial appraisal of existing technologies and trends. *Trends in Food Science & Technology, 15*(7), 330–347. https://doi.org/10.1016/j.tifs.2003.10.005.
79. Seth, D., Mishra, H. N., & Deka, S. C., (2017). Effect of microencapsulation using extrusion technique on viability of bacterial cells during spray drying of sweetened yoghurt. *International Journal of Biological Macromolecules, 103*, 802–807. https://doi.org/10.1016/j.ijbiomac.2017.05.099.
80. Dimitrellou, D., Kandylis, P., Lević, S., Petrović, T., Ivanović, S., Nedović, V., & Kourkoutas, Y., (2019). Encapsulation of *Lactobacillus casei* ATCC 393 in alginate capsules for probiotic fermented milk production. *LWT-Food Science and Technology, 116*, 108501. https://doi.org/10.1016/j.lwt.2019.108501.
81. Panghal, A., Jaglan, S., Sindhu, N., Anshid, V., Sai, C. M. V., Surendran, V., & Chhikara, N., (2019). Microencapsulation for delivery of probiotic bacteria. In: Prasad, R., Kumar, V., Kumar, M., & Choudhary, D., (eds.), *Nanobiotechnology in Bioformulations* (pp. 135–160). Nanotechnology in the Life Sciences. Springer International Publishing: Cham. https://doi.org/10.1007/978-3-030-17061-5_6.

82. Tiebackx, F. W., (1911). Gleichzeitige ausflockung zweier Kolloide. *Zeitschrift für Chemie und Industrie Der Kolloide, 8*, 198–201.
83. Eghbal, N., Yarmand, M. S., Mousavi, M., Degraeve, P., Oulahal, N., & Gharsallaoui, A., (2016). Complex coacervation for the development of composite edible films based on LM pectin and sodium caseinate. *Carbohydrate Polymers, 151*, 947–956. https://doi.org/10.1016/j.carbpol.2016.06.052.
84. Turgeon, S. L., Schmitt, C., & Sanchez, C., (2007). protein-polysaccharide complexes and coacervates. *Current Opinion in Colloid and Interface Science, 12*(4, 5), 166–178. https://doi.org/10.1016/j.cocis.2007.07.007.
85. Jain, A., Thakur, D., Ghoshal, G., Katare, O. P., Singh, B., & Shivhare, U. S., (2016). Formation and functional attributes of electrostatic complexes involving casein and anionic polysaccharides: An approach to enhance oral absorption of lycopene in rats *in vivo*. *International Journal of Biological Macromolecules, 93*, 746–756. https://doi.org/10.1016/j.ijbiomac.2016.08.071.
86. Da Silva, T. M., Lopes, E. J., Codevilla, C. F., Cichoski, A. J., Flores, É. M. M., Motta, M. H., Da Silva, C. B., et al., (2018). Development and characterization of microcapsules containing *Bifidobacterium* Bb-12 produced by complex coacervation followed by freeze-drying. *LWT- Food Science and Technology, 90*, 412–417. https://doi.org/10.1016/j.lwt.2017.12.057.
87. Koupantsis, T., Pavlidou, E., & Paraskevopoulou, A., (2016). Glycerol and tannic acid as applied in the preparation of milk proteins- CMC complex coavervates for flavor encapsulation. *Food Hydrocolloids, 57*, 62–71. https://doi.org/10.1016/j.foodhyd.2016.01.007.
88. Paula, D. A., Martins, E. M. F., Costa, N. A., Oliveira, P. M., Oliveira, E. B., & Ramos, A. M., (2019). Use of gelatin and gum Arabic for microencapsulation of probiotic cells from *Lactobacillus plantarum* by a dual-process combining double emulsification followed by complex coacervation. *International Journal of Biological Macromolecules, 133*, 722–731. https://doi.org/10.1016/j.ijbiomac.2019.04.110.
89. Dickinson, E., (2011). Double emulsions stabilized by food biopolymers. *Food Biophysics, 6*, 1–11. https://10.1007/s11483-010-9188-6.
90. Leal-Calderon, F., Thivilliers, F., & Schmitt, V., (2007). Structured emulsions. *Current Opinion in Colloid & Interface Science, 12*, 206–212, https://doi.org/10.1016/j.cocis.2007.07.003.
91. Yamanaka, Y., KobayashI, I., Neves, M. A., Ichikawa, S., Uemura, K., & Nakajima, M., (2017). Formulation of W/O/W emulsions loaded with short-chain fatty acid and their stability improvement by layer-by-layer deposition using dietary fibers. *LWT-Food Science and Technology, 76*, 344–350. https://doi.org/10.1016/j.lwt.2016.07.063.
92. O'Donnell, P. B., & McGinity, J. W., (1997). Preparation of microspheres by the solvent evaporation technique. *Advanced Drug Delivery Reviews, 28*(1), 25–42. https://doi.org/10.1016/s0169-409x(97)00049-5.
93. Song, H., Yu, W., Gao, M., Liu, X., & Ma, X., (2013). Microencapsulated probiotics using emulsification technique coupled with internal or external gelation process. *Carbohydrate Polymers, 96*(1), 181–189. https://doi.org/10.1016/j.carbpol.2013.03.068.
94. Schoubben, A., Blasi, P., Giovagnoli, S., Rossi, C., & Ricci, M., (2010). Development of a scalable procedure for fine calcium alginate particle preparation. *Chemical Engineering Journal, 160*(15), 363–369. https://doi.org/10.1016/j.cej.2010.02.062.

95. Raddatz, G. C., (2020). Use of prebiotic sources to increase probiotic viability in pectin microparticles obtained by emulsification/internal gelation followed by freeze-drying. *Food Research International, 130*. https://doi.org/10.1016/j.foodres.2019.108902.
96. Butstraen, C., & Salaün, F., (2014). Preparation of microcapsules by complex coacervation of gum Arabic and chitosan. *Carbohydrate Polymers, 99*, 608–616. https://doi.org/10.1016/j.carbpol.2013.09.006.
97. Babin, H., & Dickinson, E., (2001). Influence of transglutaminase treatment on the thermoreversible gelation of gelatin. *Food Hydrocolloids, 15*(3), 271–276. https://doi.org/10.1016/S0268-005X(01)00025-X.
98. Gharibzahedi, S. M. T., (2017). Ultrasound-mediated nettle oil nanoemulsions stabilized by purified jujube polysaccharide: Process optimization, microbial evaluation, and physicochemical storage stability. *Journal of Molecular Liquids, 234*, 240–248. https// doi:10.1016/j.molliq.2017.03.094.
99. Da Silva, T. M., Deus, C., Fonseca, B. S., Lopes, E. J., Cichosk, A. J., Esmerino, E. A., Da Silva, C. B., et al., (2019). The effect of enzymatic crosslinking on the viability of probiotic bacteria (*Lactobacillus acidophilus*) encapsulated by complex coacervation. *Food Research International, 125*, 1–8. https://doi.org/10.1016/j.foodres.2019.108577.
100. Sunny-Roberts, E. O., & Knorr, D., (2009). The protective effect of monosodium glutamate on the survival of *Lactobacillus rhamnosus* GG and *Lactobacillus rhamnosus* E-97800 (E800) strains during spray-drying and storage in trehalose-containing powders. *International Dairy Journal, 19*(4), 209–214. https://doi.org/10.1016/j.idairyj.2008.10.008.
101. Grigoriev, D., & Miller, R., (2009). Mano- and multilayer covered drops as carriers. *Current Opinion in Colloid & Interface Science, 14*(1), 48–59. https://doi.org/10.1016/j.cocis.2008.03.003.
102. Etchepare, M. D. A., Nunes, G. L., Nicoloso, B. R., Barin, J. S., Flores, E. M. M., De Oliveira, M. R., & De Menezes, C. R., (2020). Improvement of the viability of encapsulated probiotics using whey proteins. *LWT-Food Science and Technology, 117*, 108601. https://doi.org/10.1016/j.lwt.2019.108601.
103. Wang, L., Yang, S., Cao, J., & Zhao, W. W., (2016). Microencapsulation of ginger volatile oil cased on gelatin/sodium alginate polyelectrolyte complex. *Chemical and Pharmaceutical Bulletin, 64*(1), 21–26. https://doi.org/10.1248/cpb.c15-00571.
104. Fennema, O. R., Damodaran, S., & Parkin, K. L., (2010). Química de alimentos de fennema. *Editorial Acribia, Espanha.*
105. Phillips, G. O., Takigami, S., & Takigami, M., (1996). Hydration characteristics of the gum exudate from *Acacia Senegal*. *Food Hydrocolloids, 10*(1), 11–19. https://doi.org/10.1016/S0268-005X(96)80048-8.
106. Burgess, D. J., & Carless, J. E., (1984). Micro electrophoretic studies of gelatin and acacia for the prediction of complex coacervation. *Journal of Colloid and Interface Science, 98*(1), 1–8. https://doi.org/10.1016/0021-9797(84)90472-7.
107. Weinbreck, F., De Vries, R., Schrooyen, P., & De Kruif, C. G., (2003). Complex coacervation of whey proteins and gum Arabic. *Biomacromolecules, 4*(2), 293–303. https://doi.org/10.1021/bm025667n.
108. Yang, Z., Peng, Z., Li, J., Li, S., Kong, L., Li, P., & Wang, Q., (2014). Development and evaluation of novel flavor microcapsules containing vanilla oil using complex coacervation approach. *Food Chemistry, 145*, 272–277. https://doi.org/10.1016/j.foodchem.2013.08.074.

109. Bokkhim, H., Bansal, N., Grondahl, L., & Bhandari, B., (2016). *In-vitro* digestion of different forms of bovine lactoferrin encapsulated in alginate microgel particles. *Food Hydrocolloids, 52*, 231–242. https://doi.org/10.1016/j.foodhyd.2015.07.007.
110. Dhamecha, D., Movsas, R., Sano, U., & Menon, J. U., (2019). Applications of alginate microspheres in therapeutics delivery and cell culture: Past, present, and future. *International Journal of Pharmaceutics, 569*, 118627. https://doi.org/10.1016/j.ijpharm.2019.118627.
111. Agüero, L., Zaldivar-Silva, D., Pena, L., & Dias, M. L., (2017). Alginate microparticles as oral colon drug delivery device: A review. *Carbohydrate Polymers, 168*(15), 32–43. https://doi.org/10.1016/j.carbpol.2017.03.033.
112. Nami, Y., Lornezhad, G., Kiani, A., Abdullah, N., & Haghshenas, B., (2020). Alginate-Persian gum-prebiotics microencapsulation impacts on the survival rate of *Lactococcus lactis* ABRIINW-N19 in orange juice. *LWT-Food Science and Technology, 124*, 109190. https://doi.org/10.1016/j.lwt.2020.109190.
113. Etchepare, M. D. A., Barin, J. S., Cichoski, A. J., Jacob-Lopes, E., Wagner, R., Fries, L. L. M., Menezes, C. R. D., et al., (2015). Microencapsulation of probiotics using sodium alginate. *Ciência Rural, 45*(7), 1319–1326. https://doi.org/10.1590/0103-8478cr20140938.
114. Goula, A. M., & Adamopoulos, K. G., (2012). A method for pomegranate seed application in food industries: Seed oil encapsulation. *Food and Bioproducts Processing, 90*(4), 639–652. https://doi.org/10.1016/j.fbp.2012.06.001.
115. Labuschagne, P., (2018). Impact of wall material physicochemical characteristics on the stability of encapsulated phytochemicals: A review. *Food Research International, 107*, 227–247. https://doi.org/10.1016/j.foodres.2018.02.026.
116. Marques, G. R., Borges, S. V., De Mendonça, K. S., De Barros, F. R. V., & Menezes, E. G. T., (2014). Application of maltodextrin in green corn extract powder production. *Powder Technology, 263*, 89–95. https://doi.org/10.1016/j.powtec.2014.05.001.
117. Fernandes, R. V. D. B., Borges, S. V., & Botrel, D. A., (2014). Gum Arabic/starch/maltodextrin/inulin as wall materials on the microencapsulation of rosemary essential oil. *Carbohydrate Polymers, 101*, 524–532. https://doi.org/10.1016/j.carbpol.2013.09.083.
118. Warren, H., & Het, P. M., (2015). Highly conducting composite hydrogels from gellan gum, PEDOT: PSS, and carbon nanofibers. *Synthetic Metals, 206*, 61–65. https://doi.org/10.1016/j.synthmet.2015.05.004.
119. Chakraborty, S., Jana, S., Gandhi, A., Sen, K. K., Zhiang, W., & Kokare, C., (2014). Gellan gum microspheres containing a novel α-amylase from marine *Nocardiopsis* sp. strain B2 for immobilization. *International Journal of Biological Macromolecules, 70*, 292–299. https://doi.org/10.1016/j.ijbiomac.2014.06.046.
120. Kalogiannis, S., Iakovidou, G., Liakopoulou-Kyriakides, M., Kyriakidis, D. A., & Skaracis, G. N., (2003). Optimization of xanthan gum production by *Xanthomonas campestris* grown in molasses. *Process Biochemistry, 39*(2), 249–256. https://doi.org/10.1016/S0032-9592(03)00067-0.
121. Burgain, J., Gaiani, C., Linder, M., & Scher, J., (2011). Encapsulation of probiotic living cells: From laboratory scale to industrial applications. *Journal of Food Engineering, 104*(4), 467–483. https://doi.org/10.1016/j.jfoodeng.2010.12.031.
122. Cai, X., Du, X., Cui, D., Wang, X., Yang, Z., & Zhu, G., (2019). Improvement of stability of blueberry anthocyanins by carboxymethyl starch/xanthan gum combinations

microencapsulation. *Food Hydrocolloids, 91*, 238–245. https://doi.org/10.1016/j.foodhyd.2019.01.034.
123. Jo, W., Bak, J. H., & Yoo, B., (2018). Rheological characterizations of concentrated binary gum mixtures with xanthan gum and galactomannans. *International Journal of Biological Macromolecules, 114*, 263–269. https://doi.org/10.1016/j.ijbiomac.2018.03.105.
124. Chakraborty, S., (2017). Carrageenan for encapsulation and immobilization of flavor, fragrance, probiotics, and enzymes: A review. *Journal of Carbohydrate Chemistry, 36*(1), 1–19. https://doi.org/10.1080/07328303.2017.1347668.
125. Grumezescu, A., & Holban, A. M., (2019). *Biotechnological Progress and Beverage Consumption: The Science of Beverages* (Vol. 19). Academic Press.
126. Elzoghby, A. O., El-Fotoh, W. S. A., & Elgindy, N. A., (2011). Casein-based formulations as promising controlled release drug delivery systems. *Journal of Controlled Release, 153*(3), 206–216. https://doi.org/10.1016/j.jconrel.2011.02.010.
127. Fox, P. F., McSweeney, P. L. H., Cogan, T. M., & Guinee, T. P., (2000). *Fundamentals of Cheese Science* (1st edn.). Springer US.
128. Cruz, A. G., Buriti, F. C. A., Souza, C. H. B., Faria, J. A. F., & Saad, S. M. I., (2009). Probiotic cheese: Health benefits, technological and stability aspects. *Trends in Food Science & Technology, 20*(8), 344–354. https://doi.org/10.1016/j.tifs.2009.05.001.
129. Leuenberger, H., (2002). Spray freeze-drying-the process of choice for low water-soluble drugs. *Journal of Nanoparticle Research, 4*(1, 2), 111–119.
130. Succi, M., Tremonte, P., Pannella, G., Tipaldi, L., Cozzolino, A., Coppola, R., & Sorrentino, E., (2017). Survival of commercial probiotic strains in dark chocolate with high cocoa and phenols content during the storage and in a static *in vitro* digestion model. *Journal of Functional Foods, 35*, 60–67. https://doi.org/10.1016/j.jff.2017.05.019.
131. Todd, R. D., (1970). Microencapsulation and flavor industry. *Flavor Industry, 1*(11), 768–771.
132. Tripathi, M. K., & Giri, S. K., (2014). Probiotic functional foods: Survival of probiotics during processing and storage. *Journal of Functional Foods, 9*, 225–241. https://doi.org/10.1016/j.jff.2014.04.030.
133. Gandomi, H., Abbaszadeh, S., Misaghi, A., Bokaie, S., & Noori, N. (2016). Effect of chitosan-alginate encapsulation with inulin on survival of *Lactobacillus rhamnosus* GG during apple juice storage and under simulated gastrointestinal conditions. Lebensmittel-Wissenschaft + Technologie, 69, 365-371. http://dx.doi.org/10.1016/j.lwt.2016.01.064 (accessed 10 August 2021).
134. King, A. H. (1988). Flavor encapsulation with alginates. *ACS Symposium Series, 370*, 122–125.
135. Vieira da Silva, B., Barreira, J. C. M., & Oliveira, M. B. P. P. (2016). Natural phytochemicals and probiotics as bioactive ingredients for functional foods: Extraction, biochemistry and protected-delivery technologies. Trends in Food Science & Technology, 50, 144–158.
136. Xu, C., Lei, C., Meng, L., Wang, C., & Song, Y. (2012). Chitosan as a barrier membrane material in periodontal tissue regeneration. *J Biomed Mater Res Part B Appl Biomater 100*, 1435–1443.

CHAPTER 10

Nutraceuticals-Based Nano-Formulations: An Overview Through Clinical Validations

SHELLY SINGH and SHILPA SHARMA

Department of Biological Sciences and Engineering,
Netaji Subhas University of Technology, Dwarka, New Delhi, India

ABSTRACT

Due to surge in diseases and awareness about health, the requirement for modified food products, known as "nutraceuticals" has also increased. In order to formulate nutraceuticals, food products are fortified with nutrients and essential elements to treat various disorders and diseases. But due to certain disadvantages like poor solubility, low bioavailability, poor adsorption, low stability, low permeability *in vivo*, etc., the potential of nutraceuticals has not been utilized fully. Nanotechnology is increasingly being used to address such issues. Nanotechnology has been used to enhance the quality of nutraceuticals, for detection and sensing of chemical and biological contaminants and for preservation and packaging of nutraceuticals, thereby increasing their shelf life. Nanotechnology has also been used to encapsulate nutraceuticals to form nano-nutraceuticals which have enhanced therapeutic activities, better solubility and increased bioavailability. The chapter discusses various nano-nutraceuticals with their applications and therapeutic outcomes, commercially available nano-nutraceuticals and nutraceuticals based nano- delivery systems through their clinical validations.

10.1 INTRODUCTION

Nutraceuticals are defined as "designer food" products, fortified by nutrients such as essential elements, minerals, amino acids, vitamins, etc., which not only act as dietary supplements but are also used in the prevention and treatment of various diseases [1]. The word 'nutraceutical' was coined in 1989 by Stephen L. DeFelice, Founder and Chairman of the Foundation of Innovative Medicine (New York), by combining two words *viz.* 'nutrition' and 'pharmaceutical' and hence is defined as hybrid of food and drug [2]. The nutraceuticals may function as immunomodulators and provide health benefits against various diseases and disorders like cancer, neurological diseases, cardiovascular disorders, respiratory disorders, diabetes, obesity, etc. Nutraceutical products have historical aspect as well because in the ancient times, many civilizations and indigenous tribes used to depend on natural herbs and minerals. In today's world, much of the information is acquired from plants and herbs used during ancient times. Some examples of different nutraceuticals and their therapeutic outcomes are presented in Table 10.1.

TABLE 10.1 Some Examples of Nutraceuticals and Their Therapeutic Outcomes

Nutraceuticals	Therapeutic Outcomes	References
Carotenoids	Antioxidant, pro-vitamin A activity, cholesterol, cataract, and other chronic diseases.	[3]
Flavonoids	Antioxidant, prevent enzymatic oxidation of ascorbic acid	[4]
Omega 3-PUFA	Cardiovascular diseases, prevention of atherosclerosis	[5]
Anthocyanins	Neuro-protective effects, liver health improvement, and anti-inflammatory effect	[6]
Theobromine	Antioxidant and psychoactive effects	[7]
Terpenoids	Therapeutic agent for liver cancer and chemopreventive agent	[8]
Caffeine	Sleep therapeutics	[9]
Eucalyptol	Antioxidant, anti-inflammatory, and gastroprotective	[10]
Curcumin	Anti-bacterial, anti-inflammatory, anti-diabetic, anti-viral, antioxidant, anti-venom, anti-obesity, anti-arthritis, anti-depressant, and wound healing	[11]
Resveratrol	Antioxidant, anti-inflammatory, immunomodulatory, glucose, and lipid regulatory, neuroprotective, and cardio-vascular protective effect	[12]
Quercetin	Anti-diabetic, anti-inflammatory, and anti-cancerous	[13]
Lutein	Effective against age-related macular degeneration, cardiovascular diseases, cataracts, and certain types of cancers	[14]

TABLE 10.1 *(Continued)*

Nutraceuticals	Therapeutic Outcomes	References
Co-enzyme Q10	Anti-inflammatory, antioxidant, anti-hyperlipidemic, anti-hyperglycemic, cardioprotective, and neuroprotective	[15]
Lycopene	Effective against certain cancers like colon, skin, and prostate cancer; cardiovascular diseases and is a strong antioxidant	[16]
Phytosterols	Lowers ratio of low density to high density lipoprotein bound cholesterol in serum and lowers blood cholesterol	[17]
Gallic acid	Antimicrobial, antioxidant, anti-cancerous, anti-hypertensive, anti-inflammatory, anticoagulant, hypolipidemic, and hypoglycemic	[18]
Polyphenols	Anti-diabetic, modulate lipid and carbohydrate metabolism, improve adipose tissue metabolism, cardiovascular diseases, neuropathy, and retinopathy	[19]
Caffeine	Antioxidant and Central nervous system stimuli	[20]
Tangeretin	Anti-inflammatory and anti-cancerous	[21]
Phosphatidyl Serine	Improves brain metabolism, memory, and brain activity in early stages of Alzheimer's disease	[22, 23]
Capsanthin	Chemopreventive effects	[24]
B Lapachone	Targets colon and lung cancer cells	[25]
Toxifolin	For liver health	[26]
Gambogic acid	Acts against lymphoma cells	[27]
Probiotics	Gastrointestinal diseases	[28]
Nucleic acid	Anti-cancerous	[29]
Functional yogurt	Anti-cancerous	[30]
Piperine	Gastrointestinal diseases	[31]

Source: Adapted from Ref. [74].

In 2014, the global market of nutraceuticals was at US $165.62 billion and according to the report published by transparency market research in September 2015, the market is growing at a compound annual growth rate (CAGR) of 7.3% from 2015 to 2021 and by the end of 2021 the market is expected to reach US $278.96 billion [32]. Some examples of companies operating in the global nutraceuticals market are Royal DSM N.V., Archer Daniels Midland Company, BASF SE, Cargill, Incorporated, Groupe Danone S.A., E. I. du Pont de Nemours, Nestle S.A., and Company, General Mills, Inc, etc.

TABLE 10.2 Some Commercial Examples of Nano-Nutraceutical Products

Nano-Nutraceutical Product	Active Component	Manufacturer	Nanomaterial Used	Benefits
Lipimed	Lactosorb complex	Lactonova®	Encapsulated in liposomes	Regulates cholesterol level
Glutasolve	Glutamine	Nestle healthcare nutrition	Delivery using gold nanoparticles	Treats deficiency of glutamine caused because of injury or illness
Nano curcumin	Curcumin	Neurvana®	Polymeric nanoparticles	Wound healing
Nano resveratrol	Resveratrol	Neurvana ®	Solid lipid and polymeric nanoparticles	Anti-cancer, anti-inflammatory, and anti-diabetes.
Mi-omega NF	Omega-3 polysaturated fatty acid, Folic acid	Midlothian laboratories	Nanoemulsions	Reduces risks of heart diseases and promotes healthy skin
Glutagut powder	L-glutamine	Bionova®	Encapsulated in nanocarriers	Dietary supplement which promotes gut and brain health
Nevical forte soft gel	Calcium carbonate, Folic acid	Bionova®	Nano calcium	Treatment of osteoporosis, joint inflammation, and arthritis
Casein hydrolysate	Casein	Chaitanya Agrobiotech	Casein nanoparticles	Used in production of minimal media for sporulation by resuspension
Promilk	Calcium	Chaitanya Agrobiotech	Nano calcium	In bone health and body cell functioning.
Soya concentrate	Soy protein	Chaitanya Agrobiotech	Nano aggregates	Protein supplement
Cinnamon extract	Essential oil	Lactonova ®	Encapsulation in chitosan nanoparticles	Lowering of blood sugar level and reduces heart disease risks.
Mulberry leaf extract	Chlorogenic acid	Lactonova®	Nano crosslinking particles	Has an antioxidant property.

TABLE 10.2 *(Continued)*

Nano-Nutraceutical Product	Active Component	Manufacturer	Nanomaterial Used	Benefits
Chondroitin sulfate	Glucosamine	Lactonova ®	Chondroitin sulfate nanoparticles	Helps in the treatment of osteoporosis.
Sydlife-D	Lozenges	Sydler ®	Nanoencapsulation	Helps in weight loss.
Nutrisyd	Biotin	Sydler ®	Nanoemulsions	For skin and bone health.
Vitabuz	Multivitamin	Zeon Lifesciences Ltd.	Lipid-based nanoparticles	Dietary supplement (multivitamin)
Biotrex	Biotin	Zeon Lifesciences Ltd.	Lipid nanoparticles	Helps to form red blood cells, has antioxidant property, and maintains energy level.
Co-enzyme Q-10	Ubiquinone	Agati Healthcare Ltd.	Carried by lipid nanoparticles	For growth and maintenance of cells.
Fracpro	*Cissus quadrangularis*	Neiss Wellness®	Lipid nanoparticles	Prevents chronic diseases, anti-aging, and improves health.
Nano tea	Selenium antioxidant	Qinhuangdao Taiji Ring Nanoproducts Company Ltd.	Nanoparticles	Absorbs cholesterol, fat, viruses, and free radicals and good supplement of selenium
Nanoceuticals™ artichoke nanoclusters	Artichoke	RBC Lifesciences®	Nanoclusters	Immunity booster, balance body pH and provides hydration
Canola active oil	Phytosterols	Shemen Industries, Israel	Nanodrops (nanosized lipid micelles)	Inhibits uptake and transportation of cholesterol

TABLE 10.2 *(Continued)*

Nano-Nutraceutical Product	Active Component	Manufacturer	Nanomaterial Used	Benefits
OilFresh®	Oil conditioning device	OilFresh Corporation, USA	Nanoparticles	Enhances heat conductivity of oil for faster cooking at low temperature.
LifePak®	Vitamins, minerals, and fatty acids	Pharmanex®	Nanoparticles	Boost's immunity, helps in cardiovascular and brain health
Novasol® ADEK-Q10	Coenzyme Q 10, Vitamin A, D, K, and E	Aquanova®	Nano micelles	Provides stability to nanomicelles with respect to pH and temperature
Nano C	Vitamin C	Neurvana®	Nanoparticles	Enhances bioavailability of quercetin and α-lipoic acid

Source: Adapted from Ref. [74].

The use of majority of nutraceuticals is limited by poor bioavailability, poor adsorption, low stability, low solubility, safety, ineffective targeting, and low permeability *in vivo*. Therefore, efforts are underway to increase the efficacy, metabolism, and prevent the physical and chemical degradation of nutraceutical products in order to achieve improved therapeutic effects. In this context, nanotechnology is increasingly being used to target the above-mentioned drawbacks. Nanotechnology is defined as the understanding and control of matter at dimensions of roughly 1–100 nm, where unique phenomena enable novel applications [33]. The properties of nanoparticles (NPs) change drastically from their bulk counterparts as the surface-to-volume ratio increases tremendously. The nutraceuticals formulated using nanotechnology are called nano-nutraceuticals. Table 10.2 represents some examples of commercial nano-nutraceuticals. They have improved pharmacokinetic and physicochemical properties. The nano-dimension and large surface area per unit mass of nanomaterials lead to higher mucoadhesive possibility within the small intestine and also higher chances of interaction with the enzymes and metabolic factors in gastrointestinal tract (GIT) thereby leading to enhancement in the biological activity, bioavailability, and solubility of encapsulated nutraceuticals [34, 35]. This enables increased therapeutic effect of nutraceuticals at low dose and hence reduced possible risk of associated toxicity as compared to nutraceuticals alone. However, safety, and quality of nano-nutraceuticals need to be tested before they reach the market. This chapter gives an overview of nutraceuticals being formulated using nanotechnology through their clinical validations.

10.2 APPLICATIONS OF NANOTECHNOLOGY IN THE NUTRACEUTICAL INDUSTRY

Over the years, nanotechnology is being increasingly used in the food industry. Figure 10.1 showcases various applications of nanotechnology in the nutraceutical industry. Table 10.3 represents some patents of nano-formulations of nutraceuticals.

10.2.1 NANOTECHNOLOGY FOR ENHANCING QUALITY OF NUTRACEUTICALS

With increasing population literacy and improved lifestyle of people, demand for nutritive products is escalating day by day. People are becoming aware about their health and the need to maintain it. There is an increasing demand for products which have positive outcome in wellbeing and are able to provide extra nourishment. For this to happen, nanotechnology is being used to develop novel nutraceutical products which can provide nourishment, fight diseases and also are low in toxicity. Nanotechnology is being used to improve the structure, texture, flavor, and fat content of the already existing product. For example, companies like Unilever and Nestle have reportedly developed nanoemulsion-based ice creams with low-fat content with retention of fatty texture and flavor like their full-fat alternatives, thereby providing a healthier option to the consumer [46]. In one study, paprika oleoresin NPs were used for increasing the marinating performance, i.e., color of the surface, color penetration, saltiness, paprika flavor, toughness, and juiciness of poultry meat, suggesting use of NPs for improvement in marinating performance and sensory acceptability of marinated meat products [47]. Similarly, a German company, Aquanova has employed 30 nm micelles named as "NovaSol" to encapsulate nutraceuticals such as vitamin E, vitamin C and fatty acids that have improved potency and bioavailability of active ingredients [48]. Products such as breads, beverages, dairy products, and cereals are fortified with NPs of probiotics, minerals, vitamins, plant sterols, antioxidants, and bioactive peptides [49].

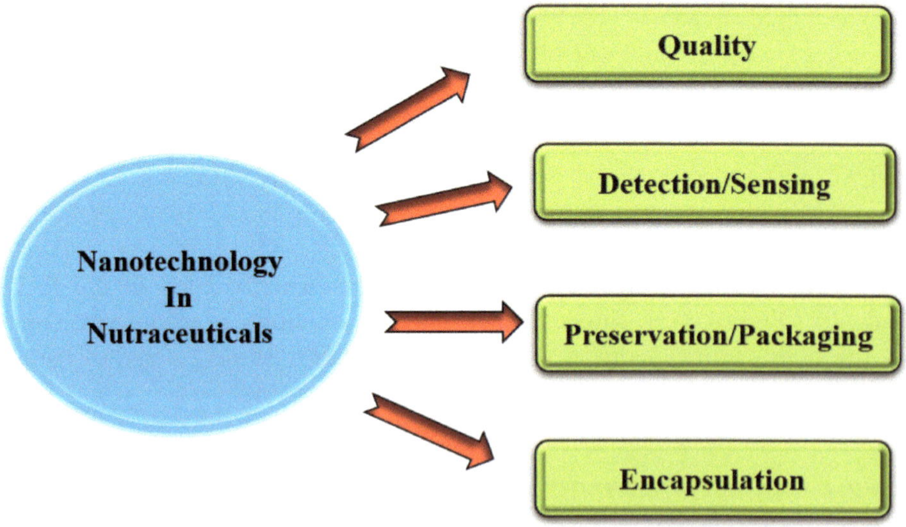

FIGURE 10.1 Applications of nanotechnology in the nutraceutical industry.

TABLE 10.3 Some Patents on Nano-Nutraceuticals

Patent Title	Patent Number	Nutraceutical Nanoformulation	Nutraceutical Active Ingredients	Year	References
Composition comprising curcumin captured ginsenoside and phospholipid-based lipid nanoparticle as effective ingredient for preventing or treating Helicobacter pylori infection	WO20181359 12A2	Curcumin was encapsulated within controlled ginsenoside or phospholipid-based nanoparticles.	Curcumin	2018	[36]
Novel nutraceutical compositions comprising epigallocatechin gallate	WO2004/041257 A3	Epigallocatechin gallate (EGCG) comprising Co-enzyme Q 10, phytanic acid, lipoic acid, etc.	EGCG	2004	[37]
Formulations containing omega-3 fatty acids or esters thereof and maqui berry extract and therapeutic uses thereof	US201802432 53A1	The active ingredients were loaded into soft gel capsules. Additionally, it was claimed that the active ingredients could be loaded into nanoparticles and other sustained released, novel formulations	Omega-3 fatty acids or esters Maqui berry extract	2018	[38]
Novel nutraceutical compositions containing Stevia extract or Stevia extract constituents and uses thereof	AU2008333570B2	A composition comprising 5 Stevia extract enhance cognitive function	Stevia extract	2008	[39]

TABLE 10.3 *(Continued)*

Patent Title	Patent Number	Nutraceutical Nanoformulation	Nutraceutical Active Ingredients	Year	References
Diabetes preventing and treating nutritional formula nanoparticles and preparing and processing method thereof	CN106174011 A	The natural ingredients were loaded into nanoparticles.	Chinese yams, Radix astragal, *Rhizoma anemarrhenae*, chicken's gizzard membranes, *Radix puerariae*, raw gypsum, *Rhizoma alismatis* and/or others.	2016	[39]
Pharmaceutical, cosmetic or food products and use of nanoparticles containing vitamin d	WO2020019043A1	The present invention relates to pharmaceutical, cosmetic or food products containing vitamin D in nanoparticles, particularly vitamin D3.	Vitamin D3	2020	[41]
Molecular particle superior delivery system	US20190289895A1	A nano solid-liquid H_2O concentrate and method, the concentrate containing molecules of H_2O each encapsulating a composite nanoparticle including selected nutrient particles of nanoscale	Encapsulation of particles, e.g., foodstuff additives	2019	[42]

TABLE 10.3 *(Continued)*

Patent Title	Patent Number	Nutraceutical Nanoformulation	Nutraceutical Active Ingredients	Year	References
Bioactive substance or composition for protein delivery and use thereof	JP6426288B2	Intimate drug-carrier mixtures characterized by the carrier, e.g., ordered mixtures, adsorbates, solid solutions, co-dried, co-solubilized, co-kneaded, co-milled, co-ground products, co-precipitates; drug nanoparticles with adsorbed surface modifiers with organic compounds	Protein (Immunoglobulins, immune serum, cyclodextrins, etc.)	2015	[43]
Combination of bioenergy and nutra-epigenetic metabolic regulators, nutraceutical compounds in conventional and nanotechnology-based combinations, for reversing and preventing cellular senescence-accelerated by chronic damage caused by diabetes and other complex chronic degenerative diseases	WO20172134 86A2	Nanoparticles were used for formulation.	Amino acids (glycine, arginine, and cysteine) resveratrol.	2017	[44]
Modified resveratrol composition and use thereof	WO20180423 24A1	Resveratrol nanoparticles were coated with tree fat.	Resveratrol	2018	[45]

Source: Adapted by Ref. [1].

The quality of nutraceuticals can be increased by increasing the shelf life of the product so that its freshness is maintained. Nanotechnology has been used to increase the shelf life of nutraceutical containing products. For instance, shelf life of tomato was increased by encapsulating quercetin in biodegradable poly-D,L-lactide NPs [34]. Similarly, the shelf life of guava was increased by application of zinc oxide NPs -containing nano structured coatings of chitosan and alginate [50].

Additionally, a lot of research is being done to increase the freshness of nutraceuticals by incorporating them within smart and biodegradable nano-packagings. The nano-packagings are made using polymers like starch, polylactic acid (PLA), polyhydroxybutyrate, and polycaprolactone (PCL). These formulated biodegradable nanopackagings are highly compatible with various food products such as dairy, fresh meats, and beverages; prevents oxidation of products, thereby maintaining their freshness for a longer duration [51]. Omega-3 unsaturated fatty acids which are naturally found in seed oils, fish oil, and some plants have been incorporated into a wide range of products such as breads, milk, fruit juices, meat, etc., using microencapsulation technology that prevents oxidative deterioration of unsaturated fatty acids and also extends their shelf life [52]. One of the bakeries in Western Australia has a successful top-selling product 'Tip-Top' Up bread which has tuna fish oil (a source of omega-3 fatty acids) incorporated in nanocapsules which break open only in stomach, thus avoiding the unpleasant taste of the fish oil [53].

10.2.2 NANOTECHNOLOGY FOR DETECTION AND SENSING OF CHEMICAL AND BIOLOGICAL CONTAMINANTS

Nanotechnology has immense applications in the fabrication of biosensors for the quantification of food constituents, detection of pathogens in the processing plants and also to alert consumers, manufacturers, and distributors about the safety standards of the product. Numerous studies have reported the development of nanosensors using NPs (gold, iron oxide), nanofibers, nanotubes, nanorods, and nanowires and their applications in detection of pathogens, pesticides, contaminants, adulterants, toxins, and nutrients with high sensitivity and quick response [54–61]. The nanosensors also act as indicators that give response when the environmental conditions such as temperature, humidity, microbial contamination are changed or there is degradation of products [62]. Highly sensitive nanotechnology-based immunosensors where specific antibodies, antigens, proteins, etc., are used for the detection of microbial cells or substances in food have been developed. For example, a hybrid nanosensor based on magnetic resonance and fluorescence for detection of *E. coli* O157:H7 could sense different concentrations of bacteria in milk in less than an hour [63]. Similarly, in a recent report, zeolitic imidazolate framework encapsulated cadmium sulfide (CdS) quantum dots were used for sandwich-type electrochemical immunodetection of *E. coli* O157:H7 in milk samples using anti-*E. coli* O157:H7 antibody, with the detection limit of 3 CFU mL^{-1} [64].

10.2.3 NANOTECHNOLOGY FOR PRESERVATION AND PACKAGING OF NUTRACEUTICALS

Nanotechnology has gained widespread attention in the preservation and packaging of nutraceuticals. The nanotechnology-based food packaging has been classified as active packaging and smart/intelligent packaging systems. Active packaging means the use of nanomaterials that are moisture-regulating agents, carbon dioxide scavengers and emitters, oxygen scavengers and antimicrobials, for providing protection and hence increasing the shelf life of the food product. Silver NPs and nanocomposites have been widely used as antimicrobials in the food industry [65]. In a study, various deposition processes and chemical modifications were explored for attaching silver NPs on the surface of plastic substrates, which facilitated the slow release of silver ions to inhibit their inclusion in food [66]. Besides, iron, silver, carbon, zinc oxides, titanium oxides, magnesium oxides, and silicon dioxide NPs have also been employed as effective antimicrobial agents in packaging. Natural antimicrobial substances encapsulated in nanoemulsions (NEs) can be adhered to via covalent, electrostatic, and hydrogen bonding interactions to develop antimicrobial packaging systems. Chemical giant Bayer (Leverkusen, Germany) has developed a transparent film in which clay NPs are dispersed uniformly on a plastic film that prevents oxygen, carbon dioxide, and moisture from reaching fresh meats and other foods [67]. A number of patents on the utilization of nanoclays and nanosilver in food packaging have been filed in the USA, Europe, and Asia [68]. Nanocomposites containing NPs of silicon dioxide, clay, titanium dioxide, nanocellulose, nanofibrillated cellulose, carbon nanotubes, etc., in polymer matrix enhance their mechanical and gas barrier properties [69].

The smart packaging system is designed to detect and alert the consumer of any biochemical or microbial changes in the food products. Nanosensors have been incorporated in packaging material to detect chemicals, toxins, gases, aromas, food pathogens, products of microbial metabolism, etc., during storage and transport. They are being increasingly used as they can provide real-time status regarding freshness of food, thereby eradicating the need of estimated expiration dates in the consumables. For instance, Timestrip developed a detection system based on gold nanoparticles (AuNPs) for chilled foods [70]. The system appeared red above freezing temperature, but when accidental freezing occurred, the AuNPs aggregated, leading to loss of red color.

Nanolaminates are another category of modification where nanotechnology is used to enhance the quality and preservative value in nutraceuticals. Edible nanofilms that protect nutraceuticals from lipids, gases, and moisture, improve their texture and serve as carriers for nutrients, antioxidants, antimicrobials, colors, and flavors come under the category of nanolaminates. Nano-laminates are basically nano dimensional thin food-grade films of two or more layers of a material, i.e., 1 nm to 100 nm each of every layer bonded chemically or physically with each other [71]. Besides protecting foods from gases and humidity, the nanolaminates can also improve the texture of food and serve as carriers of nutrients, colors, antioxidants, flavors, antimicrobials, etc. Nano laminates are used in products such as fruits, vegetables, meats, baked goods, chocolates, and candies to increase their quality as well as their shelf life [72].

10.2.4 NANOTECHNOLOGY FOR ENCAPSULATION OF NUTRACEUTICALS

A majority of nutraceuticals have low aqueous solubility and low permeability across membrane, short half-life, low stability, and fast metabolism thereby making them poorly bioavailable. For example, curcumin, which has anti-inflammatory, antioxidant, chemopreventive and anti-neoplastic properties is lipophilic and hence water-insoluble. Some nutraceuticals have stability issues *in vivo*, they get degraded or oxidized in the GIT. Others may form complexes with gastrointestinal (GI) fluid constituents (like bile salts, phospholipids, proteins, dietary fibers, surfactants, etc.), thereby decreasing their availability in systemic circulation. For instance, lutein, a xanthophyll known to be effective at retarding the development of age-related macular degeneration, gets degraded in the acidic environment and by enzymes present in GI fluid. Similarly, lipids, which are major constituent of vitamins are susceptible to oxidation, leading to bad taste and degradation. Hence, the full therapeutic potential of nutraceutical product is not utilized. High dosage may be associated with toxicity. Therefore, achieving maximum therapeutic outcome of nutraceutical at a dose which causes minimal/negligible toxicity is a challenge. In this context, nano-encapsulation has emerged as an effective delivery system of nutraceuticals.

Nano-encapsulation is a process of encapsulation of bioactive compounds such as vitamins, antioxidants, proteins, lipids, carbohydrates, aromas, etc., inside a material at the size of nanoscale in order to provide more stability and thus increase shelf life of the nutraceutical formulation [73]. It also increases bio-availability of encapsulated nutraceuticals. Nanoencapsulation increases protection against high temperatures during processing of nutraceutical products so that they are able to retain their nutritional properties. Additionally, nano-encapsulated nutraceuticals can be easily incorporated in clear and transparent foods, because of their size, which is much smaller than the wavelength of light, without causing problems of colors. Liposomes, NEs, solid lipid nanoparticles (SLNs), nanostructured lipid carriers (NLCs), polysaccharide, and protein-based NPs, nanosuspensions, etc., are different nanoencapsulation technologies used. These nanocarriers release a controlled amount of bioactive compounds at the right time and at right place.

NutraLease, an Israel based nutraceutical company has commercial beverages and food products in the market which contain encapsulated functional compounds like omega-3, β-carotene, lutein, lycopene, coenzyme Q10, phytosterols, isoflavones, and vitamins A, D, and E in self-assembled NEs [74]. The NEs increase the encapsulation rate and bioavailability of these nutraceuticals which have poor water solubility and low oral bioavailability. Similarly, nanoencapsulation is employed for the delivery of probiotics as it provides a protective coating on the probiotic bacteria separating it from the surrounding environment, thereby enhancing the viability rate of probiotic bacteria. Novel nano-formulations of nutraceuticals having therapeutic properties are under development to target cancer and cardiovascular diseases (CVD). For instance, using antioxidants in cherry extract for treatment of CVD has issue of low bioavailability because of oxidation and less absorption in GIT. This issue was addressed by encapsulating cherry extract in NPs based on chitosan derivatives that increases the residence time in GI lumen, and improves the intestinal absorption of cherry antioxidants, thereby enhancing their antioxidant and anti-inflammatory activity [75].

10.2.4.1 AN OVERVIEW OF NANOTECHNOLOGY-BASED DELIVERY SYSTEMS FOR NUTRACEUTICALS

The nutraceuticals exist naturally in a range of different molecular structures having a range of polarities, conformations, and molecular weights leading to distinct physicochemical properties such as solubility, chemical, and physical stabilities. These parameters influence their bioavailability and absorption in GIT. Therefore, a number of different types of nanotechnology-based delivery systems (Figure 10.2) have been developed with different physicochemical properties and functional attributes that improve the factors influencing the bioavailability of nutraceuticals. Depending on the specific physicochemical requirement for a specific nutraceutical, specific nanocarrier can be employed for developing nutraceutical nano-formulation. The advantages of using nanotechnology for nutraceuticals are:

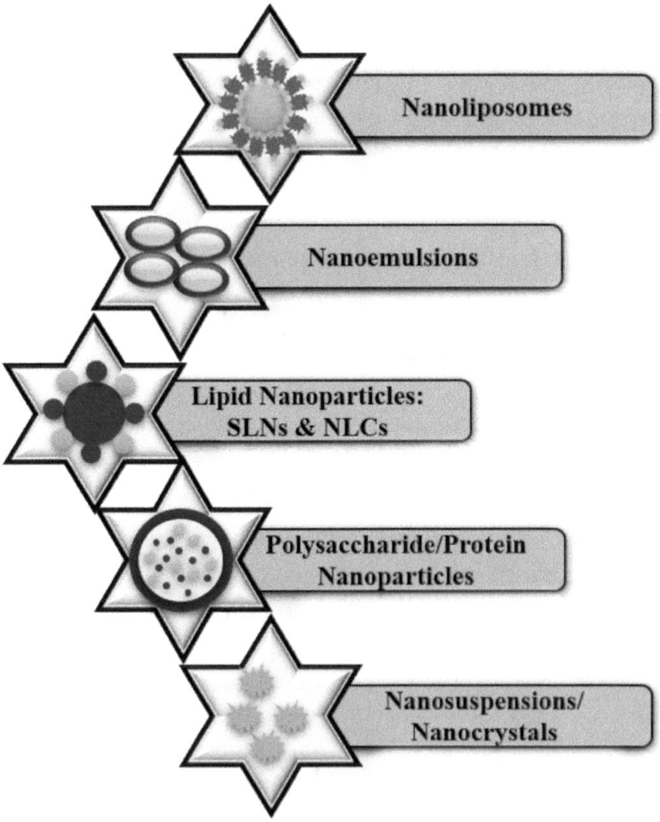

FIGURE 10.2 Some nanotechnology-based delivery systems for nutraceuticals.

- Efficient encapsulation;
- Protection of labile nutraceuticals from degradation due to environmental stresses;
- Better physicochemical stability in GIT;
- Increased gastric retention time;
- Controlled release;
- Improved pharmacokinetic properties (aqueous solubility, stability, etc.);
- Enhanced bioavailability and hence better therapeutic benefits;
- Reduced dose, hence minimal side effects.

Some of the design considerations for development of nano-nutraceuticals are as under:

- The formulation should use legally approved ingredients and processing methods.
- It should be stable, able to withstand different types of environmental stresses (pH changes, cooling, heating, dehydration, and mechanical agitation, etc.), during production, storage, and transportation.
- It should protect the encapsulated nutraceutical from chemical degradation as well as in the human digestive system and maintain its bioaccessibility.
- For nutraceuticals that are insoluble or less soluble in aqueous media, the nano delivery system should be able to increase its solubility in aqueous solution.
- It should not compromise with the quality of food product such as texture, appearance, flavor, rheology, and shelf life, i.e., must be compatible with the food medium.
- Cost of development should not be high.

In the following section, different nanotechnology-based delivery vehicles for nutraceuticals shall be discussed:

1. **Nanoliposomes:** Liposomes are spherical phospholipid vesicles formed by folding of lipid bilayer(s) with an aqueous core and hydrophobic tails facing each other. They can be used for encapsulating both lipophilic and hydrophilic nutraceuticals either simultaneously or alone. They are used widely for the administration of nutritional ingredients and drugs into the tissues. They are made artificially from non-toxic phospholipids for the sole purpose of transportation

of nutritional ingredients into the system. "Second-generation liposomes" are also under research which are obtained by manipulation of lipid composition, size, and charge of the vesicle. Using various molecules like sialic acid and glycolipids, surfaces of the liposomes can be modified. The release of ingredient depends upon the rigidity and permeability of the liposome, which can be modified by modifying the composition of its bilayer. For example, components such as dipalmitoylphosphatidylcholine forms a rigid bilayer, whereas egg and soybean phosphatidylcholines form a more permeable bilayer [76]. The advantages and disadvantages of liposomes in the nano-delivery system of nutraceutical ingredients are listed in Table 10.4.

TABLE 10.4 Advantages and Disadvantages of Nanoliposome as Delivery System

Advantages	Disadvantages
High stability	Low solubility
Biodegradable and biocompatible	Sometimes its phospholipids undergo oxidation
Non-toxic	Might undergo leakage
High efficiency and accuracy	High production cost
Useful for encapsulation	Short half-life

Nanoliposomes have been used as delivery vehicles for nutrients, food additives, enzymes, and antimicrobials. In a very recent report, β-carotene containing liposomes of sizes about 90–150 nm was prepared using supercritical carbon dioxide with ultrasonication, the liposomes obtained had improved stability, an important factor in nutraceuticals [77]. In another recent study, garlic extract and nisin were encapsulated in phosphatidylcholine liposomes using oleic acid and cholesterol as stabilizers for membranes, and it was shown that oleic acid stabilized liposomes showed the highest antimicrobial activity against *Salmonella enteric* and *Listeria monocytogenes* [78]. Nanoliposomes loaded with nutraceuticals have also been used for the treatment of skin diseases. Quercetin loaded-vitamin C-based aspasomes (ascorbyl palmitate vesicles), of sizes in the range of 125–184 nm was shown to have beneficial effects in the treatment of acne [79]. In a recent study, ammonium glycyrrhizinate which is a derivative of glycyrrhizic acid found in plant *Glycyrrhiza glabra* and is known to have anti-inflammatory and anti-allergic properties was entrapped in ultradeformable liposomes (transfersomes) which were demonstrated to cause decrease in skin inflammation on the human volunteers, thereby making them a potential topical drug delivery system for anti-inflammatory therapy [80].

2. **Nanoemulsions (NEs):** A mixture of two or more immiscible liquids such as oil and water stabilized by surfactants or other types of stabilizing agents to form a solution, in which one of the liquids is dispersed as spherical droplets in the other liquid is known as an emulsion [81]. An oil-in-water emulsion (O/W) has organic phase (oil) in the form of droplets in the aqueous continuous phase and vice versa. An emulsion is said to be a nanoemulsion when the size of dispersed droplets is in the range of 10–100 nm. The lipophilic nutraceuticals are encapsulated within the droplets of NEs, which serve as an effective delivery system with improved properties like the ability to modulate product texture, high optical clarity, better stability to droplet aggregation and gravitational separation, and increased bioavailability of lipophilic components in comparison to conventional emulsions [82]. Besides, the bioavailability of encapsulated nutraceutical in droplets of NEs is increased. The large surface area owing to the small size of droplets in nanoemulsion, makes their digestion rate higher, releasing their contents which are absorbed more easily. The absorption is also increased owing to increased residence time due to penetration of small droplets into mucous layer coating the epithelial cells in small intestine, and increased aqueous solubility of lipophilic components as the droplet size is decreased. Another advantage of using NEs is that they can be incorporated into clear or slightly turbid products without changing their visual appearance.

   In a very recent report, capsanthin, a nutraceutical with poor aqueous solubility, poor stability, and low/variable oral bioavailability was encapsulated in nanoemulsion (with size <50 nm) to increase its solubility without compromising its physical and chemical stability and retention of its antioxidant properties [83]. Another hydrophobic nutraceutical namely benzyl isothiocyanate having antitumor and antimicrobial properties but with low bioavailability was successfully encapsulated in rhamnolipid based nanoemulsion which provided solution to low solubility, poor stability, and diminished bioavailability of benzyl isothiocyanate [84]. The developed nanoemulsion was effective against bacterial strains *E. coli* and *S. aureus*. The cinnamon essential oil nanoemulsion incorporated in the pullulan coating on strawberries lowered the loss in fruit mass, firmness, total soluble solids, and titratable acidity of strawberries after six days of storage [85]. The developed nanoemulsion also exhibited antimicrobial activity, thereby prolonging the shelf life of strawberries during room storage.

3. **Lipid Nanoparticles (NPs):** These are used for encapsulating water-insoluble nutraceuticals. The lipid NPs are classified as SLNs and NLCs. SLNs are made of lipids that are solid at room temperature like paraffins, triacylglycerols. The aqueous dispersions of SLNs usually have 0.1% and 30% w/w solid lipids and are stabilized by a 0.5%–5% w/w surfactant. They are in the size range of 50–1000 nm and can be prepared via hot homogenization and cold homogenization methods depending on the thermostability of nutraceutical encapsulated. The SLNs used for encapsulating nutraceuticals that are heat stable are prepared by hot homogenization technique, and the ones used for entrapping nutraceuticals that are thermolabile, are prepared by cold homogenization method [86]. They are easy to synthesize, their small size gives them high surface area that improves bioavailability of encapsulated nutraceutical and have higher loading capacity. In a very recent report, α-tocopherol acetate was successfully loaded on SLN, which was prepared using stearic acid as solid lipid, phosphatidylcholine as stabilizer and coated by chitosan [87]. The nanoformulation was stable with high entrapment efficiency of $90.58 \pm 1.38\%$ with a no-burst slow release up to 10 days tested, indicating its potential as a promising drug delivery system for vitamin E. In another report, α-bisabolol loaded SLNs were synthesized through hot homogenization method and exhibited improved therapeutic efficacy and bioavailability of α-bisabolol for combating Alzheimer's disease [88].

   NLCs are another type of lipid NPs which use liquid lipid or a mixture of liquid lipids to form NPs. They have greater stability that SLNs. They have been used for delivering nutraceuticals with slow-release profile and also provide protection to encapsulated nutraceutical from degradation. Many nutraceuticals like lutein, quercetin, etc., have been incorporated into NLCs to achieve their slow-release pattern and enhanced bioavailability [89, 90].

4. **Polysaccharide and Protein-based Nanoparticles (NPs):** These have also been explored for encapsulation/entrapment of bioactive

ingredients for formulation of novel nano-nutraceuticals. The advantages of using polysaccharide NPs are enhanced bioavailability, efficiency, sustained release, and a higher control of drug targeting [91]. Polymeric NPs provide protection to the entrapped nutraceuticals from degradation in the GIT. Diffusion, swelling, and erosion are some of the mechanisms of release of active ingredients from polymeric NPs into the gastrointestinal tract [92]. Biodegradable and smart polymers are good option for encapsulation as they can be degraded in the body by biological or chemical processes and can release encapsulated bioactive ingredient in response to particular environmental conditions, respectively. Polysaccharide NPs are composed of polymeric matrices which can be synthetic or natural in nature. Natural polysaccharides include chitosan, alginate, dextran, etc. Synthetic polymers hold distinctive properties because of their chemical structure, the method of synthesis, type of functional groups present in the molecule and the degree of polymerization. Most explored synthetic polymers are aliphatic polyesters like poly-lactic acid, poly-ε-caprolactone and their copolymers [93]. In one study, nutraceutical lycopene was encapsulated in thermosensitive PNIPAAM-PEG-based co-polymeric NPs that demonstrated stronger antioxidant and anti-cancerous activity as compared to free lycopene *in vivo* [94]. In another study, resveratrol was loaded on to poly (dl-lactide-co-glycolide) (PLGA) NPs, which showed improved bioavailability in male Wistar rats as compared to pure drug and marketed product [95].

Nutraceuticals have also been incorporated in protein NPs. The proteins commonly used are from animal origin like casein, gelatin, whey proteins and albumin (serum albumin and ovalbumin). Nanocarriers arising from self-assembly of some milk proteins have been successfully used for the delivery of hydrophobic nutraceuticals such as ω-3 polyunsaturated fatty acids (PUFAs) and vitamin D in casein micelles, resveratrol, and curcumin in β-lactoglobulin nano delivery systems [96–99]. They have also been used for the delivery of hydrophilic nutraceuticals such as tea polyphenols [100]. The polyphenols (catechin and epicatechin) were nanoencapsulated in BSA NPs in tea to enhance their stability and bioavailability [101]. In a very recent report, curcumin was encapsulated in insect mealworm protein NPs that were uncoated or coated with chitosan [102]. Curcumin bound to the hydrophobic core of the insect protein NPs was more stabilized in the coated nano-complexes and around 90% of it was released after exposure to model GI conditions. Plant proteins, namely zein, soy proteins, wheat gliadins, and barley proteins, are also increasingly being used as delivery vehicles for nutraceuticals [103].

5. **Nanocrystals/Nanosuspensions:** These are colloidal dispersions (nanoparticles) of pure bioactive/nutraceutical compounds which are carrier-free and contain very little amount of surfactant and/or polymer for its stabilization [104]. Nanocrystals/nanosuspensions can be synthesized using a combination of top-down and bottom-up techniques like supercritical fluid methods, aerosol solvent extraction method, precipitation-lyophilization-homogenization technique, spray freezing into liquids and solution enhanced dispersion by the supercritical fluids [105]. Nanosupsension of nutraceuticals like α-tocopherol, quercetin, curcumin, β-amyrin, etc., have been reported which had increased solubility, stability, dissolution rate and bioavailability [106–108]. In a very recent report, the technique rapid expansion of supercritical solution into air (RESS) process was successfully used for the production of nanosuspensions of nutraceutical antioxidants namely α-tocopherol and β-amyrin to increase their efficacy and bioavailability [109].

10.3 CLINICAL VALIDATION OF NUTRACEUTICALS

With rising preference for foods with high content of nutraceuticals, it becomes imperative to evaluate the efficiency, safety, and toxicity of nutraceuticals. Also, due to increasing customer awareness, the demand for information regarding the biological efficiency of nutraceutical products is growing. Therefore, many companies are now investing in the research and development to pass the clinical trials of their product. According to Jay Udani, MD, CEO, and medical director of Medicus research at Northridge CA, though the clinical trials conducted by the nutraceuticals industry has grown over the last few years, it is still less than the products launched into the market. Due to the repositioning of dietary supplements into drugs, companies are taking care of clinical as well as pre-clinical research as there are more regulations for the drug development industries to follow. This trend has increased the standard and quality of the nutraceutical product because companies want to create their USPs (unique selling products), USPs can be in the terms of nutrition provided, flavor, image improvement, health benefits or health claims. For a clinical trial, the first step is data collection, i.e., assembling all the existing data about the product or its ingredients. Sometimes enough data is obtained to jump into the clinical trial, but when data is unavailable, *in-vitro* testing is done, which is followed by a pilot study or a proof-of-concept study. Good extraction of basic data helps in maximizing the chances of getting robust data which is later used for marketing and regulatory purposes. After this *in-vivo* studies are done to determine the safety of the product. AIBMR is a United States (US) based company which performs a thorough literature survey to get the information regarding the product and its ingredients. AIBMR is known to maintain one of the biggest nutraceutical research libraries in the world. After analyzing all the required data, the product is then sent for clinical validation; after which the product is launched into the market.

10.4 SAFETY AND REGULATIONS

Inclusion of nanotechnology into the nutraceutical and food industry poses arrays of risks. There is no denial in the fact that NPs and nano-foods can cause serious health issues. Research is being done to identify the problems caused by nano products and the ways to tackle them. Reactive oxygen species (ROS) generation leading to oxidative stress, which causes degeneration of mitochondria and induces apoptosis is one of the crucial mechanisms of toxicity by nanomaterials. There have been reports according to which these products can be labeled as toxic and dangerous to health. This could be due to the fact that these products in nano form have higher chemical reactivity than their larger counterparts; they have higher bioavailability which may lead to their toxic behavior; they readily cross the membrane barriers and capillaries leading to different toxicokinetic and toxicodynamic properties; they may undergo changes in the body and may not be same as originally administered; our immune system can be compromised by them and its possible that they might have long term pathological effects. They can also cause oxidative cell damage by translocating into the skin, liver, and brain cells. They are also linked with escalating levels of immune dysfunction and inflammation in the gastrointestinal tract, causing inflammatory bowel disease (IBD), known as Crohn's disease (CD) [40]. They can also cause lesions in kidneys and liver, clots, cancer, and granulomas due to build up toxicity when used in access. They can be taken up by damaged skin and brain cells due to their minute size. Impairment of DNA replication and transcription can also occur in some cases because particles having size less than 70 nm are able to enter cell nuclei [40]. The use of new nanotechnology-based food products is still a challenge as they need to undergo safety assessment before being commercialized for human use. There are a number of regulatory bodies working currently such as Food and Drug Administration (FDA), Environmental Protection Agency (EPA), European Food and Safety Authority (EFSA), National Institute for Occupational Safety and Health (NIOSH), Consumer Product Safety Commission (CPSC), US Patent and Trademark Office (USPTO), US Department of Agriculture (USDA) and Occupational Safety and Health Administration (OSHA). The safety regulations are not clear for nanofoods because the fate and toxicity of the NPs is still less understood by the researchers. There needs to be a widely accepted international regulatory framework for the regulation of the use of nanomaterials in the food industry. Proper government guidelines and directives and rigorous toxicological screening methods are the need of hour for the commercialization of nanofoods.

10.5 CONCLUSION

The fortification of food products with nutraceuticals has gained increasing significance for preventing and improving the health of people suffering from various diseases. Nanotechnology has a wide potential in addressing challenges presently faced by nutraceuticals like limited solubility, stability, shelf life and bioavailability, which compromise their therapeutic effectiveness. Nanotechnology has been used to improve the quality of nutraceuticals, for detection and sensing of chemical and biological contaminants, in preservation and packaging of nutraceuticals and for nanoencapsulation of nutraceuticals. Several nanosystems like nanoliposomes, SLNs, NLCs, polysaccharide NPs, protein-based NPs, NEs, and nanocrystals/nanosuspensions have been used for effective delivery of nutraceuticals *in vivo* and also to enhance their efficiency. There are a number of nanotechnology-based nutraceutical products commercially available in the market. However, proper evaluation of the efficacy and safety of nutraceutical nano-formulations is still needed. Incomplete knowledge about the fate of NPs once they enter organs, tissues, and cells and associated toxicity of nanomaterials is still a concern while using these commercial nano-formulations of nutraceuticals.

KEYWORDS

- **nutraceuticals**
- **nano-formulation**
- **nanocarrier**
- **nanostructured lipid carriers**
- **nanosuspension**
- **solid lipid nanoparticles**

REFERENCES

1. Helal, N. A., Eassa, H. A., Amer, A. M., Eltokhy, M. A., Edafiogho, I., & Nounou, M. I., (2019). Nutraceuticals' novel formulations: The good, the bad, the unknown and patents involved. *DDF, 13*(2), 105–156. https://doi.org/10.2174/187221131366619050311204 0.
2. Andlauer, W., & Fürst, P., (2002). Nutraceuticals: A Piece of history, present status and outlook. *Food Research International, 35*(2), 171–176. https://doi.org/10.1016/S0963-9969(01)00179-X.
3. Cardoso, L. A. C., Karp, S. G., Vendruscolo, F., Kanno, K. Y. F., Zoz, L. I. C., & Carvalho, J. C., (2017). Biotechnological production of carotenoids and their applications in food and pharmaceutical products. In: Cvetkovic, D. J., & Nikolic, G. S., (eds.), *Carotenoids*. InTech, https://doi.org/10.5772/67725.
4. Marín, F. R., Frutos, M. J., Pérez-Alvarez, J. A., Martinez-Sánchez, F., & Del, R. J. A., (2002). Flavonoids as nutraceuticals: Structural related antioxidant properties and their role on ascorbic acid preservation. In: Atta-Ur-Rahman, (ed.), *Studies in Natural Products Chemistry: Bioactive Natural Products* (Vol. 26, pp. 741–778). Elsevier. https://doi.org/10.1016/S1572-5995(02)80018-7.
5. Massaro, M., Scoditti, E., Carluccio, M. A., & Caterina, R. D., (2010). Nutraceuticals and prevention of atherosclerosis: Focus on ω-3 polyunsaturated fatty acids and Mediterranean diet polyphenols. *Cardiovascular Therapeutics, 28*(4), e13–e19. https://doi.org/10.1111/j.1755-5922.2010.00211.x.
6. Ghafoor, K., & Fahad, Y. A. J., (2014). Effects of anthocyanins as nutraceuticals. *Agro Food Industry Hi-Tech, 25*(4), 10.
7. Martínez-Pinilla, E., Oñatibia-Astibia, A., & Franco, R., (2015). The relevance of theobromine for the beneficial effects of cocoa consumption. *Front Pharmacol., 6*. https://doi.org/10.3389/fphar.2015.00030.
8. Thoppil, R. J., & Bishayee, A., (2011). Terpenoids as potential chemopreventive and therapeutic agents in liver cancer. *World J. Hepatol.,* 3(9), 228–249. https://doi.org/10.4254/wjh.v3.i9.228.
9. Yurcheshen, M., Seehuus, M., & Pigeon, W., (2015). *Updates on Nutraceutical Sleep Therapeutics and Investigational Research.* https://www.hindawi.com/journals/ecam/2015/105256/ (accessed on 10August 2021). https://doi.org/10.1155/2015/105256.

10. Nair, H. B., Sung, B., Yadav, V. R., Kannappan, R., Chaturvedi, M. M., & Aggarwal, B. B., (2010). Delivery of anti-inflammatory nutraceuticals by nanoparticles for the prevention and treatment of cancer. *Biochemical Pharmacology, 80*(12), 1833–1843. https://doi.org/10.1016/j.bcp.2010.07.021.
11. Rathore, S., (2020). *Curcumin: A Review for Health Benefits, 1*, 18.
12. Meng, X., Zhou, J., Zhao, C. N., Gan, R. Y., & Li, H. B., (2020). Health benefits and molecular mechanisms of resveratrol: A narrative review. *Foods, 9*(3), 340. https://doi.org/10.3390/foods9030340.
13. Khursheed, R., Singh, S. K., Wadhwa, S., Gulati, M., & Awasthi, A., (2020). Enhancing the potential preclinical and clinical benefits of quercetin through novel drug delivery systems. *Drug Discovery Today, 25*(1), 209–222. https://doi.org/10.1016/j.drudis.2019.11.001.
14. Spinola, M. V., & Díaz-Santos, E., (2020). Microalgae nutraceuticals: The role of lutein in human health. In: Alam, M. A., Xu, J. L., & Wang, Z., (eds.), *Microalgae Biotechnology for Food, Health and High Value Products* (pp. 243–263). Springer: Singapore. https://doi.org/10.1007/978-981-15-0169-2_7.
15. Chiu, H. F., Venkatakrishnan, K., & Wang, C. K., (2020). The role of nutraceuticals as a complementary therapy against various neurodegenerative diseases: A mini-review. *Journal of Traditional and Complementary Medicine*, S2225411020301425. https://doi.org/10.1016/j.jtcme.2020.03.008.
16. Singh, S., & Gaur, S., (2020). Lycopene: Chemistry, biosynthesis, health benefits and nutraceutical applications. In: Swamy, M. K., (ed.), *Plant-Derived Bioactives: Chemistry and Mode of Action* (pp. 251–263). Springer: Singapore. https://doi.org/10.1007/978-981-15-2361-8_11.
17. Mohammadi, M., Jafari, S. M., Hamishehkar, H., & Ghanbarzadeh, B., (2020). Phytosterols as the core or stabilizing agent in different nanocarriers. *Trends in Food Science & Technology, 101*, 73–88. https://doi.org/10.1016/j.tifs.2020.05.004.
18. Tavares, L., & Zapata, N. C. P., (2019). Encapsulation of garlic extract using complex coacervation with whey protein isolate and chitosan as wall materials followed by spray drying. *Food Hydrocolloids, 89*, 360–369. https://doi.org/10.1016/j.foodhyd.2018.10.052.
19. Bahadoran, Z., Mirmiran, P., & Azizi, F., (2013). Dietary polyphenols as potential nutraceuticals in management of diabetes: A review. *J. Diabetes Metab. Disord., 12*(1), 43. https://doi.org/10.1186/2251-6581-12-43.
20. Tan, S., Ebrahimi, A., & Langrish, T., (2019). Controlled release of caffeine from tablets of spray-dried casein gels. *Food Hydrocolloids, 88*, 13–20. https://doi.org/10.1016/j.foodhyd.2018.09.038.
21. Chen, J., Zheng, J., McClements, D. J., & Xiao, H., (2014). Tangeretin-loaded protein nanoparticles fabricated from zein/β-lactoglobulin: Preparation, characterization, and functional performance. *Food Chemistry, 158*, 466–472. https://doi.org/10.1016/j.foodchem.2014.03.003.
22. Josmi, P. J., Divya, P. D., & Rosemol, J. M., (2019). Role of nutraceuticals in Alzheimer's disease. *The Pharma Innovation Journal, 8*(4), 1129–1132.
23. Vayeda, D. R., & Mukherjee, N., (2017). *Nutraceuticals: The New Generation Therapeutics for Alzheimer's Disease, 4*, 9.

24. Fernández-Bedmar, Z., & Alonso-Moraga, A., (2016). *In vivo* and *in vitro* evaluation for nutraceutical purposes of capsaicin, capsanthin, lutein and four pepper varieties. *Food and Chemical Toxicology, 98*, 89–99. https://doi.org/10.1016/j.fct.2016.10.011.
25. Jeong, S. Y., Park, S. J., Yoon, S. M., Jung, J., Woo, H. N., Yi, S. L., Song, S. Y., et al., (2009). Systemic delivery and preclinical evaluation of Au nanoparticle-containing β-lapachone for radiosensitization. *Journal of Controlled Release, 139*(3), 239–245. https://doi.org/10.1016/j.jconrel.2009.07.007.
26. Shikov, A. N., Pozharitskaya, O. N., Miroshnyk, I., Mirza, S., Urakova, I. N., Hirsjärvi, S., Makarov, V. G., et al., (2009). Nanodispersions of taxifolin: Impact of solid-state properties on dissolution behavior. *International Journal of Pharmaceutics, 377*(1), 148–152. https://doi.org/10.1016/j.ijpharm.2009.04.044.
27. Yq, L., Ba, C., Ww, W., F, G., Gh, X., Zy, S., J, C., et al., (2010). Effects of magnetic Nanoparticle of Fe_3O_4 on apoptosis induced by gambogic acid in U937 leukemia cells. *Zhongguo Shi Yan Xue Ye Xue Za Zhi, 18*(1), 67–73.
28. Ghibaudo, F., Gerbino, E., Copello, G. J., Campo, D. O. V., & Gómez-Zavaglia, A., (2019). Pectin-decorated magnetite nanoparticles as both iron delivery systems and protective matrices for probiotic bacteria. *Colloids and Surfaces B: Biointerfaces, 180*, 193–201. https://doi.org/10.1016/j.colsurfb.2019.04.049.
29. Palmerston, M. L., Pan, J., & Torchilin, V. P., (2017). Dendrimers as nanocarriers for nucleic acid and drug delivery in cancer therapy. *Molecules, 22*(9), 1401. https://doi.org/10.3390/molecules22091401.
30. Esther, L. D., Khusro, A., Immanuel, P., Esmail, G. A., Al-Dhabi, N. A., & Arasu, M. V., (2020). Photo-activated synthesis and characterization of gold nanoparticles from *Punica granatum* L. Seed oil: An assessment on antioxidant and anticancer properties for functional yoghurt nutraceuticals. *Journal of Photochemistry and Photobiology B: Biology, 206*, 111868. https://doi.org/10.1016/j.jphotobiol.2020.111868.
31. Chen, S., Li, Q., McClements, D. J., Han, Y., Dai, L., Mao, L., & Gao, Y., (2020). Co-delivery of curcumin and piperine in zein-carrageenan core-shell nanoparticles: Formation, structure, stability and *in vitro* gastrointestinal digestion. *Food Hydrocolloids, 99*, 105334. https://doi.org/10.1016/j.foodhyd.2019.105334.
32. Nutraceuticals Product Market, (2015). *Asia Pacific Market Size, Segment and Country Analysis and Forecasts.* http://www.transparencymarketresearch.com/global-nutraceuticals-product-market.html (accessed on 21 June 2021).
33. *National Nanotechnology Initiative*, (2006). Available from: https://www.nano.gov/sites/default/files/pub_resource/nni_06budget.pdf. (accessed on 10 August 2021).
34. Yadav, S. K., (2017). Realizing the potential of nanotechnology for agriculture and food technology. *J. Tissue Sci. Eng., 08*(01). https://doi.org/10.4172/2157-7552.1000195.
35. Elham, A., & Seid, M. J., (2018). A systematic review on nanoencapsulation of food bioactive ingredients and nutraceuticals by various nanocarriers. *Critical Reviews in Food Science and Nutrition.*
36. Kyu, Y. B., Vijayakumar, A., Won, J. K., & Won, C. J., (2018). *Composition Comprising Curcumin-Captured Ginsenoside and Phospholipid-Based Lipid Nanoparticle as Effective Ingredient for Preventing or Treating Helicobacter Pylori Infection.* KR20180085947A.
37. Raedderstoff, D., Teixeira, S. R., & Weber, P., (2004). *Novel Nutraceutical Compositions Comprising Epigallocatechin Gallate.* WO2004/041257 A3.

38. Witham, P. H., & Paul, E. L., (2018). *Formulations Containing Omega-3 Fatty Acids or Esters Thereof and Maqui Berry Extract and Therapeutic uses Thereof.* US20180243253A1.
39. Ann, F., Regina, G., Claus, K., Mayne-Mechan, A. O., Hasan, M., Bernd, M., & Adrian, W., (2008). *Novel Nutraceutical Compositions Containing Stevia Extract or Stevia Extract Constituents and Uses Thereof.* AU2008333570B2.
40. Chaudhry, Q., Scotter, M., Blackburn, J., Ross, B., Boxall, A., Castle, L., Aitken, R., & Watkins, R., (2008). Applications and implications of nanotechnologies for the food sector. *Food Additives & Contaminants: Part A, 25*(3), 241–258. https://doi.org/10.1080/02652030701744538.
41. Andrè, M. C., (2020). *Pharmaceutical, Cosmetic or Food Products and use of Nanoparticles Containing Vitamin D.* WO2020019043A1.
42. Edward, V., (2019). *Molecular Particle Superior Delivery System.* US20190289895A1.
43. Lemone, X., (2015). *Bioactive Substance or Composition for Protein Delivery and Use Thereof.* JP6426288B2.
44. De La Vega, H. A., (2017). *Combination of Bioenergy and Nutra-Epigenetic Metabolic Regulators, Nutraceutical Compounds in Conventional and Nanotechnology-based Combinations, for Reversing and Preventing Cellular Senescence Accelerated by Chronic Damage Caused by Diabetes and Other Complex Chronic Degenerative Diseases.* WO2017213486A2.
45. Vinaykumar, T., (2018). *Modified Resveratrol Composition and Use Thereof.* WO2018042324A1 ().
46. Alfadul, S. M., & Elneshwy, A. A., (2010). Use of nanotechnology in food processing, packaging and safety: Review. *African Journal of Food, Agriculture, Nutrition and Development, 10*(6). https://doi.org/10.4314/ajfand.v10i6.58068.
47. Yusop, S. M., O'Sullivan, M. G., Preuß, M., Weber, H., Kerry, J. F., & Kerry, J. P., (2012). Assessment of nanoparticle paprika oleoresin on marinating performance and sensory acceptance of poultry meat. *LWT-Food Science and Technology, 46*(1), 349–355. https://doi.org/10.1016/j.lwt.2011.08.014.
48. Dasgupta, N., & Ranjan, S., (2018). An introduction to food grade nanoemulsions. *An Introduction to Food Grade Nanoemulsions.*
49. Thiruvengadam, M., Rajakumar, G., & Chung, I. M., (2018). Nanotechnology: Current uses and future applications in the food industry. *3 Biotech., 8*(1), 74. https://doi.org/10.1007/s13205-018-1104-7.
50. Arroyo, B. J., Bezerra, A. C., Oliveira, L. L., Arroyo, S. J., Melo, E. A. D., & Santos, A. M. P., (2020). Antimicrobial active edible coating of alginate and chitosan add ZnO nanoparticles applied in guavas (*Psidium guajava* L.). *Food Chemistry, 309*, 125566. https://doi.org/10.1016/j.foodchem.2019.125566.
51. Kuswandi, B., (2017). Environmental friendly food nano-packaging. *Environ. Chem. Lett., 15*(2), 205–221. https://doi.org/10.1007/s10311-017-0613-7.
52. Kaushik, P., Dowling, K., Barrow, C. J., & Adhikari, B., (2015). Microencapsulation of omega-3 fatty acids: A review of microencapsulation and characterization methods. *Journal of Functional Foods, 19*, 868–881. https://doi.org/10.1016/j.jff.2014.06.029.
53. Ameta, S. K., Rai, A. K., Hiran, D., Ameta, R., & Ameta, S. C., (2020). Use of nanomaterials in food science. In: Ghorbanpour, M., Bhargava, P., Varma, A., & Choudhary, D. K., (eds.), *Biogenic Nano-Particles and their Use in Agro-Ecosystems* (pp. 457–488). Springer: Singapore. https://doi.org/10.1007/978-981-15-2985-6_24.

54. Liu, X., Marrakchi, M., Xu, D., Dong, H., & Andreescu, S., (2016). Biosensors based on modularly designed synthetic peptides for recognition, detection and live/dead differentiation of pathogenic bacteria. *Biosensors and Bioelectronics, 80*, 9–16. https://doi.org/10.1016/j.bios.2016.01.041.
55. Sanvicens, N., Pastells, C., Pascual, N., & Marco, M. P., (2009). Nanoparticle-based biosensors for detection of pathogenic bacteria. *TrAC Trends in Analytical Chemistry, 28*(11), 1243–1252. https://doi.org/10.1016/j.trac.2009.08.002.
56. Yan, X., Li, H., & Su, X., (2018). Review of optical sensors for pesticides. *TrAC Trends in Analytical Chemistry, 103*, 1–20. https://doi.org/10.1016/j.trac.2018.03.004.
57. Wu, D., Du, D., & Lin, Y., (2016). Recent Progress on nanomaterial-based biosensors for veterinary drug residues in animal-derived food. *TrAC Trends in Analytical Chemistry, 83*, 95–101. https://doi.org/10.1016/j.trac.2016.08.006.
58. Wang, C., Hu, L., Zhao, K., Deng, A., & Li, J., (2018). Multiple signal amplification electrochemiluminescent immunoassay for Sudan I using gold nanorods functionalized graphene oxide and palladium/aurum core-shell nanocrystalline as labels. *Electrochimica Acta, 278*, 352–362. https://doi.org/10.1016/j.electacta.2018.05.061.
59. Anirudhan, T. S., Athira, V. S., & Chithra, S. V., (2018). Electrochemical sensing and nanomolar level detection of bisphenol-A with molecularly imprinted polymer tailored on multiwalled carbon nanotubes. *Polymer, 146*, 312–320. https://doi.org/10.1016/j.polymer.2018.05.052.
60. Neethirajan, S., Weng, X., Tah, A., Cordero, J. O., & Ragavan, K. V., (2018). Nano-biosensor platforms for detecting food allergens-new trends. *Sensing and Bio-Sensing Research*, 13–30. https://doi.org/10.1016/j.sbsr.2018.02.005.
61. Lin, X., Ni, Y., & Kokot, S., (2013). Glassy carbon electrodes modified with gold nanoparticles for the simultaneous determination of three food antioxidants. *Analytica Chimica Acta, 765*, 54–62. https://doi.org/10.1016/j.aca.2012.12.036.
62. Bouwmeester, H., Dekkers, S., Noordam, M. Y., Hagens, W. I., Bulder, A. S., De Heer, C., Ten, V. S. E. C. G., et al., (2009). Review of health safety aspects of nanotechnologies in food production. *Regulatory Toxicology and Pharmacology, 53*(1), 52–62. https://doi.org/10.1016/j.yrtph.2008.10.008.
63. Banerjee, T., Sulthana, S., Shelby, T., Heckert, B., Jewell, J., Woody, K., Karimnia, V., et al., (2016). Multiparametric magneto-fluorescent nanosensors for the ultrasensitive detection of *Escherichia coli* O157:H7. *ACS Infect. Dis., 2*(10), 667–673. https://doi.org/10.1021/acsinfecdis.6b00108.
64. Zhong, M., Yang, L., Yang, H., Cheng, C., Deng, W., Tan, Y., Xie, Q., & Yao, S., (2019). An Electrochemical immunobiosensor for ultrasensitive detection of *Escherichia coli* O157:H7 using CdS quantum dots-encapsulated metal-organic frameworks as signal-amplifying tags. *Biosensors and Bioelectronics, 126*, 493–500. https://doi.org/10.1016/j.bios.2018.11.001.
65. He, X., & Hwang, H. M., (2016). Nanotechnology in food science: Functionality, applicability, and safety assessment. *Journal of Food and Drug Analysis, 24*(4), 671–681. https://doi.org/10.1016/j.jfda.2016.06.001.
66. Morris, M. A., Padmanabhan, S. C., Cruz-Romero, M. C., Cummins, E., & Kerry, J. P., (2017). Development of active, nanoparticle, antimicrobial technologies for muscle-based packaging applications. *Meat Science, 132*, 163–178. https://doi.org/10.1016/j.meatsci.2017.04.234.

67. Nile, S. H., Baskar, V., Selvaraj, D., Nile, A., Xiao, J., & Kai, G., (2020). Nanotechnologies in food science: Applications, recent trends, and future perspectives. *Nano-Micro Lett., 12*(1), 45. https://doi.org/10.1007/s40820-020-0383-9.
68. Drew, R., & Hagen, T., (2016). *Nanotechnologies in Food Packaging: An Exploratory Appraisal of Safety and Regulation*. Prepared for Food Standards Australia New Zealand. Science Media Centre New Zealand: New Zealand.
69. Mihindukulasuriya, S. D. F., & Lim, L. T., (2014). Nanotechnology development in food packaging: A review. *Trends in Food Science & Technology, 40*(2), 149–167. https://doi.org/10.1016/j.tifs.2014.09.009.
70. Robinson, D. K. R., & Morrison, M. J., (2010). *Nanotechnologies for Food Packaging: Reporting the Science and Technology Research Trends*. Report for the observatory NANO. Available from: https://nanopinion.archiv.zsi.at/sites/default/files/full_report_nanotechnology_in_agrifood_may_2009.pdf (accessed on 10 August 2021).
71. Ravichandran, R., (2010). Nanotechnology applications in food and food processing: Innovative green approaches, opportunities and uncertainties for global market. *International Journal of Green Nanotechnology: Physics and Chemistry, 1*(2), P72–P96. https://doi.org/10.1080/19430871003684440.
72. Pavlath, A. E., & Orts, W., (2009). Edible films and coatings: Why, what, and how? In: Huber, K. C., & Embuscado, M. E., (eds.), *Edible Films and Coatings for Food Applications* (pp. 1–23). Springer: New York, NY. https://doi.org/10.1007/978-0-387-92824-1.
73. Sekhon, B. S., (2010). Food nanotechnology: An overview. *Nanotechnol. Sci. Appl., 3*, 1–15.
74. Ali, A., Ahmad, U., Akhtar, J., Badruddeen, & Khan, M. M., (2019). Engineered nano scale formulation strategies to augment efficiency of nutraceuticals. *Journal of Functional Foods, 62*, 103554. https://doi.org/10.1016/j.jff.2019.103554.
75. Beconcini, D., Felice, F., Fabiano, A., Sarmento, B., Zambito, Y., & Di Stefano, R., (2020). Antioxidant and anti-inflammatory properties of cherry extract: Nanosystems-based strategies to improve endothelial function and intestinal absorption. *Foods, 9*(2), 207. https://doi.org/10.3390/foods9020207.
76. Sahoo, S. K., & Labhasetwar, V., (2003). Nanotech approaches to drug delivery and imaging. *Drug Discovery Today, 8*(24), 1112–1120. https://doi.org/10.1016/S1359-6446(03)02903-9.
77. Tanaka, Y., Uemori, C., Kon, T., Honda, M., Wahyudiono, Machmudah, S., Kanda, H., & Goto, M., (2020). Preparation of liposomes encapsulating β-carotene using supercritical carbon dioxide with ultrasonication. *The Journal of Supercritical Fluids, 161*, 104848. https://doi.org/10.1016/j.supflu.2020.104848.
78. Pinilla, C. M. B., Reque, P. M., & Brandelli, A., (2020). Effect of oleic acid, cholesterol, and octadecyl amine on membrane stability of freeze-dried liposomes encapsulating natural antimicrobials. *Food Bioprocess Technol., 13*(4), 599–610. https://doi.org/10.1007/s11947-020-02419-8.
79. Amer, S. S., Nasr, M., Abdel-Aziz, R. T. A., Moftah, N. H., El Shaer, A., Polycarpou, E., Mamdouh, W., & Sammour, O., (2020). Cosm-nutraceutical nanovesicles for acne treatment: Physicochemical characterization and exploratory clinical experimentation. *International Journal of Pharmaceutics, 577*, 119092. https://doi.org/10.1016/j.ijpharm.2020.119092.
80. Barone, A., Cristiano, M. C., Cilurzo, F., Locatelli, M., Iannotta, D., Di Marzio, L., Celia, C., & Paolino, D., (2020). Ammonium glycyrrhizate skin delivery from ultra

deformable liposomes: A novel use as an anti-inflammatory agent in topical drug delivery. *Colloids and Surfaces B: Biointerfaces, 193*, 111152. https://doi.org/10.1016/j.colsurfb.2020.111152.
81. McClements, D. J., (2004). *Food Emulsions: Principles, Practices, and Techniques*. CRC.
82. McClements, D. J., & Rao, J., (2011). Food-grade nanoemulsions: formulation, fabrication, properties, performance, biological fate, and potential toxicity. *Critical Reviews in Food Science and Nutrition, 51*(4), 285–330. https://doi.org/10.1080/10408398.2011.559558.
83. Kulkarni, M., Goge, N., & Date, A. A., (2019). Development of nanoemulsion preconcentrate of capsanthin with improved chemical stability. *ASSAY and Drug Development Technologies, 18*(1), 34–44. https://doi.org/10.1089/adt.2019.916.
84. Uppal, S., Sharma, P., Kumar, R., Kaur, K., Bhatia, A., & Mehta, S. K., (2020). Effect of benzyl isothiocyanate encapsulated biocompatible nanoemulsion prepared via ultrasonication on microbial strains and breast cancer cell line MDA MB 231. *Colloids and Surfaces A: Physicochemical and Engineering Aspects, 596*, 124732. https://doi.org/10.1016/j.colsurfa.2020.124732.
85. Chu, Y., Gao, C., Liu, X., Zhang, N., Xu, T., Feng, X., Yang, Y., et al., (2020). Improvement of storage quality of strawberries by pullulan coatings incorporated with cinnamon essential oil nanoemulsion. *LWT, 122*, 109054. https://doi.org/10.1016/j.lwt.2020.109054.
86. Pardeshi, C., Rajput, P., Belgamwar, V., Tekade, A., Patil, G., Chaudhary, K., & Sonje, A., (2012). Solid lipid based nanocarriers: An overview. *Acta Pharmaceutica, 62*(4), 433–472. https://doi.org/10.2478/v10007-012-0040-z.
87. Nasiri, F., Faghfouri, L., & Hamidi, M., (2020). Preparation, optimization, and in-vitro characterization of α-tocopherol-loaded solid lipid nanoparticles (SLNs). *Drug Development and Industrial Pharmacy, 46*(1), 159–171. https://doi.org/10.1080/03639045.2019.1711388.
88. Sathya, S., Shanmuganathan, B., Manirathinam, G., Ruckmani, K., & Devi, K. P., (2018). *α-Bisabolol Loaded Solid Lipid Nanoparticles Attenuates Aβ Aggregation and Liquids, 264*, 431–441. https://doi.org/10.1016/j.molliq.2018.05.075.
89. Lacatusu, I., Mitrea, E., Badea, N., Stan, R., Oprea, O., & Meghea, A., (2013). Lipid nanoparticles based on omega-3 fatty acids as effective carriers for lutein delivery. Preparation and *in vitro* characterization studies. *Journal of Functional Foods, 5*(3), 1260–1269. https://doi.org/10.1016/j.jff.2013.04.010.
90. Liu, L., Tang, Y., Gao, C., Li, Y., Chen, S., Xiong, T., Li, J., et al., (2014). Characterization and biodistribution *in vivo* of quercetin-loaded cationic nanostructured lipid carriers. *Colloids and Surfaces B: Biointerfaces, 115*, 125–131. https://doi.org/10.1016/j.colsurfb.2013.11.029.
91. Froiio, F., Lammari, N., Tarhini, M., Alomari, M., Louaer, W., Meniai, A. H., Paolino, D., et al., (2020). Chapter 16-polymer-based nanocontainers for drug delivery. In: Nguyen-Tri, P., Do, T. O., & Nguyen, T. A., (eds.), *Smart Nanocontainers: Micro and Nano Technologies* (pp. 271–285). Elsevier. https://doi.org/10.1016/B978-0-12-816770-0.00016-2.
92. Dima, C., Assadpour, E., Dima, S., & Jafari, S. M., (2020). Bioavailability of nutraceuticals: Role of the food matrix, processing conditions, the gastrointestinal tract,

and nanodelivery systems. *Comprehensive Reviews in Food Science and Food Safety, 19*(3), 954–994. https://doi.org/10.1111/1541-4337.12547.
93. Andonova, V., (2017). Synthetic polymer-based nanoparticles: Intelligent drug delivery systems. In: Reddy, B. S. R., (ed.), *Acrylic Polymers in Healthcare*. InTech, https://doi.org/10.5772/intechopen.69056.
94. Bano, S., Ahmed, F., Khan, F., Chand, C. S., & Samim, M., (2020). Targeted delivery of thermoresponsive polymeric nanoparticle-encapsulated lycopene: *In vitro* anticancer activity and chemopreventive effect on murine skin inflammation and tumorigenesis. *RSC Advances, 10*(28), 16637–16649. https://doi.org/10.1039/C9RA10686C.
95. Singh, G., & Pai, R. S., (2014). Optimized PLGA nanoparticle platform for orally dosed trans-resveratrol with enhanced bioavailability potential. *Expert Opinion on Drug Delivery, 11*(5), 647–659. https://doi.org/10.1517/17425247.2014.890588.
96. Semo, E., Kesselman, E., Danino, D., & Livney, Y. D., (2007). Casein micelle as a natural nano-capsular vehicle for nutraceuticals. *Food Hydrocolloids, 21*(5), 936–942. https://doi.org/10.1016/j.foodhyd.2006.09.006.
97. Zimet, P., Rosenberg, D., & Livney, Y. D., (2011). Re-assembled casein micelles and casein nanoparticles as nano-vehicles for ω-3 polyunsaturated fatty acids. *Food Hydrocolloids, 25*(5), 1270–1276. https://doi.org/10.1016/j.foodhyd.2010.11.025.
98. Liang, L., Tajmir-Riahi, H. A., & Subirade, M., (2008). Interaction of β-lactoglobulin with resveratrol and its biological implications. *Biomacromolecules, 9*(1), 50–56. https://doi.org/10.1021/bm700728k.
99. Sneharani, A. H., Karakkat, J. V., Singh, S. A., & Rao, A. G. A., (2010). Interaction of curcumin with β-lactoglobulin—stability, spectroscopic analysis, and molecular modeling of the complex. *J. Agric. Food Chem., 58*(20), 11130–11139. https://doi.org/10.1021/jf102826q.
100. Shpigelman, A., Israeli, G., & Livney, Y. D., (2010). Thermally-induced protein-polyphenol co-assemblies: Beta lactoglobulin-based nanocomplexes as protective nanovehicles for EGCG. *Food Hydrocolloids, 24*(8), 735–743. https://doi.org/10.1016/j.foodhyd.2010.03.015.
101. Yadav, R., Kumar, D., Kumari, A., & Yadav, S. K., (2014). Encapsulation of catechin and epicatechin on BSA NPS improved their stability and antioxidant potential. *EXCLI J., 13*, 331–346.
102. Okagu, O. D., Verma, O., McClements, D. J., & Udenigwe, C. C., (2020). Utilization of insect proteins to formulate nutraceutical delivery systems: Encapsulation and release of curcumin using mealworm protein-chitosan nano-complexes. *International Journal of Biological Macromolecules, 151*, 333–343. https://doi.org/10.1016/j.ijbiomac.2020.02.198.
103. Wan, Z. L., Guo, J., & Yang, X. Q., (2015). plant protein-based delivery systems for bioactive ingredients in foods. *Food Funct., 6*(9), 2876–2889. https://doi.org/10.1039/C5FO00050E.
104. Rabinow, B. E., (2004). Nanosuspensions in drug delivery. *Nature Reviews Drug Discovery, 3*(9), 785–796. https://doi.org/10.1038/nrd1494.
105. Junyaprasert, V. B., & Morakul, B., (2015). Nanocrystals for enhancement of oral bioavailability of poorly water-soluble drugs. *Asian Journal of Pharmaceutical Sciences, 10*(1), 13–23. https://doi.org/10.1016/j.ajps.2014.08.005.

106. Karadag, A., Ozcelik, B., & Huang, Q., (2014). Quercetin nanosuspensions produced by high-pressure homogenization. *J. Agric. Food Chem., 62*(8), 1852–1859. https://doi.org/10.1021/jf404065p.
107. Aditya, N. P., Yang, H., Kim, S., & Ko, S., (2015). Fabrication of amorphous curcumin nanosuspensions using β-lactoglobulin to enhance solubility, stability, and bioavailability. *Colloids and Surfaces B: Biointerfaces, 127*, 114–121. https://doi.org/10.1016/j.colsurfb.2015.01.027.
108. Chaharband, F., Kamalinia, G., Atyabi, F., Mortazavi, S. A., Mirzaie, Z. H., & Dinarvand, R., (2018). Formulation and *in vitro* evaluation of curcumin-lactoferrin conjugated nanostructures for cancerous cells. *Artificial Cells, Nanomedicine, and Biotechnology, 46*(3), 626–636. https://doi.org/10.1080/21691401.2017.1337020.
109. Yekefallah, M., & Raofie, F., (2020). Preparation of potent antioxidant nanosuspensions from olive leaves by rapid expansion of supercritical solution into aqueous solutions (RESSAS). *Industrial Crops and Products, 155*, 112756. https://doi.org/10.1016/j.indcrop.2020.112756.

CHAPTER 11

Growth Patterns, Emerging Opportunities, and Future Trends in Nutraceuticals and Functional Foods

ASAD UR REHMAN,[1,2] SALMAN AKRAM,[1] and THIERRY VANDAMME[1]

[1]*University of Strasbourg, CNRS 7199, Faculty of Pharmacy, 74 Route du Rhin, CS – 60024, 67401 ILLKIRCH CEDEX, France*

[2]*University of Paris Descartes, UTCBS CNRS UMR 8258-INSERM U1267, Faculty of Pharmacy, 4 Avenue de l'Observatoire, Paris – 75006, France*

ABSTRACT

This chapter will first provide information about the background of the nutraceuticals. This background includes the historical perspective of using food items for healing purposes, gap in literature for stringent regulatory definitions for these products. Then, growing market trends in the domain of nutraceuticals and functional foods and their potential use in the prevention and treatment of diseases are discussed. In later parts, recent developments in their industrial manufacturing, novel dosage forms for their delivery into the body, including nanotechnology-based dosage forms, will be discussed in detail. Finally, some of the prominent clinical trials both completed and in the process will be discussed in detail to throw light on progress in their use as a potential alternative treatment.

11.1 BACKGROUND AND GROWTH PATTERNS IN NUTRACEUTICALS AND FUNCTIONAL FOODS OVER THE YEARS

Historically for a very long time, there was no distinction between the food and drugs in terms of their use for medical purposes. The drugs were presumed to be some sort of special food items needed to heal injury, to replenish energy and to antagonize the effects of accidental or intentional poisoning. During this era of human history, there were no scientific methods established to check the authenticity of the reports of medicinal use of different food items in different cultures. Despite this history accounts various statements on the importance of the food items for health purpose. The famous Greek physician Hippocrates mentioned "...difference of diseases depends on the nutriment" [1]. From that time to dawn of modern science the idea of use of food elements for medical purpose was popular in all cultures.

Dr. Stephen DeFelice was the first person to use the term "nutraceutical" from the combination of words "nutrition" and "pharmaceutical" in 1989 [2]. He described them as food or part of food that provides medicinal and health effects. FDA covers most of the nutraceuticals under the definition of "Food supplements" [3]. According to European Nutritional Association, the term nutraceutical is defined as "Nutritional products that provide health and medical benefits, including the prevention and treatment of disease" (ENA 2016). Whereas the "Functional food" term is defined by Hardy as "Any food or ingredient that has a positive impact on an individual's health, physical performance, or state of mind, in addition to its nutritive value" [4]. A lot of people have tried to define these terms for clarification of their meaning and scope. According to literature, the definition of these terms overlaps with each other and there is no strict regulatory definition which can differentiate between the terms "food supplements," "functional foods" and "nutraceuticals" [2, 3]. For the nutraceuticals there is debate in literature either to set same safety and efficacy parameters for them as pharmaceuticals or establish special safety protocols for them. The work on establishing proper regulatory guidelines on their use is in progress and clinical trials of the many nutraceuticals have been done to demonstrate the safety and efficacy of these products.

The global market of nutraceutical was \$106 billion in 2004 [5] and \$128 in 2008 [6]. The global market of nutraceuticals was projected to cross \$171.8 billion in 2014 and \$241.1 billion by 2019 [7]. It is difficult to estimate current global market value of the nutraceutical due to overlapping between nutraceuticals and functional foods and other simple food supplements as they are marketed under the umbrella of the same group name [8]. This substantial growth in the global market is attributed to the following factors [9]:

- Recent developments in science and technology;
- Higher proportion of aging population;

- Increase in lifestyle related diseases;
- Awareness on the medical benefits of the food items;
- Exploration of alternative treatment options;
- Biotechnology and genetic engineering modified plants and food products.

In addition to above-mentioned factors, this recent growth in the market is also influenced by various other factors, for instance, in Canada whey waste in millions of tons is produced each year. This whey was considered as waste byproduct produced during the industrial processing [10]. Recently, the industry has developed processes to process this whey waste to produce essential healthy nutrients. Important proteins [11] obtained by biotechnology from the whey proteins are now on market. Another factor driving the growth in this domain is lesser regulatory constraints and less product development time required [12].

11.2 EMERGING OPPORTUNITIES AND FUTURE TRENDS IN NUTRACEUTICALS AND FUNCTIONAL FOODS

11.2.1 PREVENTION AND TREATMENT OF DISEASES

The nutraceuticals are emerging as an alternative option in prevention and therapy of number of diseases. By each passing day, their scope is getting broader and they are finding their applications in new pharmacotherapy areas. Following is the some of the key fields in pharmacotherapy in which they are finding their applications.

11.2.1.1 NEURODEGENERATIVE DISEASES

Neurogenerative diseases is a term used mainly for number of diseases related to peripheral or central nervous systems such as Parkinson's diseases, Alzheimer's disease, and such others. In these diseases, the neuronal integrity or neuronal transmission is compromised, which ultimately leads to the loss of sensory or motor activities [13, 14]. This leads to life-threatening conditions, death, and a huge financial burden on public health systems [15, 16]. They also lead to the psychological and physical stress of the affected families [17]. Below are mentioned some of the representative nutraceuticals currently under investigation and use for the management and cure of these group of diseases:

1. **Curcumin:** The active ingredient of turmeric known as curcumin is one of the most widely used nutraceutical used owing to its beneficial therapeutic effects [18]. Historically it has been reported for its use as flavoring and coloring agent and as traditional remedy for various diseases [19]. Besides its effects as anti-inflammatory and antioxidant effects it is well reported for its use for neuro-protective functions. Curcumin can cross the blood brain barrier and so can be exploited for treatment of various neurodegenerative disorders [20]. The major limitation in its use as therapeutic agent is its poor bioavailability due to its limited absorption and immediate clearance from body [21]. Clinical trial results indicate its beneficial effects for improving cognitive functions and as antidepressant [22, 23].
2. **Coenzyme Q10:** The role of Coenzyme Q10 in ATP cycle and as anti-oxidant makes it one of essential micronutrients for the human body. Despite its many other functions as antidiabetic, antihyperlipidemic, cardioprotective agent it can also cross the blood-brain barrier and has positive impact on different neurodegenerative diseases [24, 25]. It exhibits its neuroprotective effects by inhibiting ACE activity and through improving activity of endogenous antioxidant system [26]. This suffers from low bioavailability due to its high lipophilicity. Clinical trial data supports its beneficial role in depression and Parkinson's disease [27, 28].
3. **Resveratrol:** This polyphenol is present mostly in red grapes, cherries, and berries. It has multiple beneficial effects and its anticancer, antidiabetic, and anti-inflammatory effects are frequently reported in literature. It can effectively cross the blood brain barrier and also shows neuroprotective effects [29, 30]. Its safety is well established as every age group has well tolerance for it. By upregulating Nrf2/HO-1 and PI3K/Akt signaling pathway it exhibits its antioxidant and hence neuroprotective properties [31, 32]. However, its major neuroprotective function is due to its ability to act as an anti-protein aggregation agent which overall improves cognitive function. This also suffers from low bioavailability due to its rapid metabolization rate and rapid clearance from body [33]. Clinical trial data suggests its beneficial role in improving cognitive functions [34].

4. **Polyunsaturated Fatty Acids (PUFAs):** These are group of essential fats as they cannot be synthesized by human body. The most important of these fatty acids are normally found in salmon fish and walnuts. These important nutritional components show antioxidant, antihyperlipidemic, and cardioprotective properties. They perform their neuroprotective function by blocking microglia/astrocytes via JNK and PPAR-y signaling pathway [35, 36]. It also enhances neurotransmission and neurogenesis and hence improve overall brain function activity [36–38]. Clinical trial data shows its beneficial effects for its use for treatment of Parkinson's and Alzheimer's diseases [38, 39].

11.2.1.2 DIABETES

Diabetes is non-communicable, multifactorial, lifestyle chronic disorder characterized by hyperglycemia [39]. In diabetes the metabolism of several nutritional components is impaired due to inadequate amount or inactivity of insulin [40]. Due to its prevalence as pandemic state and enormous burden on the public health system by causing lifetime morbidity and mortality diabetes is major challenge of today's world [41]. The diabetes can cause number of secondary health problems or aggravate them and hence complicate the therapy of the individual elements. Several various therapeutic options have been proposed for the prevention and cure of this public health problem but still there are huge gaps in its effective treatment. This gap leads to poor patient compliance and hence decreases the chances of the effectiveness of overall pharmacotherapy. Various nutraceuticals have been explored in traditional medicine for the prevention and cure of this metabolic disorder as an alternative option for prevention, management, or therapy of the diabetes [42]:

1. **Antioxidants:** These can potentially play a vital role in the management of diabetes. Various animal studies suggest that antioxidants can delay the onset of the diabetes induced complications by relieving oxidative stress [43]. Vitamin C reduces protein glycation, sorbitol accumulation and lipid peroxides [44]. All these impacts on reduction of diabetes induced complications. Similarly, vitamin E also leads to the reduction of peroxides and aldehydes as well as increases the platelets activation [44, 45]. Similarly, vitamin D intake by infants helps in reduction in development of type 1 diabetes [46].

While in adult patients the intake of vitamin D led to the decrease of the amount of insulin needed for the diabetes management [47]. α-Lipoic acid is also an important antioxidant that protects the retina against ischemic injury [43]. Ischemic injury occurs in diabetes and is one of the most important causes of vision loss. Clinical trial data suggests that supplementation by α-Lipoic acid helps in the treatment of diabetic neuropathy [48].

2. **Minerals:** Chromium in an important trace element needed for the healthy function of the body. It has been shown that chromium intake leads to the insulin sensitivity and glucose tolerance [48]. Another important mineral that increases the insulin sensitivity is magnesium [49]. Similarly, zinc supplementation reduced the oxidative stress over the retina and decreases the chances of the age-related eye diseases [50].
3. **Plant Derived Nutraceuticals:** The plant-based nutraceuticals act by inhibiting the key enzymes involved in the starch degradation into the smaller carbohydrate units and ultimately their absorption into the blood circulation [51, 52]. The two key enzymes involved in the glucose metabolism in the body are α-amylase and α-glucosidase [52]. Limonoids from *A. indica* (Neem), myricetin from the guava, aqueous, and ethanolic extracts from brown seeds and several other plant extracts are reported in literature to effectively inhibit the activity α-amylase and α-glucosidase [53–56]. In addition to that, plant-based nutraceuticals are also effective against glucose regulatory enzymes such as DPP4 (dipeptidyl peptidase-IV), GLP-1 glucagon-like peptide-1 (GLP-1), and GIP (glucose-dependent insulinotropic polypeptide). Essential oil from Cymbopogon citratus, active ingredients from Inonotus obliquus, Rhizophora mucronate, ethanolic extract of Urena lobate are the prominent plant nutraceuticals inhibiting the activity of these enzymes and hence aids in diabetes management [57–59]. In addition to this many phenolic derivatives are also responsible for aldose reductase enzyme which is responsible for the sorbitol accumulation and hence aggravation of diabetic complications [60]. The peroxisome proliferator-activated receptors (PPARγ), whom activation leads to the increase of cellular glucose uptake and reduces plasma glucose level is an important target for diabetes treatment. Catechin from green tea, activates them and hence decreases the chances of the diabetes type 2 [61].

11.2.2 TECHNOLOGY FOR DESIGN AND DEVELOPMENT OF NUTRACEUTICALS AND FUNCTIONAL FOODS

There has been a lot of innovation in design and development of nutraceuticals and functional foods over the past few decades. Modern pharmaceutical technology, biotechnology, and genetic engineering techniques have been employed to manufacture these products. Below is a summary of production of nutraceuticals by two different key techniques over the past few years, i.e., biotechnology and pharmaceutical nanotechnology:

1. **Biotechnology:** It is playing a major role in the development of this industry. Biologically active non-nutritive components of food and food items are being utilized to manufacture nutraceuticals by biotechnology approaches [12]. To produce enzymes and recombinant microorganism used in nutraceuticals there is immense need to explore new food elements. The biotechnology is key for exploration of these new food items for production of novel nutraceuticals [62]. Although the biotechnology is playing a vital role in development of this new industry, but the nutraceuticals manufactured by this way suffer from technology constraints and consumer distrust [63] and regulatory limitations [64]. Despite this, there are some incentives for companies exploiting biotechnology to produce the nutraceuticals. For instance, a successful food product life cycle is around 21 years [12]. By employing the biotechnology and genetic engineering exploring new advantages of this food product and mentioning some health claims to it can increase the market value and life cycle of this food product [12].
2. **Pharmaceutical Nanotechnology:** It is one of keyway used in recent times to produce nutraceuticals and functional foods for research lab scale level and at industrial scale. Various nanoscale dosage forms have been developed over the years, such as nanoemulsions, liposomes, solid lipid nanoparticles (SLNs), and others. Details on these dosage forms and their contribution in field of nutraceuticals will be provided in the next sections of this chapter. Here, we will briefly use different strategies employed to produce these dosage forms. The technology has huge impact on the properties of the final delivery system produced and on its capacity to effectively encapsulate and release the nutraceuticals [65]. The amount of energy supplied has impact on the final particle size of the formulation which ultimately defines the stability of the product. The comparison of two different

high energy methods, i.e., mircrofluidization, and high-pressure homogenization were compared in a study. Results showed that low particle size produced in mircrofluidization due to higher energy employed in mircrofluidization have a long stability [66]. Other studies compared the ultrasonication with mircrofluidization and mircrofluidization and have reported better results with mircrofluidization [67, 68]. Although these high energy methods are very effective, but they suffer from limitations of easy scale up and use with sensitive food components in some nutraceuticals such as proteins. The excessive high energy can have negative impact on the stability of the formulation [68, 69]. The low energy methods are interesting for sensitive components [70]. The two most important methods in this domain are spontaneous emulsification and phase inversion method [71]. There are other various methods used in domain for production of nano scale dosage forms and although they are mainly affected by process parameters, formulation parameters also play an important part in determining the final particle size, stability, and effectiveness of the nutraceuticals produced [72, 73].

11.2.3 DEVELOPMENT OF VARIOUS DELIVERY SYSTEMS FOR NUTRACEUTICALS AND FUNCTIONAL FOODS

Nutraceuticals include a wide range of compounds such as carotenoids, bioactive peptides, lipids, phenolic compounds, vitamins, essential minerals, etc., and provide physiological or therapeutic benefits beyond the basic nutritional needs. In the recent years, the incorporation of nutraceuticals in food products has provided a very simple and efficient way of developing novel functional foods. However, the nutraceuticals often show low aqueous solubility, instability during food processing or storage, chemical transformation, or/and poor absorption within the gastrointestinal tract (GIT), thus resulting in low bioavailability and reduced health benefits of the nutraceuticals [74]. Some of the nutraceuticals (e.g., carotenoids, Vitamins), when directly added into the food products, can show unwanted interactions with other food components, which may affect the texture, appearance, bioavailability, and stability of those components. All these factors limit their direct incorporation into the food products. Over the past couple of decades, the efforts to overcome these limitations and an increased research interest in the formulation of food products enriched with nutraceuticals, have led to the development of the delivery systems for the nutraceuticals and functional foods. The major classes of the nutraceuticals and functional food components, which have shown great therapeutic potential and can benefit from their encapsulation into the delivery systems, are shown in Table 11.1. However, the development of a delivery system for nutraceuticals is very challenging process because the components which are used to form delivery systems must be of food grade (so only a limited number of approved materials can be used) as well as the method to be used for the development of delivery system must be very economical, robust, and reproducible [75].

11.2.3.1 IDEAL PROPERTIES OF NUTRACEUTICAL DELIVERY SYSTEMS

The development of nutraceutical delivery system is a very challenging process and it requires a thorough knowledge of the properties of nutraceuticals to be encapsulated as well as the use of the suitable materials and techniques. The knowledge of the properties of an ideal delivery system significantly assist to understand the main parameters to be considered during the selection or development of a delivery system for nutraceuticals. Following are the main properties of an ideal delivery system for nutraceuticals,

11.2.3.1.1 HIGH ENCAPSULATION EFFICIENCY (EE)

The ideal deliver system should provide a very high encapsulation of the nutraceuticals, and effectively retain the encapsulated nutraceutical until it reaches the desired site of action, providing a high bioavailability [75, 76]. The lipophilic/hydrophobic nature of the delivery system also helps to increase the bioavailability of encapsulated lipophilic active ingredients (e.g., nutrients, vitamins, and other bioactive agents). However, the selection of the encapsulation technique is very important to keep the nutraceuticals in their bioactive form, e.g., use of heat producing encapsulation techniques for the nutraceuticals which are sensitive to heat can result in the loss of the bioactivity of the nutraceuticals.

11.2.3.1.2 PROTECTION OF THE LOADED NUTRACEUTICALS

The delivery system should provide protection to the encapsulated nutraceuticals against the chemical degradation (e.g., degradation resulting from oxidation or hydrolysis) as well as protection from the adverse physiological factors (e.g., harsh gastrointestinal (GI) pH, digestive enzymes) while it is carrying the loaded components to the site of action. It should also provide protection to nutraceuticals during production, storage, and transport against the factors such as temperature, light, pH, etc. As a result, the nutraceuticals are safely delivered at the desired site of action in their bioactive form and provide maximum health benefits.

11.2.3.1.3 TARGETED AND CONTROLLED DELIVERY OF THE FUNCTIONAL COMPONENTS

The delivery system should carry the loaded nutraceuticals to the desired site of action and provide either a controlled release or a release under the influence of an environmental trigger (e.g., temperature, pH, enzyme activity, ionic strength) of the loaded functional components. The components to be used to develop a delivery system are selected based on the environment at the site of action as well as the mechanism of the delivery or release of the loaded components.

11.2.3.1.4 COMPATIBLE WITH THE COMPONENTS OF THE FOOD PRODUCT

The delivery system should be compatible with the other components of the food product (or matrix) and it should not affect the texture, appearance, flavor, aroma, and stability of the final product. The food matrix may be composed of single or multiple components (e.g., water, lipids, proteins, surfactants, polysaccharides, etc.). Delivery systems can be incorporated into different kinds of food matrices, e.g., trapped inside a biopolymer matrix (sauces, yogurts) or a solid matrix (cereal products) or it can be dispersed in an aqueous solution (drinks, beverages) [77].

11.2.3.1.5 MASKING THE BITTER TASTE OR OFF-FLAVORS

The bitter taste and unwanted flavors are major contributors of non-compliance in the consumers. However, this bitter taste as well as the off-flavor can be masked by inhibiting the direct interaction of bioactive molecule with the oral mucosal surface. The lipophilic delivery systems prevent the encapsulated bitter components and unwanted flavors, from interacting directly with the taste receptors, thus improving the consumer compliance.

Table 11.1 Different Classes of Nutraceuticals which can Benefit from Encapsulation

Class	Health Benefits	Encapsulation Benefits
Bioactive proteins and peptide (e.g., immunoglobulin, lactoferrin, casein phosphopeptides)	Have antimicrobial, anticancer, antioxidant, antihypertensive, and cholesterol lowering properties Immune system mediators	No interactions with other food components Resistance against denaturation and conformational changes Improved bioavailability
Vitamins (e.g., Vitamin A, B complex, C, D, E, and K)	Antioxidant properties Play vital role in the cellular functions, e.g.: o The production of red blood cell and insulin o The metabolism of proteins, fats, and carbohydrates o The formation of teeth and bones o Clotting and wound healing Prevention and treatment of cardiovascular diseases, cancer, and diabetes	Easy incorporation into food products without losing their bioactivity Masking the off-flavors Enhanced chemical stability
Carotenoids (e.g., β-carotene and lutein)	Antioxidant activity Promote good vision and protect from eye disease Treatment of Cancer and coronary heart disease	Protection from the environmental factors, e.g., oxygen, temperature, and light Ideal solution for their limitations to be used directly, e.g., low aqueous solubility, crystalline structure, etc.
Phytosterols (e.g., Stigmasterol, β-Sitosterol)	Treatment of coronary heart disease Regulate the cholesterol level	Protection from the external factors, e.g., oxygen, temperature, and light Potential solution for their low aqueous solubility
Essential minerals (e.g., zinc, iron, chromium, calcium)	Play an important role in the cellular functions, e.g., o Protein function o Growth and development o Keep the bones healthy o Enhance insulin activity	Protection of the food products from adverse effects Improved bioavailability Masking the unpleasant taste

Table 11.1 *(Continued)*

Class	Health Benefits	Encapsulation Benefits
Fatty acids (e.g., butyric acid, ω-3 fatty acids, conjugated linoleic acid)	Provide protection against various chronic diseases, e.g., arthritis, coronary heart disease and cancer Keep the bones and brain healthy	Enhance their bioavailability Easy incorporation into food products without losing their bioactivity Mask the unpleasant taste
Polyphenols (e.g., curcumin)	Possess antioxidant, anticancer, antimicrobial, and anti-inflammatory properties	Easy incorporation into food products without losing their bioactivity Masking the unpleasant taste Enhanced chemical stability
Fibers (e.g., pectin, cellulose)	Regulation of cholesterol and blood glucose Prebiotic effects To treat constipation	Easy incorporation into food products without affecting the texture and flavor of the food products

Source: Refs. [74, 75, 78–83].

11.2.3.1.6 FACILITATE THE LYMPHATIC UPTAKE

The food components and nutraceuticals which suffer from hepatic first pass metabolism can be benefitted by their encapsulation into a lipid-based delivery system. The lymphatic uptake is an efficient alternative for systematic transport of lipophilic compounds which helps them to bypass the hepatic metabolism and results in an increase in the bioavailability of those nutraceuticals [84]. As the extent of lymphatic uptake is directly related to the ability of the bioactive compounds to associate with the lipoproteins within the enterocyte [85], the lipid-based delivery systems (having nano-sized particles) provide an effective way to improve the direct intestinal lymphatic uptake of the lipophilic bioactive compounds.

11.2.3.2 THE EMERGING NUTRACEUTICAL DELIVERY SYSTEMS

Different types of delivery systems have been developed to encapsulate and deliver the nutraceuticals, depending upon the molecular and physicochemical properties of the nutraceuticals and food components to be encapsulated. These delivery systems usually differ from one another in terms of their cost, biocompatibility, ease of development, ease of use, biodegradability, and their capacity to encapsulate, protect, and deliver the nutraceuticals. As mentioned earlier, the materials which are used for the development of these delivery systems to encapsulate nutraceuticals and food components should be of food grade. That is why the bio-based materials are usually used as the encapsulating materials for nutraceuticals because of their biocompatibility, biodegradability, and non-toxicity, e.g., proteins, lipids, surfactants, polysaccharides, etc. The most important types of delivery systems which are currently used as well as the ones having potential to be used as delivery systems for nutraceuticals are summarized here.

11.2.3.2.1 LIPOSOMES

The liposomes are the spherical structures consisting of an internal aqueous compartment enclosed by one or more phospholipidic bilayers, known as lamellae. Liposomes were first described by Bangham et al. as spherical bilayer structures composed of phospholipid and cholesterol [86], known as the classical or conventional liposomes. When the phospholipids are hydrated, they form bilayers (by self-association) surrounding an aqueous interior. The cholesterol improves the fluidity and stability of the bilayers, which prevents the leakage of the active payload. However, incorporation of cholesterol in food products should be avoided in certain health conditions (e.g., hypercholesterolemia), in that case some other compounds can be used instead of cholesterol to help maintain the integrity of the liposome membrane (e.g., phytosterols). Among all the emerging systems developed to encapsulate nutraceuticals, liposomes have gained much importance due to their unique bilayer structure which enables them to encapsulate both lipophilic and hydrophilic nutraceuticals. The general structure of the classical liposomes is shown in Figure 11.1.

Based on the lamellarity and size, liposomes can be classified into the following types:

- Small unilamellar vesicles (SUVs) with a size < 100 nm;
- Large unilamellar vesicles (LUVs) with a size > 100 nm;
- Multilamellar vesicles (MLVs) with a size > 0.5 μm.

The classical liposomes are generally cleared from the blood very rapidly, that's why scientists have developed the advanced liposomes to deal with this problem. The advanced liposomes can be categorized into the following main types, stealth liposomes, cationic liposomes, fusogenic liposomes, and ligand targeted liposomes, having a wide range of applications in food, pharmaceutical, and cosmetic industry.

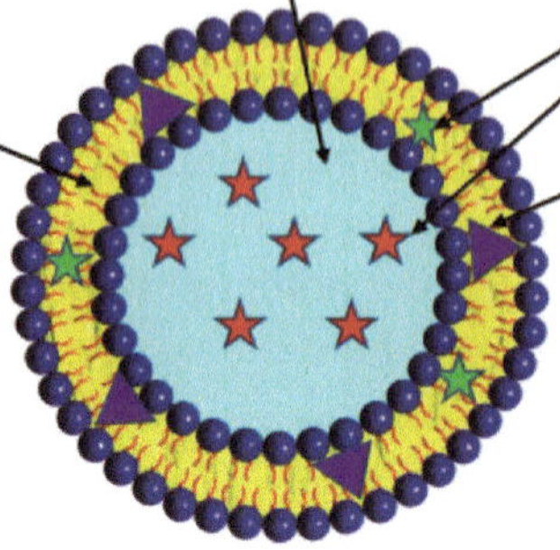

FIGURE 11.1 Structure of liposome.

Liposomes can be prepared by a number of different methods with each method influencing the properties of the resulting liposomes. The most widely used methods to prepare liposomes are:

- Thin film hydration or Bangham's method;
- Polycarbonate membrane extrusion technique;
- Ultrasonication;
- Microfluidization;
- Reverse phase evaporation method;
- Solvent injection method;
- Detergent removal method.

The liposomes are most commonly formed by using either phospholipids (from milk, soy, and eggs) or non-ionic surfactants [87]. The development of liposomes is an effective strategy for encapsulation and controlled release of bioactive molecules in food matrices. The desired particle size, cost, reproducibility, type of aqueous medium to form the liposomes, physicochemical properties of the encapsulating material as well as of the nutraceuticals to be encapsulated are the most important factors to be considered to choose the appropriate method to prepare liposomes. Each method has its own significance, as far as achieving the desired objectives is concerned. For example, Alexander et al. developed ascorbic acid encapsulated liposomes using soy phospholipids by different methods and studied the effect of incorporation of phytosterol on the encapsulation efficiency (EE) and stability of the liposomes obtained [88]. They found that the liposomes obtained by the ethanol method showed the highest EE and better stability as compared to the liposomes obtained by other methods and that the incorporation of phytosterols in the liposomes caused an increase in the average size of the formed liposomes. It was concluded that the method of preparation and the ratio of phospholipids to phytosterol affect the properties of the formed liposomes (e.g., average size, EE, etc.), and that by carefully controlling these two factors, liposomes having limited aggregation, high EE and stability upon dilution can be obtained. A few examples of the nutraceuticals and food components benefitted from their incorporation/encapsulation into the liposomes are shown in Table 11.2. Liposomes have many advantages over other delivery systems like biodegradability, biocompatibility, possibility of encapsulation of hydrophilic, lipophilic as well as multi-compounds, targeted delivery, and very low toxicity [74, 89–91]. However, they have a few disadvantages as well, e.g., low storage stability, short release time and sensitivity to external factors such as temperature, oxygen, and light [14]. But these shortcomings can be avoided or improved by using suitable components based on the desired application, for example, Li *et al.* have developed chitosan coated liposomes for entrapment of antidiabetic peptides and showed that the chitosan coating improved greatly the EE and stability of the liposomes and it prolonged the release of peptides in simulated gastric fluids as well [12].

11.2.3.2.2 CONVENTIONAL EMULSIONS

The emulsions are the heterogeneous dispersions of two immiscible liquids, with the one dispersed as small droplet (called dispersed phase) into other liquid (called continuous phase) stabilized by an emulsifying agent. The emulsifying agents are the amphiphilic molecules, which arrange themselves at the oil/water interface and stabilize the emulsions. Emulsions are generally classified as O/W, when oil droplets are dispersed in continuous aqueous phase and water-in-oil type, when water droplets are dispersed in continuous oil phase. For the encapsulation and delivery of lipophilic bioactive compounds the O/W type emulsions are used. Encapsulation of bioactive compounds into emulsion-based delivery systems is an effective approach to overcome the issues associated with incorporation of bioactive compounds into the food product, like low aqueous solubility, interaction with other food components, physicochemical, and physiological degradation, low bioavailability, etc. The stability and droplet size of the emulsions mainly depends on the composition of the oil phase, co-solvents used and surfactant to oil ratio. The commonly used encapsulating materials in emulsion systems include lipids, proteins, polymers, low molecular weight surfactants and polysaccharides [92]. The emulsions are widely used in the development of food products like meat products, beverages, bakery products, dressings, butter, etc. The conventional Emulsions have many advantages over other systems (even more sophisticated ones), such as, the possibility of encapsulation of both the lipophilic and hydrophilic nutraceuticals, which can be easily developed from food-grade ingredients in different dosage forms based on the desired application (e.g., pastes, viscous liquids, elastic solids, etc.), and finally, they can be prepared by low cost and very simple methods, like homogenization and mixing [75]. However, there are some limitations of conventional emulsions, for example, they have large particle size (500 nm to 1000 μm) as compared to other advanced delivery systems which causes aggregation of the droplets and associated instability problems (flocculation, coalescence, etc.), and secondly, they have poor stability towards environmental stresses, such as pH extremes, freezing, heating, drying, and high mineral concentrations.

These limitations have led to the shift of scientific interest to encapsulate nutraceuticals and processed foods in more advanced types of emulsions (e.g., nanoemulsions (NEs)).

11.2.3.2.3 NANOEMULSIONS (NES)

Nanoemulsions (NEs) are very fine, kinetically stable, and optically clear and translucent dispersions of two immiscible liquids (either oil-in-water (O/W) or water-in-oil), generally stabilized by an amphiphilic surfactant or emulsifying agent (e.g., polysorbate 80), with average droplet radii < 100 nm (Figure 11.2) [70, 76, 108]. Apart from oil, water, and emulsifier, some other components are also added either to modify the physicochemical properties or to enhance the stability of NEs, e.g., co-surfactants, thickening agents, pH-stabilizers, ripening inhibitors, gelling agents, preservatives, etc. Based on constituents and the relative distribution of continuous phase and dispersed phase, NEs can be classified into two types, which are biphasic NEs and double (or multiple) NEs. As the name indicates, biphasic nanoemulsion consists of two phases, either O/W or water-in-oil, whereas double or multiple nanoemulsion is a complex colloidal system consisting of more than two phases, i.e., oil-in-water-in-oil (O/W/O) or water-in-oil-in-water (W/O/W) (Figure 11.3). Contrary to conventional emulsions, NEs have very small droplet sizes, faster absorption following oral administration, show higher and long-term stability against coalescence and flocculation and are optically clear and translucent systems. NEs are prepared either by high energy methods (like high pressure homogenization, ultrasonication, and microfluidization) or by low energy methods (like spontaneous emulsification, phase inversion method).

Table 11.2 The Emerging Nutraceutical Delivery Systems

Delivery System	Functional Foods and Nutraceuticals	Health Benefits	Objective of Encapsulation	References
Liposomes	Catechin	Antioxidant and neuroprotective	Protection against degradation To improve bioavailability and distribution to brain	[17]
	Curcumin	Antioxidant, anticancer, anti-HIV, and anti-inflammatory	Protection from GI-degradation To improve the solubility and enhance bioavailability	[18]
	Silymarin	Hepatoprotective, antioxidant, cardioprotective, anticancer, and anti-inflammatory	Protection from GI-degradation To improve the solubility and enhance bioavailability	[93]
	Vitamin C	Antioxidant, repairs bones, teeth, and cartilage, protects skin, delays aging	To enhance encapsulation and bioavailability	[94]
	Bioactive peptides	Antioxidant, ACE-inhibitors	To improve encapsulation and stability	[95]
Nanoemulsions	Vitamin E	Antioxidant, prevention of chronic diseases, e.g., cardiovascular diseases and cancer	To improve solubility and bioavailability	[96]
	Fish oil	Prevention of heart and kidney diseases	To improve the absorption	[97]
	Resveratrol	Antioxidant, chemoprotective, and cardioprotective effects	Protection against degradation and to achieve sustained release	[98]
	Quercetin	Anti-cancer, antioxidant, and anti-inflammatory activity	To improve solubility and bioavailability	[94]

Table 11.2 *(Continued)*

Delivery System	Functional Foods and Nutraceuticals	Health Benefits	Objective of Encapsulation	References
Microemulsions	Berberine	Anti-bacterial, anticancer, improves cerebral ischemia	To improve bioavailability	[99]
	Simvastatin and phytosterol	Hypercholesterolemia	To improve solubility and absorption	[100]
	Puerarin	Antioxidant and cardio protective	To improve the absorption	[101]
Solid lipid nanoparticles (SLN)	β-carotene	Acts as pro-vitamin A	To improve physicochemical stability and to prevent exclusion from the matrix	[102]
	Triptolide	Anti-inflammatory, anti-neoplastic, immunosuppressive	To decrease the toxicity	[103]
Nanostructured lipid carriers (NLCs)	Hesperetin	Antioxidant, inhibits the chemically induced colon carcinogenesis and heart attack	To improve the solubility and to mask the bitter taste	[104]
	Ascorbyl palmitate	Antioxidant	To enhance the chemical stability	[105]
Biopolymer microgels and nanogels	Curcumin	Antioxidant, anticancer, anti-HIV, and anti-inflammatory	To control the release through the GI-tract	[26]
	Omega-3 fatty acids	Prevention of heart and kidney diseases	Protection against degradation	[106]
Polyelectrolyte complexes (PECs)	Resveratrol	Antioxidant, chemoprotective, and cardioprotective effects	To provide controlled release and to improve therapeutic index	[107]

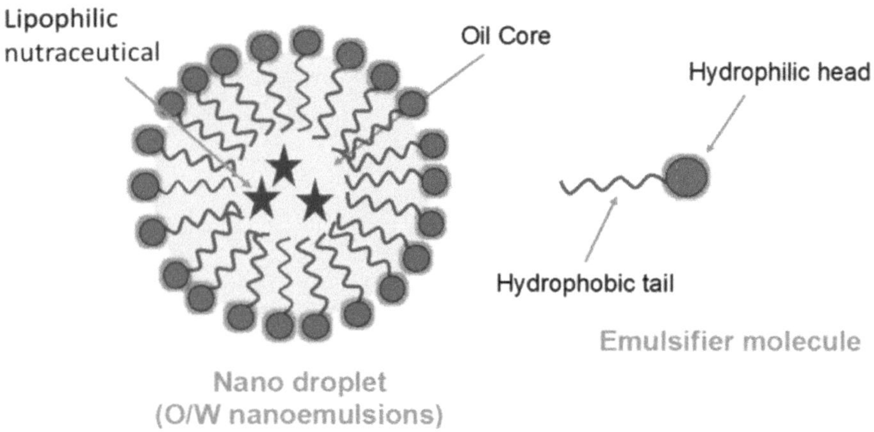

FIGURE 11.2 Structure of nano-droplet and emulsifier molecule.

O/W NEs are widely used to encapsulate lipophilic nutraceuticals and food components by solubilizing them in oil phase prior to the addition of the surfactant and exposure to mechanical disruption, resulting in the successful entrapment of the nutraceuticals. NEs protect functional properties of the bioactive compounds (for example, aromas, vitamins, food colorants, antioxidants, antimicrobial agents, etc.), by preventing their interaction with other food components and by providing protection against oxidation, hydrolysis, and harsh GI environment [75, 95, 109]. They also improve the bioavailability of lipophilic bioactive compounds by improving their aqueous solubility and rate of passive diffusion as well as by facilitating their direct uptake by intestinal lymphatic system. A number of studies have shown that the nanoemulsion encapsulation of the lipophilic nutraceuticals and food components enhances their bioavailability, bio-accessibility, and bioactivity, a few of the examples are shown in the Table 11.2.

Although NEs are the most commonly used delivery system for bioactive compounds, but they have some limitations as well, for example, they are thermodynamically unstable and have the tendency to destabilize over long period of time through a variety of destabilization mechanisms, such as Ostwald ripening and gravitational separation [76]. This destabilization can be avoided by using suitable stabilizers, for example ripening inhibitors, co-surfactants, texture modifiers and weighting agents, etc.

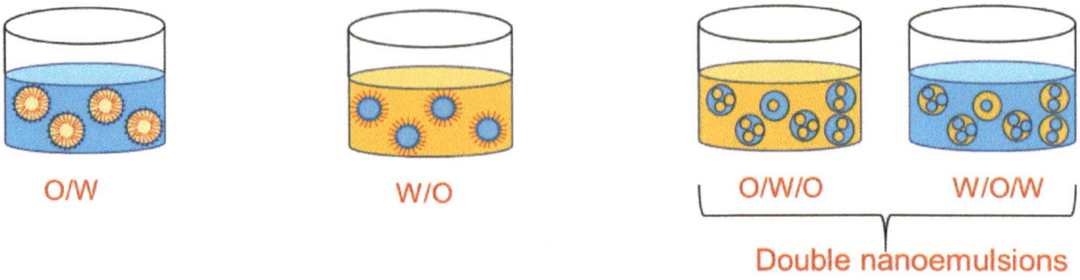

FIGURE 11.3 Types of nanoemulsions.

11.2.3.2.4 *MICROEMULSIONS*

The term "microemulsion" generally refers to thermodynamically stable isotropic liquids formed by mixing surfactants, oil, and water together [76]. Microemulsions are sometimes mistakenly considered as the conventional emulsions because of the term "micro," but these are two completely different systems. Firstly, microemulsions are thermodynamically stable, whereas both the conventional emulsions as well as NEs are thermodynamically unstable systems and secondly, the average droplet diameter for microemulsions is very small (< 100 nm) as compare to the conventional emulsions, which have much bigger droplets (in μm range) and lastly, microemulsions are kinetically unstable whereas both the conventional emulsions and NEs are kinetically stable systems. Microemulsions can be classified into O/W and water-in-oil types. O/W microemulsions are thermodynamically stable colloidal dispersions of small spheroid particles (composed of surfactant, oil, and sometimes co-surfactant as well) in an aqueous medium. In o/w microemulsions systems, the non-polar tails of the surfactant molecules form the hydrophobic core by associating with each other, which minimizes the thermodynamically unfavorable contact area between aqueous phase and the non-polar groups, whereas the hydrophilic/polar head groups protrude into the aqueous phase. Oil molecules usually get incorporated in-between the non-polar tails or as a separate core into the hydrophobic interior of a micelle. For the formation of microemulsion the concentration of the surfactant, in the dispersion medium, is very important. As soon as the concentration of certain types of the polymer or amphiphilic surfactant exceeds a fixed limit (called critical micelle concentration or CMC) the microemulsions are formed spontaneously. Like other emulsion-based systems, natural surfactants are preferred to prepare microemulsions (e.g., lecithin, and lecithin derivatives, glycerol fatty acids esters, etc.), [87] because synthetic surfactants have the tendency to cause toxicity. Microemulsions have some advantages over other emulsion-based systems used for the encapsulation of bioactive compounds, for example, they can be prepared by very simple processing operations (like blending or stirring), thus easy to scale up for industrial production, have long shelf-life because of their thermodynamic stability and they are optically transparent, so can be incorporated into clear food products intended for oral consumption. The delivery systems based on microemulsions have high potential for the solubilization as well as for transport of hydrophobic bioactive compounds having weak absorption and provide protection against chemical oxidation and phase separation of the food components during digestion. Like other emulsion-based systems, both the lipophilic as well as hydrophilic bioactive compounds can be encapsulated in the microemulsions; few examples are mentioned in the Table 11.2. Along with numerous advantages, microemulsions have some disadvantages as well, such as, they have a relatively low loading capacity because of very small hydrophobic interiors, usually require higher levels of surfactants for their formation (20% or even higher) [110].

11.2.3.2.5 SOLID LIPID NANOPARTICLES (SLNS) AND NANOSTRUCTURED LIPID CARRIERS (NLCS)

Solid lipid nanoparticles (SLNs) were first introduced as drug carrier systems in early 1990s, however, over the years they have found numerous applications in food industry as well (Figure 11.4). SLNs have properties of the lipid-based systems (emulsions and liposomes) as well as those of polymeric nanoparticle systems (PNPs). The average particle size of SLNs ranges from 40 to 1000 nm [111]. SLNs are developed by substituting the oil phase (liquid lipid) of an o/w emulsion by a solid lipid. The commonly used lipids are glycerides, fatty acids, waxes, paraffin, and triacylglycerol. The method of development involves preparation of a NEs at high temperature (i.e., above the melting point of the lipid used), followed by rapid drop of the temperature of the system which induces lipid crystallization. The system is usually stabilized by an emulsifying agent (e.g., poloxamers, polysorbates, sodium glycolate, etc.). SLNs successfully overcome the limitations associated with the emulsions, liposomes, and polymeric nanoparticles (NPs) by, providing protection to the encapsulated bioactive molecules against degradation, retention of the bioactive molecules, targeted, and/or controlled/sustained release of bioactive compounds, biocompatibility, biodegradability, high stability and lastly the possibility of an affordable and easy to scale-up preparation methods [111, 112]. However, there are some limitations associated with SLNs as well, which include, low free space for the entrapment of bioactive molecules, aggregation of particles, high aqueous content of the particle suspension and lastly, expulsion of bioactive components after liquid-to-solid phase transition during storage. In the late 1990s, nanostructured lipid carriers (NLCs) were developed to overcome these limitations. The average size of the NLCs (also known as second generation SLNs) ranges from 100 to 500 nm and are fabricated using blends of liquid lipids and solid lipids (preferably in a ratio of 30:70) by the same two step method as described for SLNs [111, 112]. Contrary to SLNs, NLCs have improved stability, lower aqueous content of the particle suspension and show higher loading capacity and no/minimized leakage of bioactive compounds during storage because of higher available space for bioactive molecules as well as due to the presence of many amorphous crystals within their structure. The rest of the properties and applications are the same as those of SLNs. SLNs as and NLCs have been widely used as delivery system for both hydrophilic and lipophilic bioactive compounds, to enhance their solubility, absorption, bioavailability, stability, and activity few examples are mentioned in Table 11.2.

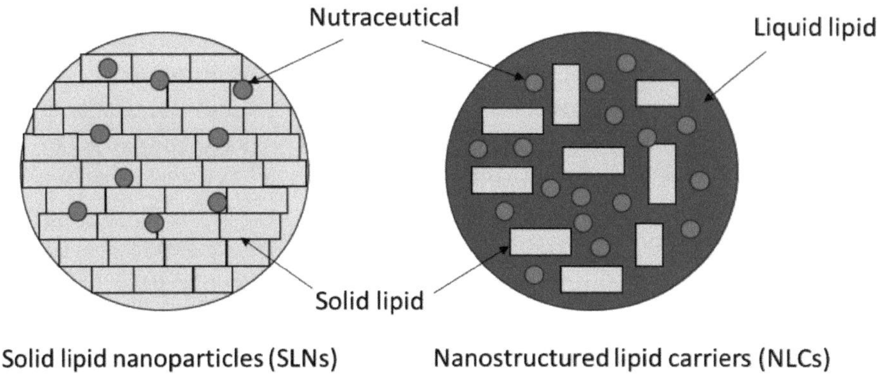

FIGURE 11.4 Structure of SLNs and NLCs.

11.2.3.2.6 BIOPOLYMER MICROGELS AND NANOGELS

Biopolymer microgels and nanogels consist of very small particles having networks of cross-linked biopolymer molecules inside, (usually food grade polysaccharides or proteins). Microgels have average particle size ranging from 100 nm to 1000 μm, whereas nanogels have very small particles <100 nm. Both the hydrophilic as well as hydrophobic nutraceuticals can be entrapped in these delivery systems. The hydrophilic nutraceuticals can be entrapped either before gel preparation, by mixing them with the biopolymer in the gelling solution or can be loaded after gel preparation. The hydrophobic nutraceuticals are first dissolved in a suitable oil phase, followed by the formation of an o/w emulsion, and then these nutraceuticals loaded oil droplets are blended with biopolymer solution, before the fabrication of gels. These gel systems can be produced by different methods, such as phase separation, extrusion, templating methods and anti-solvent precipitation [106, 113]. The functional attributes of these gel systems can be tailored by varying their dimensions, compositions, morphologies, surface characteristics and pore-sizes, to achieve the desired applications. These delivery systems have numerous advantages, such as, biodegradability, biocompatibility, protect the loaded nutraceuticals and provide targeted or triggered release of the loaded bioactive compounds. Few examples of these delivery systems, used for the encapsulation, protection, and release of nutraceuticals, are mentioned in Table 11.2.

11.2.3.2.7 POLYELECTROLYTE COMPLEXES (PECS)

Polyelectrolyte complexes (PECs) are biopolymer-based delivery systems, which are formed by an electrostatic interaction between two oppositely charged biopolymers (e.g., protamine, and carrageenan). The nutraceuticals are entrapped in the PECs during their formation, i.e., the nutraceuticals are first solubilized either in negatively or positively charged biopolymer followed by the addition of the oppositely charged biopolymer and then they are mixed together to from nutraceutical entrapped PECs. The food biopolymers (such as, starch, amylose, pectin, chitosan, etc.), are an abundant source for the development of PECs. PECs are a very promising delivery system for nutraceuticals because of their biocompatibility, biodegradability, ease of manufacturing, loading capacity and tendency to improve the bioavailability of nutraceuticals, for example, Resveratrol loaded Gelatin PEC showed improved bioavailability and anti-proliferative efficacy than free resveratrol after intravenous administration in mice [107]. PECs are not very well exploited as nutraceutical delivery systems so far, but research interest in PECs shows that they are one of the most important emerging delivery systems for nutraceuticals.

11.2.4 FUTURE TRENDS IN NUTRACEUTICALS AND FUNCTIONAL FOODS

Since the last few decades, functional foods or nutraceuticals appear as an emerging food category defined as novel term gaining increasing interest to provide medical or health benefits such as the prevention and treatments of diseases. Among these functional foods or nutraceuticals, applications allowing to improve the well-being can be found from certain groups of foods or "nutrients" containing bioactive phyto-constituents of potential health benefits [114, 115]. Among the different examples, we can cite carotenoids in most fruits and vegetables [116] allicin in garlic [117] phytosterols in the non-saponifiable fraction of plant oils [118] and certainly the most used polyphenols (e.g., anthocyanins, flavonoids, isoflavonoids, stilbenoids, ellagic acid, etc.), for treatment of various diseases related to the blood circulation system and improvement of the performance of different organs of the body [119–121]. The bioactive molecules are susceptible to degradation in the GI environment and/or in the external environment such as topical applications on skin.

More recently, it has been shown that these bioactive compounds can be more optimized in using as micro- or nanocarriers. Therefore, the great potential of micro- or nanoscale delivery systems in the nutraceutical and functional food industry has rapidly established, especially for the encapsulation, protection, and delivery of nutraceuticals [122]. As described above, various micro- or nanoscale delivery systems (micro- and NEs, liposomes, proteins, and biopolymers, nanostructured, and solid lipid nanocarriers, and nanosuspension) have been designed for these applications including their advantages and disadvantages for the preparation and characterization for industrial applications.

Just like pharmaceutical industry, nanotechnology plays a major role in the food sector through the food production of these carriers intended for advanced food processing, the manufacture of smart packaging, and providing long-term storage as well as improving quality of foods by improving their flavor and texture. Nanomaterials and nanosensors help consumers to provide information about the condition of the food itself and its nutritional status with enhanced security thanks to the detection of pathogens. In addition, they improve solubilization of hydrophobic food bioactive substances against various diseases, therefore improving their lower bioavailability and their stability. Nevertheless, the design of foods based on nanotechnology lead to major challenges for both the legislator and the industry, in order to ensure consumer confidence and acceptance of nanofoods available on the market. The active use of nano colloidal particles in different branches of the food industry must necessarily lead to an improvement in quality, safety, nutrition. In addition, their uses in packaging have been widely reported recently as well. Currently, NPs are manufactured all over the world. However, very few countries have standard regulatory rules for the use of nanotechnology in food products. Therefore, insufficient scientific research of nanosystems for food can create difficulties in reaching conclusions about their effectiveness. Among the applications of NPs, it is obvious that their industrial applications in food packaging will present less harmful effects than the use of these same NPs as food ingredients. In question, there are still the risk of an interaction by the food chain with DNA, a disruption of the cell membrane and cell death. So far, very few *in vivo* studies have been conducted on the effects of nanofoods in human and animal health. This is the reason why there should be appropriate labeling and the establishment of recommended regulations for marketing nanofoods. These studies and this legislation should help to increase consumer acceptability. Therefore, the use of these nanotechnology, if managed and regulated properly, should play an important role in improving food processing and ensuring food products to have a better quality of effects beneficial to human and animal health [123].

The purpose of packaging food, sensitive to degradation (such as meat and similar products), is to suppress or slow down deterioration and allow the food to retain its original characteristics and the color red. It is the same for fruits and vegetables, so that they can keep their original properties. The use of nanosensors makes it possible to alert the consumer following an alteration or contamination of food by detection of toxins generated by degradation. The same is true for the detection of pesticides and microbial contamination in food products, based on the production of flavor and color formation. Most of the NPs used for smart packaging design in the food industry have potential antimicrobial activity. Their mechanisms of action are based on their properties of antimicrobial polypeptide carriers and of being able to provide protection against microbial spoilage. From a practical standpoint, future trends will include packaging materials made using a coating of starch colloids containing the antimicrobial agent with an objective to act as a barrier against microorganisms by the release of antimicrobials from packaged material.

NPs are used as containers to encapsulate enzymes, antioxidants, anti-browning agents, flavors, and other ingredients to improve shelf-life even after opening the package. Metals and metal oxides formulated as inorganic nanocontainers such as iron, silver, zinc oxides, carbon, oxides of magnesium, oxides of titanium, and silicon dioxide NPs, are widely used as antimicrobials and under certain conditions as food ingredients. TiO_2 has the property of producing reactive oxygen species (ROS). These have the capacity to harm microorganisms therefore allowing an effective antimicrobial role to be played.

Nanocontainers are commonly used for packaging design, most often for coating purposes as well. Many other NPs like SiO_2, clay silicate, carbon nanotubes, graphene or even natural or semi-synthetic organic substances such as chitosan or chitin can also be used. In the latter case, they will be formulated in the form of nanofibers based on cellulose and other inorganic substances. The formulator will take care to add polymers to make the polymer matrix lighter and ensure low gas permeability.

Future trends in improving the preservation of nutraceuticals and functionalized foods will support the inclusion of charged nanocontainers in the polymer matrix to increase the performance of food packaging. These nanocontainers will provide functional attributes such as antioxidants, antimicrobial, and toxin scavenger that will allow a longer shelf life of packaged food products.

11.3 NUTRACEUTICALS AND FUNCTIONAL FOODS PRODUCTS IN THE CLINICAL TRIALS AND IN THE MARKET

In addition to the feasibility of industrial scale-up, the transposition of formulations is not only challenged by technological issues and strategies but also by regulatory affairs. In the field of drugs and food in general, this can be easily managed. However, in the field of nutraceuticals and functional foods it remains more complicated because it is a very recent field for which the legislation only partially exists. In fact, carrying out preclinical studies turns out to be very complex in certain cases, as the models and clinical parameters have in certain cases not been defined beforehand. At the same time, paradoxically, translational medicine operates in the opposite direction, integrating clinical needs and observations with scientific hypothesis and innovative technological proposals. The real difficulty lies mainly in proving the efficacy of the extracted biomolecules of marine, plant, or inorganic origin without however being considered as drugs. Moreover, the optimization and interest to be formulated in micro-or nanoscale must be proven. Consequently, the innovation does not lie only in the definition of new biomolecules but also in the implementation of clinical studies which can serve as a support.

As reported in a plethora of *in vitro* and *in vivo* studies, undoubtedly, bioactive molecules such as polyphenols, flavonoids, lipids, enzymes, peptides, etc., exert numerous biological activities. In the literature, several systematic reviews and meta-analyses of observational and intervention studies describe a reduced risk for numerous chronic diseases. As it was reported about 10 years ago, now at present as well, only few, and conflicting results are available for total polyphenols. Therefore, it remains difficult to establish a standard or a reference and/or prudent intake of total polyphenols, although an approximate mean intake of about 900 mg/day is given in the literature. On the contrary, some studies suggest an inverse association between high total flavonoid intake (generally higher 500 mg/day) and cardiovascular events and/or mortality. Meanwhile, the values published in the scientific literature should be considered as an approximative level due to the heterogeneity of the experimental or clinical studies and the numerous limitations associated with the evaluation and estimation of polyphenols or flavonoids intake. In this context, it is important to consider that polyphenols intake can be modified if these ones are administered as a diverse food sources or formulated into micro- or nanocarriers. For this reason, it is better to argue in terms of dietary patterns or formulated at micro- or nanoscale and then focusing on single contributions. Therefore, a distinction must be established if the biomolecules as polyphenol-rich dietary pattern are administered regularly to exert health benefits and therefore should be considered as a valid tool for the prevention of numerous chronic diseases.

Similar conclusions can be drawn from studies including other biomolecules and other routes of administrations. For example, one can cite the clinical trials lead by Abd-Allah [124] on chitosan NPs on their topical use for acne treatment and by Bonfigli [125] on effects of a novel nutraceutical combination (BruMeChol™) in subjects with mild hypercholesterolemia or alternative therapeutic options to antibiotics for the treatment of urinary tract infections [126]. Pellegrini [127] studied the role of nutrition and nutritional supplements in ocular surface diseases whereas Talebi [128] studied the beneficial effects of nutraceuticals and natural products on small dense LDL levels, LDL particle number and LDL particle size and Bumrungpert [129] determined the clinical conditions to evaluate how the nutraceuticals can improve the glycemic control, insulin sensitivity, and oxidative stress in hyperglycemic subjects by using a randomized, double-blind, placebo-controlled clinical trial.

Further investigation is highly recommended in order to optimize the use of biomolecules such as: (1) evaluation of the diet in terms of the need for the human or animal organism; (2) establishment of standardized and validated analytical procedures for the analysis of biomolecules and their related subclasses in foods; (3) the establishment of databases of biomolecules contained in food products and the interest of formulating them at the micro- or nanoscale; (4) validation of specific biomarkers to assess the added value of particular biomolecule forums. Although it is obvious that the data from observational studies are important to identify the potential intake of food compounds, well-controlled dietary studies such as in the field of drugs by setting up control and specifically targeted groups (determination of dose-response effects) are essential to identify the reference biomolecule (s). It is also necessary to assess the justified or unjustified contribution to health for bioactive molecules from food such as polyphenols, flavonoids, lipids, peptides, etc., intended for the general population or for particular, more vulnerable groups such as infants, children or even the elderly.

At present, several nanofoods products are commercialized, as well as nanomaterial-based biosensors are used in food science and food nanotechnology. Different types of nanoformulations are already applied in food industry. Shivraj [123] reviewed systematically and completely all these marketed systems and those which are still in development for industrial applications.

KEYWORDS

- **encapsulation efficiency**
- **glucagon-like peptide-1**
- **large unilamellar vesicles**
- **nanoemulsions**
- **oil-in-water-in-oil**
- **peroxisome proliferator-activated receptors**
- **small unilamellar vesicles**
- **water-in-oil-in-water**

REFERENCES

1. Hippocrates, et al., (1923). Heraclitus of Ephesus. London: Heinemann; New York: Putnam, Cambridge, Mass.: Harvard University Press, 1923–1931.
2. Kalra, E. K., (2003). Nutraceutical-definition and introduction. *AAPS PharmSci., 5*(3), E25.
3. Santini, A., et al., (2018). Nutraceuticals: Opening the debate for a regulatory framework. *British Journal of Clinical Pharmacology, 84*(4), 659–672.
4. Hardy, G., (2000). Nutraceuticals and functional foods: Introduction and meaning. *Nutrition, 16*(7, 8), 688, 689.
5. Jain, N., & Ramawat, K. G., (2013). Nutraceuticals and antioxidants in prevention of diseases. In: Ramawat, K. G., & Mérillon, J. M., (eds.), *Natural Products: Phytochemistry, Botany and Metabolism of Alkaloids, Phenolics and Terpenes* (pp. 2559–2580). Springer Berlin Heidelberg: Berlin, Heidelberg.
6. Ruchi, S., (2017). Role of nutraceuticals in health care: A review. *International Journal of Green Pharmacy (IJGP), 11*(03).
7. Wang, J., et al., (2016). Microbial production of value-added nutraceuticals. *Current Opinion in Biotechnology, 37*, 97–104.
8. Williamson, E. M., Liu, X., & Izzo, A. A., (2020). Trends in use, pharmacology, and clinical applications of emerging herbal nutraceuticals. *British Journal of Pharmacology, 177*(6), 1227–1240.
9. Rahul Dev, Sunil Kumar, Jagbir Singh, & Bhupendra Chauhan. (2011). Potential role of nutraceuticals in present scenerio: A review. *Journal of Applied Pharmaceutical Science 1*(4), 26–28.
10. Ben-Hassan, R., & Ghaly, A., (1994). Continuous propagation of *Kluyveromyces* fragilis in cheese whey for pollution potential reduction. *Applied Biochemistry and Biotechnology, 47*(1), 89–105.
11. Belem, M. A. F., & Lee, B. H., (1998). Production of bioingredients from *Kluyveromyces marxianus* grown on whey: An alternative. *Critical Reviews in Food Science and Nutrition, 38*(7), 565–598.
12. Belem, M. A. F., (1999). Application of biotechnology in the product development of nutraceuticals in Canada. *Trends in Food Science & Technology, 10*(3), 101–106.
13. Tahereh, F., Hanieh, S. Y., & Saeed, S., (2019). The protective effects of green tea catechins in the management of neurodegenerative diseases: A review. *Current Drug Discovery Technologies, 16*(1), 57–65.
14. Adefegha, S. A., (2018). Functional foods and nutraceuticals as dietary intervention in chronic diseases; novel perspectives for health promotion and disease prevention. *Journal of Dietary Supplements, 15*(6), 977–1009.
15. Agnihotri, A., & Aruoma, O. I., (2020). Alzheimer's disease and Parkinson's disease: A nutritional toxicology perspective of the impact of oxidative stress, mitochondrial dysfunction, nutrigenomics and environmental chemicals. *Journal of the American College of Nutrition, 39*(1), 16–27.
16. Wimo, A., et al., (2013). The worldwide economic impact of dementia 2010. *Alzheimer's & Dementia, 9*(1), 1–11. e3.
17. Qiu, C., De Ronchi, D., & Fratiglioni, L., (2007). The epidemiology of the dementias: An update. *Current Opinion in Psychiatry, 20*(4), 380–385.

18. Venkatakrishnan, K., Chiu, H. F., & Wang, C. K., (2019). Extensive review of popular functional foods and nutraceuticals against obesity and its related complications with a special focus on randomized clinical trials. *Food & Function, 10*(5), 2313–2329.
19. Mantzorou, M., et al., (2018). Effects of curcumin consumption on human chronic diseases: A narrative review of the most recent clinical data. *Phytotherapy Research, 32*(6), 957–975.
20. Perrone, L., et al., (2019). The autophagy signaling pathway: A potential multifunctional therapeutic target of curcumin in neurological and neuromuscular diseases. *Nutrients, 11*(8), 1881.
21. Tiwari, S. K., et al., (2014). Curcumin-loaded nanoparticles potently induce adult neurogenesis and reverse cognitive deficits in Alzheimer's disease model via canonical Wnt/β-catenin pathway. *ACS Nano, 8*(1), 76–103.
22. Teter, B., et al., (2019). Curcumin restores innate immune Alzheimer's disease risk gene expression to ameliorate Alzheimer pathogenesis. *Neurobiology of Disease, 127*, 432–448.
23. Kulkarni, S. K., Bhutani, M. K., & Bishnoi, M., (2008). Antidepressant activity of curcumin: Involvement of serotonin and dopamine system. *Psychopharmacology, 201*(3), 435.
24. Gutierrez-Mariscal, F. M., et al., (2019). Coenzyme Q10: From bench to clinic in aging diseases, a translational review. *Critical Reviews in Food Science and Nutrition, 59*(14), 2240–2257.
25. Spindler, M., Beal, M. F., & Henchcliffe, C., (2009). Coenzyme Q10 effects in neurodegenerative disease. *Neuropsychiatric Disease and Treatment, 5*, 597.
26. Yang, X., et al., (2016). Neuroprotection of coenzyme Q10 in neurodegenerative diseases. *Current Topics in Medicinal Chemistry, 16*(8), 858–866.
27. Shults, C. W., et al., (1997). Coenzyme Q10 levels correlate with the activities of complexes I and II/III in mitochondria from parkinsonian and nonparkinsonian subjects. *Annals of Neurology: Official Journal of the American Neurological Association and the Child Neurology Society*, *42*(2), 261–264.
28. Mehrpooya, M., et al., (2018). Evaluating the effect of coenzyme q10 augmentation on treatment of bipolar depression: A double-blind controlled clinical trial. *Journal of Clinical Psychopharmacology, 38*(5).
29. Mohammad, H. P., et al., (2019). The effect of resveratrol on neurodegenerative disorders: Possible protective actions against autophagy, apoptosis, inflammation and oxidative stress. *Current Pharmaceutical Design, 25*(19), 2178–2191.
30. Berman, A. Y., et al., (2017). The therapeutic potential of resveratrol: A review of clinical trials. *NPJ Precision Oncology, 1*(1), 1–9.
31. Moussa, C., et al., (2017). Resveratrol regulates neuro-inflammation and induces adaptive immunity in Alzheimer's disease. *Journal of Neuroinflammation, 14*(1), 1.
32. Hui, Y., et al., (2018). Resveratrol attenuates the cytotoxicity induced by amyloid-β1-42 in PC12 cells by upregulating heme oxygenase-1 via the PI3K/Akt/Nrf2 pathway. *Neurochemical Research, 43*(2), 297–305.
33. Coimbra, M., et al., (2011). Improving solubility and chemical stability of natural compounds for medicinal use by incorporation into liposomes. *International Journal of Pharmaceutics, 416*(2), 433–442.

34. Witte, A. V., et al., (2014). Effects of resveratrol on memory performance, hippocampal functional connectivity, and glucose metabolism in healthy older adults. *Journal of Neuroscience, 34*(23), 7862–7870.
35. Eckert, G. P., Lipka, U., & Muller, W. E., (2013). Omega-3 fatty acids in neurodegenerative diseases: Focus on mitochondria. *Prostaglandins, Leukotrienes and Essential Fatty Acids, 88*(1), 105–114.
36. Dong, Y., et al., (2018). Dietary eicosapentaenoic acid normalizes hippocampal omega-3 and 6 polyunsaturated fatty acid profile, attenuates glial activation and regulates BDNF function in a rodent model of neuroinflammation induced by central interleukin-1β administration. *European Journal of Nutrition, 57*(5), 1781–1791.
37. Avallone, R., Vitale, G., & Bertolotti, M., (2019). Omega-3 fatty acids and neurodegenerative diseases: New evidence in clinical trials. *International Journal of Molecular Sciences, 20*(17), 4256.
38. Morris, M. C., et al., (2003). Consumption of fish and n-3 fatty acids and risk of incident Alzheimer disease. *Archives of Neurology, 60*(7), 940–946.
39. Wu, S., et al., (2015). Omega-3 fatty acids intake and risks of dementia and Alzheimer's disease: A meta-analysis. *Neuroscience & Biobehavioral Reviews, 48*, 1–9.
40. Beverley, B., & Eschwège, E. (2003). The Diagnosis and Classification of Diabetes and Impaired Glucose Tolerance. In: Pickup, J. C. & Williams, G., (eds.), *Textbook of Diabetes,* 3rd Edition, Blackwell Publishing, UK, 2.1–2.11.
41. Ahmed, A. M., (2002). History of diabetes mellitus. *Saudi Medical Journal, 23*(4), 373–378.
42. Marra, M. V., & Boyar, A. P., (2009). Position of the American dietetic association: Nutrient supplementation. *Journal of the American Dietetic Association, 109*(12), 2073–2085.
43. Elliott, D. B., (2012). Systematic reviews of optometric interventions. *Ophthalmic and Physiological Optics, 32*(3), 173.
44. Riccioni, G., et al., (2007). Antioxidant vitamin supplementation in cardiovascular diseases. *Annals of Clinical & Laboratory Science, 37*(1), 89–95.
45. Davì, G., & Patrono, C., (2007). Platelet activation and atherothrombosis. *New England Journal of Medicine, 357*(24), 2482–2494.
46. Group, E. S. S., (1999). Vitamin D supplement in early childhood and risk for Type I (insulin-dependent) diabetes mellitus. *Diabetologia, 42*(1), 51–54.
47. Danescu, L. G., Levy, S., & Levy, J., (2009). Vitamin D and diabetes mellitus. *Endocrine, 35*(1), 11–17.
48. Ziegler, D., et al., (1999). α-Lipoic acid in the treatment of diabetic polyneuropathy in Germany: Current evidence from clinical trials. *Experimental and Clinical Endocrinology & Diabetes, 107*(07), 421–430.
49. Bo, S., & Pisu, E., (2008). Role of dietary magnesium in cardiovascular disease prevention, insulin sensitivity and diabetes. *Current Opinion in Lipidology, 19*(1), 50–56.
50. Evans, J., & Henshaw, K., (2000). Antioxidant vitamin and mineral supplementation for preventing age-related macular degeneration. *The Cochrane Database of Systematic Reviews*, (2), CD000253-CD000253.
51. Eichler, H., et al., (1984). The effect of a new specific α-amylase inhibitor on post-prandial glucose and insulin excursions in normal subjects and type 2 (non-insulin-dependent) diabetic patients. *Diabetologia, 26*(4), 278–281.

52. Tarling, C. A., et al., (2008). The search for novel human pancreatic α-amylase inhibitors: High-throughput screening of terrestrial and marine natural product extracts. *ChemBioChem, 9*(3), 433–438.
53. Ponnusamy, S., et al., (2015). Gedunin and azadiradione: Human pancreatic alpha-amylase inhibiting limonoids from neem (*Azadirachta indica*) as antidiabetic agents. *PloS One, 10*(10), e0140113.
54. Wang, H., Du, Y. J., & Song, H. C., (2010). α-Glucosidase and α-amylase inhibitory activities of guava leaves. *Food Chemistry, 123*(1), 6–13.
55. Lordan, S., et al., (2013). The α-amylase and α-glucosidase inhibitory effects of Irish seaweed extracts. *Food Chemistry, 141*(3), 2170–2176.
56. Wulan, D. R., Utomo, E. P., & Mahdi, C., (2015). Antidiabetic activity of *Ruellia tuberosa* L., role of α-amylase inhibitor: In silico, *in vitro*, and *in vivo* approaches. *Biochemistry Research International, 2015.*
57. Geng, Y., et al., (2013). Bioassay-guided isolation of DPP-4 inhibitory fractions from extracts of submerged cultured of *Inonotus obliquus*. *Molecules, 18*(1), 1150–1161.
58. Gurudeeban, S., et al., (2012). Antidiabetic effect of a black mangrove species *Aegiceras corniculatum* in alloxan-induced diabetic rats. *Journal of Advanced Pharmaceutical Technology & Research, 3*(1), 52–56.
59. Purnomo, Y., et al., (2015). Antidiabetic potential of Urena lobata leaf extract through inhibition of dipeptidyl peptidase IV activity. *Asian Pacific Journal of Tropical Biomedicine, 5*(8), 645–649.
60. Wang, Z., et al., (2009). Docking and molecular dynamics studies toward the binding of new natural phenolic marine inhibitors and aldose reductase. *Journal of Molecular Graphics and Modelling, 28*(2), 162–169.
61. Shin, D. W., et al., (2009). (−)-Catechin promotes adipocyte differentiation in human bone marrow mesenchymal stem cells through PPARγ transactivation. *Biochemical Pharmacology, 77*(1), 125–133.
62. James, J., Simpson, B. K., & Marshall, M. R., (1996). Application of enzymes in food processing. *Critical Reviews in Food Science and Nutrition, 36*(5), 437–463.
63. Katz, F., (1999). Top product development trends in Europe. *Food Technology (Chicago), 53*(1), 38–42.
64. Smith, B. L., Marcotte, M., & Harrison, G., (1997). A comparative analysis of the regulatory framework affecting functional food development and commercialization in Canada, Japan, the European union and the United States of America. *Journal of Nutraceuticals, Functional & Medical Foods, 1*(2), 45–87.
65. Sanguansri, L., Oliver, C. M., & Leal-Calderon, F., (2013). Nanoemulsion technology for delivery of nutraceuticals and functional-food ingredients. *Bio-Nanotechnology: A Revolution in Food, Biomedical and Health Sciences*, 667–696.
66. Pinnamaneni, S., Das, N. G., & Das, S. K., (2003). Comparison of oil-in-water emulsions manufactured by micro fluidization and homogenization. *Pharmazie, 58*(8), 554–558.
67. Mahdi, J. S., He, Y., & Bhandari, B., (2006). Nano-emulsion production by sonication and microfluidization: A comparison. *International Journal of Food Properties, 9*(3), 475–485.
68. Jafari, S. M., et al., (2008). Nano-particle encapsulation of fish oil by spray drying. *Food Research International (Ottawa, Ont.), 41*(2), 172–183.
69. Cheong, J. N., et al., (2008). α-Tocopherol nanodispersions: Preparation, characterization and stability evaluation. *Journal of Food Engineering, 89*(2), 204–209.

70. Anton, N., & Vandamme, T. F., (2009). The universality of low-energy nano-emulsification. *International Journal of Pharmaceutics, 377*(1, 2), 142–147.
71. Anton, N., Benoit, J. P., & Saulnier, P., (2008). Design and production of nanoparticles formulated from nano-emulsion templates: A review. *Journal of Controlled Release, 128*(3), 185–199.
72. Relkin, P., Jung, J. M., & Ollivon, M., (2009). Factors affecting vitamin degradation in oil-in-water nano-emulsions. *Journal of Thermal Analysis and Calorimetry, 98*(1), 13.
73. Cornacchia, L., & Roos, Y. H., (2011). Stability of β-carotene in protein-stabilized oil-in-water delivery systems. *Journal of Agricultural and Food Chemistry, 59*(13), 7013–7020.
74. Gonçalves, R. F. S., et al., (2018). Advances in nutraceutical delivery systems: From formulation design for bioavailability enhancement to efficacy and safety evaluation. *Trends in Food Science & Technology, 78*, 270–291.
75. McClements, D. J., et al., (2009). Structural design principles for delivery of bioactive components in nutraceuticals and functional foods. *Critical Reviews in Food Science and Nutrition, 49*(6), 577–606.
76. McClements, D. J., (2012). Requirements for food ingredient and nutraceutical delivery systems, In: *Encapsulation Technologies and Delivery Systems for Food Ingredients and Nutraceuticals* (pp. 3–18). Elsevier.
77. Yao, M., Xiao, H., & McClements, D. J., (2014). Delivery of lipophilic bioactives: Assembly, disassembly, and reassembly of lipid nanoparticles. *Annual Review of Food Science and Technology, 5*(1), 53–81.
78. Stringham, J. M., & Hammond, B. R., (2005). Dietary lutein and zeaxanthin: Possible effects on visual function. *Nutrition Reviews, 63*(2), 59–64.
79. Korhonen, H., Marnila, P., & Gill, H. S., (2000). Milk immunoglobulins and complement factors. *British Journal of Nutrition, 84*(S1), 75–80.
80. Park, Y. W., & Nam, M. S., (2015). Bioactive peptides in milk and dairy products: A review. *Korean Journal for Food Science of Animal Resources, 35*(6), 831–840.
81. Hallikainen, M. A., Sarkkinen, E. S., & Uusitupa, M. I. J., (2000). Plant stanol esters affect serum cholesterol concentrations of hypercholesterolemic men and women in a dose-dependent manner. *The Journal of Nutrition, 130*(4), 767–776.
82. White, P., & Broadley, M., (2005). Biofortifying crops with essential mineral elements. *Trends in Plant Science, 10*(12), 586–593.
83. Hibbeln, J. R., et al., (2006). Healthy intakes of n–3 and n–6 fatty acids: Estimations considering worldwide diversity. *The American Journal of Clinical Nutrition, 83*(6), 1483S–1493S.
84. Ting, Y., et al., (2014). Common delivery systems for enhancing *in vivo* bioavailability and biological efficacy of nutraceuticals. *Journal of Functional Foods, 7*, 112–128.
85. Trevaskis, N. L., Charman, W. N., & Porter, C. J. H., (2008). Lipid-based delivery systems and intestinal lymphatic drug transport: A mechanistic update. *Advanced Drug Delivery Reviews, 60*(6), 702–716.
86. Bangham, A. D., Standish, M. M., & Watkins, J. C., (1965). Diffusion of univalent ions across the lamellae of swollen phospholipids. *Journal of Molecular Biology, 13*(1), 238-IN27.
87. Dima, Ş., Dima, C., & Iordăchescu, G., (2015). Encapsulation of functional lipophilic food and drug biocomponents. *Food Engineering Reviews, 7*(4), 417–438.

88. Alexander, M., et al., (2012). Incorporation of phytosterols in soy phospholipids nanoliposomes: Encapsulation efficiency and stability. *LWT, 47*(2), 427–436.
89. Rehman, A. U., et al., (2018). Development of doxorubicin hydrochloride loaded pH-sensitive liposomes: Investigation on the impact of chemical nature of lipids and liposome composition on pH-sensitivity. *European Journal of Pharmaceutics and Biopharmaceutics, 133*, 331–338.
90. Huang, Y. B., et al., (2011). Elastic liposomes as carriers for oral delivery and the brain distribution of (+)-catechin. *Journal of Drug Targeting, 19*(8), 709–718.
91. Takahashi, M., et al., (2009). Evaluation of an oral carrier system in rats: Bioavailability and antioxidant properties of liposome-encapsulated curcumin. *Journal of Agricultural and Food Chemistry, 57*(19), 9141–9146.
92. Livney, Y. D., (2015). Nanostructured delivery systems in food: Latest developments and potential future directions. *Current Opinion in Food Science, 3*, 125–135.
93. Elsamaligy, M., Afifi, N., & Mahmoud, E., (2006). Evaluation of hybrid liposomes-encapsulated silymarin regarding physical stability and *in vivo* performance. *International Journal of Pharmaceutics, 319*(1, 2), 121–129.
94. Karadag, A., et al., (2013). Optimization of preparation conditions for quercetin nanoemulsions using response surface methodology. *Journal of Agricultural and Food Chemistry, 61*(9), 2130–2139.
95. Chay, S. Y., Tan, W. K., & Saari, N., (2015). Preparation and characterization of nanoliposomes containing winged bean seeds bioactive peptides. *Journal of Microencapsulation, 32*(5), 488–495.
96. Mayer, S., Weiss, J., & McClements, D. J., (2013). Vitamin E-enriched nanoemulsions formed by emulsion phase inversion: Factors influencing droplet size and stability. *Journal of Colloid and Interface Science, 402*, 122–130.
97. Dey, T. K., et al., (2012). Comparative study of gastrointestinal absorption of EPA & DHA rich fish oil from nano and conventional emulsion formulation in rats. *Food Research International, 49*(1), 72–79.
98. Sessa, M., et al., (2014). Bioavailability of encapsulated resveratrol into nanoemulsion-based delivery systems. *Food Chemistry, 147*, 42–50.
99. Gui, S. Y., et al., (2008). Preparation and evaluation of a microemulsion for oral delivery of berberine. *Die Pharmazie, 63*(7), 516–519.
100. Fisher, S., et al., (2013). Solubilization of simvastatin and phytosterols in a dilutable microemulsion system. *Colloids and Surfaces B: Biointerfaces, 107*, 35–42.
101. Yu, A., et al., (2011). Formulation optimization and bioavailability after oral and nasal administration in rabbits of Puerarin-loaded microemulsion. *Journal of Pharmaceutical Sciences, 100*(3), 933–941.
102. Mehrad, B., et al., (2018). Enhancing the physicochemical stability of β-carotene solid lipid nanoparticle (SLNP) using whey protein isolate. *Food Research International, 105*, 962–969.
103. Mei, Z., et al., (2005). The research on the anti-inflammatory activity and hepatotoxicity of triptolide-loaded solid lipid nanoparticle. *Pharmacological Research, 51*(4), 345–351.
104. Fathi, M., et al., (2013). Hesperetin-loaded solid lipid nanoparticles and nanostructure lipid carriers for food fortification: Preparation, characterization, and modeling. *Food and Bioprocess Technology, 6*(6), 1464–1475.

105. Teeranachaideekul, V., Muller, R., & Junyaprasert, V., (2007). Encapsulation of ascorbyl palmitate in nanostructured lipid carriers (NLC)—Effects of formulation parameters on physicochemical stability. *International Journal of Pharmaceutics, 340*(1, 2), 198–206.
106. Chen, F., et al., (2017). Inhibition of lipid oxidation in nanoemulsions and filled microgels fortified with omega-3 fatty acids using casein as a natural antioxidant. *Food Hydrocolloids, 63*, 240–248.
107. Karthikeyan, S., et al., (2013). Anticancer activity of resveratrol-loaded gelatin nanoparticles on NCI-H460 non-small cell lung cancer cells. *Biomedicine & Preventive Nutrition, 3*(1), 64–73.
108. Rehman, A. U., et al., (2019). Spontaneous nano-emulsification with tailor-made amphiphilic polymers and related monomers. *European Journal of Pharmaceutical Research, 1*(1), 27–36.
109. McClements, D. J., (2020). Advances in nanoparticle and microparticle delivery systems for increasing the dispersibility, stability, and bioactivity of phytochemicals. *Biotechnology Advances, 38*, 107287.
110. Akhavan, S., et al., (2018). Lipid nano scale cargos for the protection and delivery of food bioactive ingredients and nutraceuticals. *Trends in Food Science & Technology, 74*, 132–146.
111. Pardeike, J., Hommoss, A., & Müller, R. H., (2009). Lipid nanoparticles (SLN, NLC) in cosmetic and pharmaceutical dermal products. *International Journal of Pharmaceutics, 366*(1, 2), 170–184.
112. Muller, R., et al., (2007). Nanostructured lipid carriers (NLC) in cosmetic dermal products. *Advanced Drug Delivery Reviews, 59*(6), 522–530.
113. Matalanis, A., Jones, O. G., & McClements, D. J., (2011). Structured biopolymer-based delivery systems for encapsulation, protection, and release of lipophilic compounds. *Food Hydrocolloids, 25*(8), 1865–1880.
114. Gul, K., Singh, A., & Jabeen, R., (2016). Nutraceuticals and functional foods: The foods for the future world. *Critical Reviews in Food Science and Nutrition, 56*(16), 2617–2627.
115. Gutiérrez-del-Río, I., Fernández, J., & Lombó, F., (2018). Plant nutraceuticals as antimicrobial agents in food preservation: Terpenoids, polyphenols and thiols. *International Journal of Antimicrobial Agents, 52*(3), 309–315.
116. Eggersdorfer, M., & Wyss, A., (2018). Carotenoids in human nutrition and health. *Archives of Biochemistry and Biophysics, 652*, 18–26.
117. Touloupakis, E., & Ghanotakis, D. F., (2010). Nutraceutical use of garlic sulfur-containing compounds. In: *Bio-Farms for Nutraceuticals* (pp. 110–121). Springer.
118. Jones, P. J., & AbuMweis, S. S., (2009). Phytosterols as functional food ingredients: Linkages to cardiovascular disease and cancer. *Current Opinion in Clinical Nutrition & Metabolic Care, 12*(2), 147–151.
119. Del, B. C., et al., (2019). Systematic review on polyphenol intake and health outcomes: Is there sufficient evidence to define a health-promoting polyphenol-rich dietary pattern? *Nutrients, 11*(6), 1355.
120. Piccolella, S., et al., (2019). Nutraceutical polyphenols: New analytical challenges and opportunities. *Journal of Pharmaceutical and Biomedical Analysis, 175*, 112774.
121. Williamson, G., (2017). The role of polyphenols in modern nutrition. *Nutrition Bulletin, 42*(3), 226–235.

122. Chen, J., & Hu, L., (2020). Nanoscale delivery system for nutraceuticals: Preparation, application, characterization, safety, and future trends. *Food Engineering Reviews, 12*(1), 14–31.
123. Nile, S. H., et al., (2020). Nanotechnologies in food science: Applications, recent trends, and future perspectives. *Nano-Micro Letters, 12*(1), 45.
124. Abd-Allah, H., Abdel-Aziz, R. T., & Nasr, M., (2020). Chitosan nanoparticles making their way to clinical practice: A feasibility study on their topical use for acne treatment. *International Journal of Biological Macromolecules.*
125. Bonfigli, A. R., et al., (2020). Effects of a novel nutraceutical combination (BruMeChol™) in subjects with mild hypercholesterolemia: Study protocol of a randomized, double-blind, controlled trial. *Trials, 21*(1), 1–8.
126. Loubet, P., et al., (2020). Alternative therapeutic options to antibiotics for the treatment of urinary tract infections. *Frontiers in Microbiology, 11.*
127. Pellegrini, M., et al., (2020). The role of nutrition and nutritional supplements in ocular surface diseases. *Nutrients, 12*(4), 952.
128. Talebi, S., et al., (2020). The beneficial effects of nutraceuticals and natural products on small dense LDL levels, LDL particle number and LDL particle size: A clinical review. *Lipids in Health and Disease, 19*, 1–21.
129. Bumrungpert, A., et al., (2020). Nutraceutical improves glycemic control, insulin sensitivity, and oxidative stress in hyperglycemic subjects: A randomized, double-blind, placebo-controlled clinical trial. *Natural Product Communications, 15*(4), 1934578X20918687.

Index

A

Abnormal cells, 105, 106, 128
Active
  ingredients, 65–67, 87, 190, 284, 297, 316, 319
  transport, 46
Adhesion, 225, 240, 241, 243, 250, 265
Adipose tissue metabolism, 279
Adulteration, 79, 80, 82, 83, 87–90, 92, 94–98
  criminally motivated adulteration, 89
  economically motivated adulteration, 88
  undeclared labeling, 90
  unintentional adulteration, 89
Agro-food, 14
Agro-industrial waste, 11
Albumin, 35, 36, 73, 113, 143, 186, 264, 297
Alkaloids, 13, 17–19
Alpha-carotene, 135
Alpha-linoleic acid (ALA), 131
Alzheimer's disease, 14, 149, 150, 279, 296, 313
American
  cancer society, 106
  dietetic association (ADA), 80
Amino acids, 17, 19, 36, 45, 84–86, 126, 138, 140, 141, 146, 151, 178–180, 183–188, 190, 242, 277
  leucine, 140
  L-glutamine, 138
Amphiphilic molecules, 114, 325
Angiogenesis, 107, 129
Angiotensin
  converting enzyme (ACE), 144, 146, 147, 176, 182, 184–186, 314
  I-enzyme (ACE), 176, 191
Anthocyanidins, 32
Antibiotics anti-inflammatory drugs, 36
Anticancer, 9, 10, 17, 19, 20, 22, 23, 31, 36, 38, 73, 107, 108, 116, 127, 135, 136, 149, 178, 185, 188, 314, 321, 322
Anticarcinogenic activities, 8
Antifertility, 73
Antihypertensive
  activity, 176, 184, 185, 189
  effects, 137
Anti-inflammatory
  activities, 20, 132, 176, 179, 184–186, 291
  drugs, 36, 38, 179
  effect, 8, 140, 189, 278
  efficacy, 74
  influence, 9
  properties, 22, 73, 131, 132, 134, 137, 149, 189
Antimicrobial
  activity, 176, 177, 182–184, 294, 296, 335
  peptides, 177, 181, 182
  properties, 22, 295
Antiobesity, 10, 176, 185
Antioxidant, 14, 30, 86, 180, 184, 186, 188, 245, 284, 290, 291, 298, 315, 329, 335, 336
  activities, 20, 22, 30, 114, 176, 180, 218
  properties, 17, 136, 180, 182, 188, 295
  proteins, 36, 37
  supplements, 180
Anti-thrombotic effects, 136
Antitumor activities, 8, 14, 116, 186, 188
Anti-tumorigenic properties, 149
Apigenin, 115, 137, 138
Apocynaceae, 19
Apoptosis, 115, 128, 129, 137, 299
Aptamers, 110
Aqueous cavity, 22
Arachidonic acid (ARA), 20
Artificial
  intelligence (AI), 147
  tissue materials, 38
Ascorbic acid, 278, 324
Atherosclerosis, 129, 176, 278
ATP-binding cassette (ABC), 108
Autoimmune diseases, 180
Autoxidation reactions, 22

B

Beta-carotene, 135, 136
Beta-cryptoxanthin, 135
Betalains, 22
Bifidobacteria, 37, 175, 211
Bifidobacterium lactis, 201, 208
Bioactive
  agents, 61, 65, 67, 319
  chemical substances, 2
  components, 34, 66, 332
  compounds, 13, 20, 22, 30, 59, 61, 68, 69, 150, 187, 200, 202, 291, 322, 325, 329, 331–334
  dietary polyphenol preparations (BDPP), 149
  indole alkaloids, 19
  ingredient, 80, 297
  molecules, 36, 324, 331, 332, 334, 337, 338
  peptides, 143–145, 179–181, 183, 184, 186, 188
  polyphenols, 14
  proteins, 180
  substances, 7, 22, 68, 334
Bioactivity, 60, 134, 135, 148, 150, 319, 321, 322, 329
Biodegradable polymers, 35, 113
Biodegradation, 35
Biogenetic origin, 13
Bio-imaging applications, 60
Biological
  activity, 12, 13, 283
  contaminants, 277, 288, 300
Biologically active peptides sources, 180
Biomolecules, 173, 190, 216, 336–338
Biopolymer, 14, 16–18, 20, 47, 48, 61, 261, 262, 320, 332–334
  microgels, 332
Biosynthetic pathways, 13
Biotransformation, 37
Blood pressure, 2, 91, 131, 144, 151, 176, 182, 183, 185
Brassica oleracea, 9, 10
Breast cancer resistance protein (BCRP), 106, 108
Brucea javanica, 113
Butyric acid, 20, 322

C

Cadmium sulfide (CdS), 289
Caffeic acid, 15, 45
  phenethyl ester (CAPE), 15
Cancer, 4, 8–10, 14, 20, 22, 30, 36, 39, 73, 74, 91, 105–111, 113, 115, 116, 123, 125, 127–129, 134–137, 139, 145, 149, 151, 153–155, 175, 176, 178, 180, 186, 206, 278, 279, 291, 299, 321, 322
  cells, 73, 107, 109, 113, 129, 137, 178, 180, 279
  community, 73
  prevention and treatment, 107, 108, 116
Caprylic acid, 45
Capsaicin, 20
Carbohydrates, 13, 32, 61, 65, 175, 213, 220, 222, 242, 246, 247, 260, 291, 321
Carbon
  atoms, 20, 115
  nanotubes, 109, 115, 289, 336
Carboxylic acid, 40
Carcinogenesis, 11, 14, 149, 206
Cardioprotective potential, 8
Cardiovascular, 2, 8, 20, 22, 23, 123, 125, 127, 130, 131, 136, 137, 144, 153, 155, 174, 176, 182, 186, 187, 240, 278, 279, 291, 321, 337
  diseases (CVD), 2–4, 8, 9, 11, 14, 23, 125–127, 131, 133–135, 137, 138, 147, 149, 151, 154, 155, 176, 186, 278, 279, 291, 321
  protective effect, 278
  vascular disease, 125, 155
Carotenoids, 40, 46, 66, 86, 106, 126, 134, 135, 175, 278, 318, 321, 334
Carrageenan, 35, 189, 210, 225–227, 253, 254, 257, 263, 333
Casein, 35, 36, 143–145, 147, 178, 183, 297, 321
Caseinophosphopeptides (CCPs), 183, 191
Cationic properties, 177
Cell
  constant, 64
  membrane, 20, 46, 47, 177, 254, 335
  molecules, 176
  surface receptors, 47
  suspension, 43

Cellular
ingestion, 110
physiology, 20
tolerance, 114
Cellulose acetate phthalate, 253
Central nervous system, 279
Cerebrovascular diseases, 9
Chalcones, 136
Chemical bonds, 48
Chemokines, 132, 148
Chemometrics, 93
Chemoneutraceuticals, 113
Chemoprevention, 4, 13, 73
Chemopreventive agent, 2, 11, 278
Chemotherapeutic agents, 105, 109
Chemotherapy, 36, 116, 128, 139, 140
Chitin, 35, 262, 336
Chitosan, 35, 37, 45, 46, 69, 73, 81, 113, 210, 221, 226, 252, 253, 257, 261, 262, 288, 291, 296, 297, 325, 333, 336, 337
Cholesterin, 85
Chondroitin sulfate, 37, 84, 113
Chromatographic technique, 64
Chronic
degenerative diseases, 10
disease, 1, 3, 4, 6, 7, 9, 11, 32, 50, 80, 123, 124, 130, 132, 133, 138, 151, 178, 182, 183, 190, 278, 322, 337
health conditions, 186
human disease, 148, 149
Chronical immune diseases, 21
Clandestine, 89
Claudin, 45, 139
Clinical trials, 107, 126, 131, 148–151, 187, 212, 298, 311, 312, 336, 337
compounds, 148
bioactive dietary polyphenol preparations (BDPP), 149, 150
KB220, 151
nutrafol, 150
omega 3-PUFAS, 151
resveratrol, 148
Clusters, 68, 128
Collagen, 35, 36, 137, 189, 261
Collateral damage, 108
Colloidal
suspension, 39, 62
vesicular transporters, 110
Colonization, 211
Combinatorial techniques, 59, 74
Compound annual growth rate (CAGR), 3, 12, 23, 279
Conjugated
linoleic acid (CLA), 242, 265, 322
α-linolenic acid (CLNA), 242
Consumer product safety commission (CPSC), 49, 300
Conventional
adulteration, 88
emulsions, 42, 295, 325, 326, 330
food, 6, 7, 81, 94
suspension dosage, 110
Co-polymers, 61
Critical solution temperature (CST), 39, 51
Crohn's disease (CD), 125, 133, 139, 299
Cryoprotectants, 44, 253, 259, 260
Crystallization, 64, 72, 247, 331
Cytomodulatory Activity, 179
Cytotoxic outcomes, 107
Cytotoxicity, 59, 61, 110, 114

D

Dairy products, 199, 201
Degenerative diseases, 2, 11, 13
Detoxification, 70
Dextran, 35, 297
Dextrose equivalent (DE), 199, 239, 262, 265
Diabetes, 4, 8–11, 13, 14, 125, 130, 138, 147, 149, 155, 182, 186, 187, 189, 278, 315, 316, 321
mellitus, 125, 176
Diarrhea, 8, 95, 130, 134, 139
Dietary
agents, 30
strategies, 179
supplements, 12, 30, 31, 49, 80, 82, 84, 90, 126, 189, 278, 298
Differential
scanning calorimetry (DSC), 59, 64
thermal analysis (DTA), 59, 64
Diffusion, 47, 71, 297
coefficient, 47
Digestive stimulant action, 9
Diindolylmethane (DIM), 108
Dipeptidyl peptidase-IV (DPP4), 316

Disability-adjusted life years (DALYs), 125, 132
Disaccharides, 253
Disease management, 81, 124, 154
Diterpenes, 15, 17
Docosahexaenoic acid (DHA), 20, 130–134, 151
Dopamine, 151
Doppler anemometry, 63
Drug
  delivery, 30, 35–38, 40, 42, 49, 50, 59, 60, 68, 109, 113, 114, 295, 296
  discovery programs, 17
  molecules, 71, 110
  release, 37, 38, 48, 70–72, 107, 113

E

Ehrlich's ascites carcinoma, 115
Eicosapentaenoic acid (EPA), 20, 49, 130–134, 151, 300
Elastin, 35, 189
Electrochemical detection, 93
Electron
  donator group, 179
  receptor, 179
Ellagic acid, 73
Emulsification techniques
  emulsions, 257
  external ionic gelation, 259
  internal ionic gelation, 258
Encapsulation, 21–23, 30, 31, 36, 37, 43, 44, 59, 108, 110, 113, 114, 209–211, 221, 224–226, 239, 249, 250, 252, 253, 255–257, 259, 260, 264, 265, 291, 293, 294, 296, 297, 319, 321, 322, 324–326, 329, 331, 333, 334, 338
  efficiency (EE), 37, 59, 64, 252, 319, 324, 338
  methods, 21
  techniques, 22, 224, 250, 319
Endocytosis, 46, 61, 114
Endometrial cancer cell line, 135
Endothelia, 44
Endothelial
  cells, 44, 109
  growth, 107, 116
Enterobacteriaceae, 43, 128, 211
Environmental protection agency (EPA), 49, 300
Eosinophilia-myalgia syndrome (EMS), 86, 98
Epidermal
  cells, 115
  growth factor receptor (EGFR), 107
Epigallocatechin
  -3-gallate, 115
  gallate (EGCG), 109
Epithelial cells, 45, 109, 145, 295
Equilibrium, 44, 86, 256
Erosion, 38, 48, 72, 297
Essential
  fatty acids, 20
  oils (EOs), 17
European
  commission (EC), 5, 12, 31, 51, 49, 92, 98, 110, 124, 153
  food
    and safety authority, 12, 23, 49, 300
    law, 92
    safety authority (EFSA), 12, 49, 92, 130, 153, 300
  union, 92
Exopolysaccharides (EPS), 243, 265

F

Fatty acid, 9, 13, 20, 31, 40, 44, 46, 84, 86, 106, 109, 110, 124, 131, 132, 151, 175, 184, 186, 205, 242, 258, 284, 288, 315, 322, 330, 331
  synthase (FAS), 20
Fertilizers, 87
Ferulic acid, 73
Fish protein hydrolysates (FPHs), 187–189, 191
Flavanones, 136
Flavonoids, 32, 66, 126, 136–138, 175, 278, 334, 337, 338
Flavonols, 136
Flocculation, 41, 326
Folic acid, 110
Food
  adulteration crime, 89
  and agriculture organization (FAO), 130, 240

and drug administration (FDA), 12, 49, 65, 95, 116, 300, 312
for specified health use (FOSHU), 4, 5, 81
grade materials, 33
industry, 59, 74, 174, 177, 183, 189, 199, 208, 227, 244–246, 250, 251, 263, 264, 283, 289, 299, 300, 331, 334, 335, 338
ingestion, 66
matrices, 22, 42, 59, 65, 67, 74, 227, 242, 244–246, 265, 320, 324
characteristics, 247
proteins, 141, 177–179
standards agency (FSA), 44
supplements, 12, 22, 92, 124, 153, 312
Fortification, 7, 174, 183, 186, 190, 300
Foundation for innovation in medicine (FIM), 12, 84, 106
Fourier transform infrared spectroscopy, 93
Fragmentation, 48
Fraud
network, 89
occasion, 89
opportunity, 89
Free
cells, 210, 252, 257
radical scavenging activity, 132, 149
Freeze-drying, 44, 225, 247, 252–254
French paradox, 2, 14
Functional
components, 175, 320
food, 1–9, 11, 22, 23, 31, 32, 50, 61, 67, 79–92, 94–98, 126, 154, 173, 174, 176, 178, 179, 182–188, 191, 199–201, 204, 207, 208, 212, 218, 227, 243, 278, 311–313, 317–319, 334, 336
biogenics, 175
center (FFC), 6
conceptual definition and classification, 4
ingredients, 174
nutrients, 175
prebiotics, 175
ingredient, 9, 179

G

Gambogic acid transport, 47
Gastric
residence time, 32, 65
retention time, 68
Gastrointestinal (GI), 20, 32, 34, 35, 37, 48, 51, 61, 65, 68, 69, 109, 134, 154, 174, 175, 177, 203, 204, 206, 210, 216, 219, 220, 227, 239, 240, 242, 243, 245, 246, 252–255, 257, 260, 261, 265, 283, 290, 291, 297–299, 318, 319, 329, 334
resistance, 216, 217, 227
tract (GIT), 32, 33, 37, 42, 44, 47, 51, 67–69, 107, 200, 206, 210, 211, 213, 216–219, 224, 239–242, 248, 250, 265, 283, 290–293, 297, 299, 318
Gelatin, 35, 36, 45, 114, 210, 215, 226, 253, 254, 257, 261, 263, 297
Gellan gum, 253, 263
Gene therapy, 60
Generally recognized as safe (GRAS), 41, 250, 264, 265
Genistein, 14
Gliadin, 35–37
Global
diseases, 2
market, 3, 23, 213, 279, 312
nutraceutical market, 3, 12
Globulin, 35, 186
Glucagon-like peptide-1 (GLP-1), 316, 338
Glucose-dependent insulinotropic polypeptide (GIP), 316
Glutelins, 187
Glycinin, 36
Glycoalkaloids (GAs), 20
Glycogen, 37
Glycolic acid, 38, 45, 46, 113
Glycoproteins, 44
Goblet cells, 44
Gold (Au), 60, 115, 288, 290
nanoparticles (AuNPs), 115, 290
Gram-negative bacteria, 177
Guacamole, 9
Gut microbiota, 128, 175, 216

H

Health
claims, 5, 8, 91, 124, 153, 299, 317
indorsing substances, 84
management, 97
promoting qualities, 9
regulation, 81
service executive (HSE), 125

HeLa cell lines, 114
Hemorrhage, 8
Hepatocytes, 115
Herbal
  medicines, 80
  supplements, 31
Herbivores, 20
Heterocyclic ring, 17
Heterogeneous erosion, 48
Homogeneous matrix, 47
Hyaluronic acid, 35, 109, 113
Hybrid, 278, 289
Hydrochloric acid (HCl), 248
Hydrocolloid solution, 43
Hydrodynamic diameter, 63
Hydrogels, 32, 39, 40, 71, 114, 115, 263
Hydrolysis, 35, 38, 72, 146, 174, 184, 185, 261, 262, 319, 329
  phenomena, 35
Hydrophilic
  polymer, 59, 60
  polymeric systems, 71
Hydrophilicity, 38
Hydrophobic compounds, 62
Hydrophobicity, 178, 180
Hypersensitivity, 108
Hypertension, 125, 129, 131, 144, 146, 176, 182, 186, 188, 189

I

Ile-pro-pro (IPP), 182, 191
Immune
  cells, 139, 145, 146, 180, 207
  function, 131, 132, 139, 141, 148, 149
  system modulation, 44, 174, 240
Immunogenicity, 36–38, 41
Immunomodulators, 179, 278
Immunostimulation, 175, 180, 200
Immunosuppression, 87, 180
In vitro, 8, 11, 110, 114, 115, 128–130, 132, 134, 136–140, 146, 149, 182–184, 186, 188, 189, 210, 216, 227, 245, 246, 252, 253, 255, 257, 260, 337
  drug release, 64
In vivo, 8, 11, 20, 30, 40, 110, 130, 132, 136–138, 144, 145, 149, 182–185, 188, 189, 210, 216, 227, 277, 283, 290, 297, 300, 335, 337
  activities, 114
Indirect health claims, 91
Infant formula, 199, 201, 211
Inflammaging, 132, 137, 154
Inflammation, 8, 11, 14, 20, 37, 39, 127, 130, 132, 133, 137–140, 145, 154, 179, 180, 186, 295, 299
Inflammatory
  biomarkers, 132, 179
  bowel
    disease (IBD), 127, 128, 133, 139, 145, 147, 242, 299
    syndrome (IBS), 138, 139, 145
  cytokines, 127, 132, 134, 137, 140, 145, 148
  diseases, 8, 126, 179
  disorders, 127, 137
Influenza virus A, 17
Infrared spectroscopy, 93
Inorganic nanocarriers, 115
Insomnia, 86
Institute of food technologists (IFT), 5
Insulin-like growth factor receptor (IGFR), 107, 116
Intercellular
  adhesion, 45
  spaces, 45
Interleukin (IL), 127, 134, 138, 146, 150, 179
International
  life sciences institutes (ILSI), 4, 5
  organizations, 4, 95
Intestinal
  cells, 46, 180
  epithelium, 45, 225, 240–242, 250, 265
  membrane fluidity, 69
  microflora, 44
Intracellular metabolism, 45
Intrinsic health-promoting potential, 8
Ionic
  polymerization, 43
  strength, 48, 67, 72, 256, 320
Ionization polymers, 40
Ischemia, 41
Isoflavones, 136, 291
Isoflavonoids, 86, 334
Isopentenyl diphosphate (IPP), 16, 23, 182
Isoprene, 15, 16, 134
Isoprenoids, 16

J

Joint FAO/WHO expert committee on food additives (JECFA), 130

K

Kaempferol, 44

L

Lactic acid, 38, 43, 72, 202, 244, 250, 265, 297
  bacteria (LAB), 43, 202, 209, 215, 220, 223, 224, 244, 246, 265
Lactobacillus
  acidophilus, 201, 202, 204, 207, 221, 245, 246, 260
  plantarum, 201, 206, 212, 213, 217, 221, 246, 257
  rhamnosus, 201, 202, 204–207, 212, 217, 220, 224, 245, 253
Lactoferrin, 47, 143, 182, 264, 321
Lactulose-derived oligosaccharide (OsLu), 208
Large unilamellar vesicles (LUVs), 323, 338
Lectin, 45
Legislative regulation, 5
Leukotrienes, 20
Lipid, 20, 32, 34, 40, 41, 44, 47, 51, 61, 62, 69, 74, 105, 108–110, 129, 131, 138, 151, 177, 178, 180, 188, 223, 247, 248, 278, 279, 290, 291, 293, 294, 296, 301, 315, 317, 318, 320, 322, 323, 325, 331, 332, 334, 337, 338
  mediators, 20
Lipidic bilayer, 22
Lipophilic
  compounds, 34, 68, 69, 322
  drugs, 41, 61
  nutrients, 68
Lipophilicity, 42, 314
Lipoprotein, 9, 69, 109, 279, 322
Liposomal vesicles, 41
Liposomes, 21–23, 34, 40, 41, 50, 107–110, 291, 293–295, 317, 323–325, 331, 334
Lipoxins, 20
Loganiaceae, 19
Long
  chain molecules, 37
  term treatments, 59, 61
Low methoxyl pectin (LMP), 215, 227
Lower critical solution temperature (LCST), 39, 51
Lutein, 38, 40, 135, 290, 291, 296, 321
Lycopene, 86, 108, 134, 135, 279, 291, 297
Lymphatic uptake, 59, 69, 322

M

Macromolecules, 36, 252
Macronutrient, 65
Magnetic fields, 72
Malonate pathways, 14
Maresins, 20
Mass spectrometry, 93, 96
Matrix metalloproteinases (MMPs), 108
M-cells, 46, 47
Mechanical stress, 44
Medicinal foods, 80
Mediterranean-style diet, 2
Melissic acid, 20
Metabolic
  disease, 127, 138, 149–151
  pathways, 13, 16
  syndrome (Met-S), 9, 11, 13, 87, 125, 126, 134, 141, 150, 155
Metabolism, 69, 85, 107, 140, 149, 154, 178, 180, 186, 213, 246, 279, 283, 290, 315, 316, 321, 322
Methylerythritol phosphate (MEP), 16
Mevalonate pathway, 16
Mevalonic acid (MVA), 16, 23
Micelles, 42, 50, 107, 108, 114, 284, 297
Microbial
  agents, 87
  sources, 176
  strains, 211
Microemulsions, 330, 331
Microencapsulation, 200, 209, 219–221, 225, 239–241, 248–255, 257, 260–263, 265, 288
Micro-nutrients, 86
Microparticles, 32, 40, 43, 253, 260
Microspheres, 38, 255
Mineral binding activity, 174, 176, 179, 190
Mitochondrial tumor suppressor 1 (MTSG1), 132
Molecular
  composition, 11
  weight, 14

Monographs, 96
Monomers, 38, 114
Mononuclear phagocyte system (MPS), 41
Monoterpenes, 15
Mucoadhesion, 61
Mucoadhesive hydrophobic materials, 44
Multidrug resistance protein (MRP), 107, 108
Multifunctional crop, 186
Multilamellar vesicles (MLV), 22
Multiple inflammatory cytokines, 36
Mutagenesis, 175
Mycotoxins, 49

N

Nano cages, 44
Nano formulations, 107
Nanocapsules, 60, 113, 288
Nanocarrier, 29–38, 40–50, 67, 70, 72, 105–111, 113, 115, 116, 291, 292, 297, 301, 334, 337
  absorption mechanisms, 44
    paracellular route, 45
    transcellular route, 46
  example, 43
  matrix, 48
  nutraceuticals delivery, 32, 35, 38
    liposomes, 40
    nanoemulsions (NES), 41
    polymers, 35
    smart/stimuli-responsive polymers, 38
    solid lipid nanoparticles (SLNS), 41
  surface, 48
Nanoclusters, 44
Nano-delivery systems, 49
Nanodrop, 44
Nano-drug carrier systems, 108
Nanoemulsions (NEs), 34, 41, 42, 59, 61, 62, 284, 289, 291, 295, 296, 300, 317, 326, 329–331, 334, 338
Nanoencapsulation, 37, 50, 291, 300
Nano-fibers, 61
Nano-films, 36
Nano-formulation, 33, 107, 283, 291, 292, 300, 301, 338
Nanogels, 332, 333
Nanogold, 44
Nanomaterials, 44, 49, 115, 283, 289, 299, 300
Nanomicelles, 114
Nano-nutraceuticals, 277, 283
Nanoparticles (NPs), 32, 34, 36, 44–47, 51, 59–64, 72–75, 107–110, 113–116, 283, 284, 288–291, 296–301, 317, 331, 335–337
  characterization, 75
  dispersions, 64
  evaluation, 63
    encapsulation efficiency (EE), 37, 48, 64, 252, 254, 255, 259, 319, 324, 325
    in vitro release studies, 64
    particle size and surface charge, 63
    storage stability, 64
    surface and interior morphology, 63
  introduction, 60
  matrix, 59, 60
  preparation, 61
    nanoemulsion techniques, 62
    nanoprecipitation, 61
    reverse-phase evaporation, 62
  release mechanisms, 70
    degradation-controlled release, 71
    diffusion-controlled release, 71
    solvent-controlled release, 71
    stimuli-controlled release, 72
Nanoprecipitation, 59
Nanoscale, 29, 33, 60, 69, 291, 317, 334, 336–338
Nanosensors, 93, 288, 334, 335
Nanosilica, 115
Nanosize, 60
Nanospheres, 60, 71, 113
Nano-sponges, 36
Nanostructured lipid carriers (NLCs), 109, 110, 291, 296, 300, 301, 331, 332
Nanosuspension, 301, 334
Nano-systems, 38, 46
Nanotechnology, 21, 23, 29, 30, 32–34, 36, 38, 44, 49, 50, 59, 60, 72, 74, 277, 283, 284, 288–290, 292, 293, 299, 300, 311, 317, 334, 335, 338
  applications, 283
National institute for occupational safety and health (NIOSH), 49, 300
Natural
  carotenoids, 134
    flavonoids, 136
    lycopene, 134

B-carotene, 135
compounds, 12, 13, 30, 179
nutraceuticals, 127
curcumin, 127
omega-3 polyunsaturated fatty acids (PUFAS), 20, 130
Neoblastoma xenographs, 149
Neurodegenerative
disease, 149, 150, 313, 314
disorders, 11, 14, 314
Neurological
diseases, 278
disorders, 176, 188
Neuropathy, 279, 316
Neutraceuticals, 113
Neutralization, 176
Nicotinic acid (NA), 125, 155, 252
Nitrogen atoms, 17, 19
NMR spectroscopy, 96
Non-communicable diseases (NCDs), 125, 126, 147, 149, 153–155
Non-covalent attractions, 114
Non-dairy products probiotics, 213
Non-melanoma cancers, 136
Non-steroidal anti-inflammatory drugs (NSAIDs), 38, 51
Novel nanocarriers, 106, 109, 116
N-trimethyl chitosan (TMC), 110
Nuclear magnetic resonance, 93
Nucleic acid, 279
Nutraceutical
bioavailability, 65, 67
external factors, 66
cancer prevention and treatment, 107
clinical validation, 298
definition and regulatory aspects, 11
delivery systems, 47, 67, 319, 322, 333
ideal properties, 319
effects, 34
industry, 283, 284
molecules, 47, 65
nutraceuticals and chemical classification, 13
alkaloids, 17
fatty acids, 20
nanotechnology, 21
polyphenols, 13
terpenoids, 15
oral bioavailability, 67
aqueous solubility improvement, 68
controlled/delayed release, 68
gastric retention time delay, 68
intestinal permeability improvement, 69
labile compounds safety, 67
lymphatic uptake, 69
metabolic activities modulation, 69
release mechanisms, 47
research and education act (NREA), 82
targets, 21
Nutraceutics, 12
Nutrafol, 150
Nutrient bioavailability, 59, 75
Nutritional
functions, 6, 81, 200
pulses, 186
value, 8, 32, 50, 67, 74, 123, 126, 185, 187, 189, 190, 264

O

Occludin, 45, 139
Occupational safety and health administration (OSHA), 49, 300
Office of combination products (OCP), 49
Oil-in-water (O/W), 62, 257, 265, 295, 325, 326, 329, 330, 338
-in-oil (O/W/O), 326, 338
Oligosaccharide, 208, 220, 227
Opioid addictions, 151
Optimal nutritional value, 7
Optimum delivery system, 67
Oral bioavailability, 22, 34, 59, 109, 115, 291, 295
Organic
compounds, 13, 90
phase, 62, 295
solvent, 37, 61–63, 93
Origanum vulgare, 17
Osteoarthritic subjects, 137
Oxidative stress, 8, 132, 218, 299, 315, 316, 337

P

Palladium (Pd), 60
Passive
diffusion, 45, 46, 329
transport, 45

Pathogenic substances, 177
Pegylation, 41
Peptides, 45, 46, 60, 61, 86, 109, 126, 143–147, 173–191, 242, 244, 284, 318, 325, 337, 338
Permeation, 44
Peroxidation, 175, 180
Peroxisome proliferator-activated receptors (PPARγ), 316, 338
Pesticides, 87, 288, 335
P-glycoprotein, 106, 109
pH, 32, 37, 39, 40, 43, 44, 48, 67, 72, 108, 113, 190, 203–205, 207–209, 215, 219, 223, 224, 226, 240, 242–245, 247–250, 256, 258, 259, 261, 262, 264, 293, 319, 320, 326
Phagocytosis, 40, 46
Phagosome, 40
Pharmaceutical, 12, 30, 31, 179, 188, 251, 312
  function, 12
  sectors, 61
Pharmacodynamic, 23, 109
  profile, 107
Pharmacokinetic, 23, 30, 108, 109, 150
  profile, 107
Pharmacology, 2, 85
Phenolic
  acids, 15, 175
  compounds, 13–15, 22, 34, 218, 318
  diterpenes, 17
  metabolites, 14
Phenylalanine, 20, 146, 151
Phenylpropanoids, 17
Phosphate buffer solution (PBS), 64
Phosphatidylcholine (PC), 40, 294, 296
Phospholipid, 21, 22, 40, 44, 62, 73, 110, 132, 178, 258, 290, 293, 294, 323–325
  vesicles, 22, 293
Phosphotungstic acid, 63
Photon correlation spectroscopy, 63
Phytochemicals, 30, 42, 50, 85–88, 135, 146, 148, 213
Phytosterols, 7, 40, 86, 213, 279, 291, 321, 323, 325, 334
Pinocytosis, 46
Placebo group, 130, 135, 144, 212, 227
Plant
  polyphenols, 14
  sources, 36, 184, 185
Plastids, 16
Platelet aggregation, 10, 129, 131
Platinum (Pt), 60
Polar
  biomolecules, 46
  molecules, 45
Poly (ethylene glycol), 59, 60
Poly(acrylamide-co-butyl methacrylate) (PAAM), 39
Poly(acrylic acid) (PAA), 39, 40
Poly(cyclohexane-1,4-diyl acetone dimethylene ketal) (PCADK), 40
Poly(ethylene glycol) (PEG), 45, 60, 73, 110, 113, 114, 297
Poly(glycolic acid) (PGA), 38, 46
Poly(lactic-co-glycolic acid) (PLGA), 38, 45, 72, 73, 109, 113, 297
Poly(N,N-diethylacrylamide) (PDEAM), 39
Poly(N-isopropylacrylamide) (PNIPAM), 39
Poly[2-(dimethylamino)ethyl methacrylate] (PDMAEMA), 39
Polyacrylamide, 39
Polyamide, 38, 71
Polyanhydrides, 38
Polyanions, 40
Polycaprolactone (PCL), 72, 73, 113, 288
Polycationic polymers, 40
Polydispersity, 63
Polyelectrolyte complexes (PECs), 333
Polyesters, 71, 297
Polyethylene glycol, 45, 109, 110, 113
Polyglycolic acid, 38, 51
Polyketals (PK), 40
Polylactic acid (PLA), 38, 73, 113, 288
Polymer, 34–40, 43, 48, 50, 59–62, 69–72, 113, 114, 116, 225, 226, 249, 251, 252, 254–262, 264, 265, 288, 289, 297, 298, 325, 330, 336
  biodegradable polymers, 35
  matrix, 43, 71, 289, 336
  molecular weight, 48
  naturally occurring biodegradable polymers, 35
  pH-sensitive polymers, 40
  polysaccharides based natural polymers, 37
  proteins based natural polymers, 36
  synthetic biodegradable polymers, 37
  temperature-sensitive polymers, 38

Polymeric
  hydrogels, 39, 114
  membrane, 71
  nanoparticle systems (PNPs), 331
  network, 39, 113
Polymerization, 64, 297
Polyorthoesters, 38
Polyphenols, 13–15, 45, 46, 74, 148, 149, 188, 297, 322, 334, 337, 338
Polysaccharides, 34, 35, 37, 71, 86, 113, 188, 218, 258, 263, 297, 320, 323, 325, 332
Polyunsaturated fatty acids (PUFAs), 20, 22, 126, 130–134, 151, 154, 297, 315
Polyvinyl acetate-polyethene glycol, 114
Potential
  antiviral activity, 17
  biological activities, 6, 11
  health benefits, 6, 50, 65, 334
Precipitation, 39, 298, 333
Preclusion stage, 88
Predictive tools, 97
Primary metabolism, 13
Principal secondary metabolite, 20
Proanthocyanins, 86
Probiotics
  application, 243
    functional food, 244
    limitations, 246
    pharmaceutical formulations, 243
  bacteria, 32, 43, 44, 175, 200–206, 209, 210, 213–215, 217–221, 223, 224, 227, 240, 242, 245–247, 254, 291
  cells, 43, 44, 200, 209, 210, 216, 224–226, 243–245, 247–250, 252, 253, 255, 257–259, 265
  microencapsulation techniques, 249, 250
    complex coacervation, 255
    emulsification techniques, 257
    extrusion, 254
    freeze-drying, 252
    spray drying, 251
  microorganisms, 175, 201, 203–207, 210, 213, 222, 224, 239–248, 251, 252, 254, 264
  receiving group, 212, 227
  strains, 202, 204, 205, 218, 247
Prolamin proteins, 36, 37
Propolis, 15
Prostaglandins, 20
Prostate intraepithelial neoplasia, 135
Protectins, 20
Protein
  and bioactive peptides, 141
    casein protein, 144
    plant-based protein, 145
    whey protein, 143
  carbohydrate conjugates, 61
  functions, 176
  inhibiting catalase, 36
  kinase, 36
  sources, 180
Protozoa, 177
Public
  health
    hazards, 88, 94
    nutrition, 9
  private mutual coordination, 88
Punica granatum, 8, 10
Pyrolysis, 64

Q

Quantum dots, 109, 115, 289

R

Radical, 36, 90, 185, 186
  scavengers, 12, 14
Raman spectroscopy, 93
Reactive oxygen species (ROS), 40, 134, 136, 180, 299, 336
Resolvins, 20
Respiratory disease mortality, 9
Reticuloendothelial system (RES), 40, 41
Retinopathy, 279
Rheumatoid arthritis (RA), 125, 127, 128, 155
Ribonucleic acid (RNA), 40, 141, 252
Ribosomes, 252
Rota
  evaporation, 62
  evaporator, 62
  vapor, 63
Rutin, 45

S

Safety issues, 79, 80, 82, 83, 87, 92, 94–98
Salicylic acid, 38
Scamorza, 202, 204

Secondary
food elements, 32
metabolism, 13
metabolites, 13, 15, 17, 19, 20
Self-micro emulsifying drug delivery system (SMEDDS), 113
Sensory parameters, 22
Serotonin, 86
Sesquiterpenes, 15, 16
Short-chain aliphatic hydrocarbon, 17
Sialic acid, 110, 294
Signal transduction, 108
Silver (Ag), 44, 60, 115, 289, 336
Simvastatin, 73
Small
interfering RNA (siRNA), 40, 50
unilamellar vesicles (SUVs), 323, 338
Smart gels, 39
Sole-gel transition, 39
Solid
lipid nanoparticles (SLN), 34, 51, 74, 108, 110, 291, 296, 301, 317, 331
matrix, 203, 245, 320
phase extraction, 93
Sonication, 63
Soy protein, 185, 253
Stakeholders, 96
Staphylococci, 43
Staphylococcus aureus, 17
Steam distillation, 17
Stringent, 10
Superfruit, 8
Superoxide
dismutase (SOD), 36, 40
scavenging enzyme, 40
Surface
erosion, 48
enhanced Raman spectroscopy (SERS), 93, 98
Surfactant, 41, 42, 60–62, 108, 109, 113, 114, 258, 259, 264, 290, 295, 296, 298, 320, 323–326, 329–331
Swelling-shrinkage mechanism, 48

T

Tannins, 14
Targeted delivery systems, 116
Terpenes, 16
Terpenoids, 13, 15–17, 278
Tetraterpenes, 16
Theobromine, 278
Therapeutic
agents, 176, 241
effects, 22, 110, 126, 182, 186, 283, 314
indications, 17
qualities, 8
sectors, 180
Thermo-sensitive polymers, 39
Tight junctions (TJs), 45, 46
Tissue engineering, 35, 36
Titanium nitride, 44
Tortuosity, 47
Traditional
medicine, 8, 22, 23, 315
oriental medicine, 9
Transcellular absorption mechanism, 46
Transcription
elements, 179
factors, 179
Transmission electron microscopy (TEM), 63
Triterpenes, 15, 16
Tryptophan, 85, 86, 151
Tumor necrosis factor, 40, 125, 155
Type 2 diabetes mellitus (T2DM), 130, 138, 147

U

Ulcerative colitis (UC), 133
Ultrasonication method, 62
Unintentional adulteration, 80, 83, 89, 90
United States (US), 49, 96, 106, 125, 279, 299, 300
department of agriculture (USDA), 49, 300
of America (USA), 125, 151, 289
patent and trademark office (USPTO), 49, 300
Unsaturated fatty acids, 20, 288
Upper critical solution temperature (UCST), 39
Ursolic acid, 110

V

Val-pro-pro (VPP), 182
Vascular endothelial growth factor (VEGF), 108, 116, 129
Vegetable tannins, 14
Vinblastine, 19

Vincristine, 19
Vitamins, 7, 12, 22, 30–32, 40, 44, 50, 65, 80, 81, 84–86, 106, 124, 153, 175, 182–184, 186, 188, 213, 222, 240, 245, 246, 249, 277, 284, 291, 318, 319, 321, 329
Volatiles terpenoids, 17

W

Water-in-oil-in-water (W/O/W), 326, 338
World Health Organization (WHO), 8, 23, 30, 31, 87, 90, 124, 125, 130, 133, 155, 220, 240

X

Xenograf, 128, 149
Xenograft mouse model, 116

Y

Yeasts, 43, 85, 177, 183, 205, 241

Z

Zein, 35–37, 298
Zero-order
  drug release profile, 70
  release profile, 71